INCLUSIVE DEVELOPMENT AND MULTILEVEL TRANSBOUNDARY WATER GOVERNANCE

THE KABUL RIVER

T0303870

Shakeel Hayat

Promotiecommissie

Promotor:	Prof. dr. J. Gupta	Universiteit van Amsterdam
Copromotor:	Dr. C.L. Vegelin	Universiteit van Amsterdam

Overige leden:	Prof. dr. I.S.A Baud	Universiteit van Amsterdam
	Prof. dr. M.Z. Zwarteveen	Universiteit van Amsterdam
	Dr. U. Daxecker	Universiteit van Amsterdam
	Prof. dr. M. Zeitoun	University of East Anglia, UK
	Dr. A. Taj	IM Sciences, Pakistan

Faculteit der Maatschappij- en Gedragswetenschappen

This research was conducted under the auspices of the Graduate School for Socio-Economic and Natural Sciences of the Environment (SENSE)

INCLUSIVE DEVELOPMENT AND MULTILEVEL TRANSBOUNDARY WATER GOVERNANCE

THE KABUL RIVER

ACADEMISCH PROEFSCHRIFT

ter verkrijging van de graad van doctor

aan de Universiteit van Amsterdam

op gezag van de Rector Magnificus

prof. dr. ir. K.I.J. Maex

ten overstaan van een door het College voor Promoties ingestelde commissie,

in het openbaar te verdedigen in de Agnietenkapel

op donderdag 5 maart 2020, te 10:00 uur

door

Shakeel Hayat

geboren te Swabi

CRC Press/Balkema is an imprint of the Taylor & Francis Group, an informa business

© 2020, Shakeel Hayat

Published by:
CRC Press/Balkema
Schipholweg 107C, 2316 XC, Leiden, the Netherlands
Pub.NL@taylorandfrancis.com
www.crcpress.com – www.taylorandfrancis.com
ISBN: 978-0-367-50074-0 (Taylor & Francis Group)

Dedicated to:

"My parents- Hakim Akbar & Parveen; Wife- Surriya & Kids- Sania & Mahad"

ACKNOWLEDGMENTS

Pursuing a PhD was always a tough choice due to financial limitations, family responsibilities and busy schedules at workplace. However, my ambitions were high, support from my family was strong and my employer was outstanding and encouraged me to make my own contribution to the continuous development of scientific research. I am extremely thankful to Almighty Allah for empowering me to finish my PhD research. To overcome the financial constraints, I am thankful to the Netherlands Fellowship Programme (NFP; Project No. 104883) and Higher Education Commission of Pakistan (HEC; Project No. 1-8/HEC/HRD/2017/8435) which provided me the funding and enabled me to carry out my research and make my dream a real one. I came to know that accomplishment in the PhD journey is a collective, emotional, and even a physical endeavour. Hence, there are various individuals who have contributed directly to the research and writing of this thesis by providing ideas, resources, feedback, data and information. There are several other and important people who indirectly provided infinite encouragement, firm support and wholesome kindness. None of these unconditional support and assistances can be measured, nor can they be appropriately acknowledged. But I will make a sincere effort to do so here.

I would like to commence by thanking Professor Joyeeta Gupta and Dr. Courtney Vegelin for doing everything in their ability to make this thesis possible, often going beyond the call of duty. Joyeeta, finishing this thesis would have demonstrated an overwhelming job, if not for your regularly pushing me to make my writing better through supervision, critique, and inspiration. Thank you for seeing that this thesis was possible long before I could. Your friendly behaviour and company has endorsed my work to arrive where it is today. You are always a true inspiration for me and I will keep following your footsteps. Courtney, you have showed me that all kind of challenges can be overcome with faith and determination. Thank you for always believing I would make it through to the end and making sure I _kept on smiling'. I appreciate your time, support, willingness and enthusiasm for coming into this process mid-stream. Dr. Hameedullah Jamali, thank you for your endless support and help throughout my PhD process and even before that. You are a true friend and a wonderful human-being.

The ultimate love and gratitude go to my family and friends in Pakistan as well as in the Netherlands. You always bothered me in both emotional and funny ways with your questions of _when am I going to finish'? Yet, through all the reservations and sufferings, you continued to be faithful, kept me in your prayers, and today I can proudly say that the _PhD Journey' is over with these words of gratefulness for all of you. My beautiful wife Surriya, thank you for your support and for being confident I would finish, even when I was not. You showed me that imperceptible development amounts to big success. My marvellous kids Sania and Mahad, thank you for cheering me up with your funny and naughty acts when I was depressed and down. My loving and respected parents, thank you for teaching me to enlarge and harness my creativity. You empowered me to include a bit of art into everything, including this thesis. My siblings Yasmin, Rukhsana, Faqir

Hayat, Nasira, and Tanzeela, thank you for listening to all of my doubts, fears, and tears without judgement and putting me back together every time I fell apart. Thanks to my aunts, uncles and cousins for helping me learn what is best for me and how to achieve it.

Dr. Khaista, Adil, Haris, Waleed, Munir and Yasir Iqbal, thank you for your prayers and best wishes, they kept me going even when I was losing ground. To my Director, Dr. Muhammed Mohsin Khan, Joint Director Dr. Usman Ghani, Deputy Director Dr. Iftikhar-ul-Amin, and colleagues at HR, finance and accounts departments of IMSciences, Mian Mabood Gul, Wajeeh, Haseeb, Ali Abbas Mirza, Waqar, Muhammad Imran Khan, Muhammad Irshad, Gulzar Ali, Qaiser Ayub, Adil, Zafar, Abdul Khaliq, thank you for making IMSciences a dream place to work. You all made administration and financial complications easier for me and it was your support that never made me worried about management and finance-related issues. To my amazing colleagues at IMSciences Javed Iqbal, Zafar Habib, Sikandar Tangi, Shabana Gul, Aamer Taj, Waseef Jamal, Atta ur Rehman, Fakhr-i-Alam Khan, Abdul Ghaffar, Muhammed Rafiq, Mukhtar Ahmed, Sajid Iqbal, Mukamil Shah, Farman Ali, Bakhtawar, Asif, Sartaj, and Nazir Khan, thank you, your presence always encouraged me to finish the PhD in time and come back to IMSciences as soon as possible.

I would also like to acknowledge my friends in Delft, Imran Ashraf, Fawad, Hussam, Nauman, Qasim, Hussam, Tanvir, Irfan, and Saad as well as my colleagues at IHE for your feedback, inputs and encouragement over the years. Raquel, Kirstin, Pedi, Mohan, Shahnoor, Eva, Jonatan, Nadine, Gabriela, Marmar, Jakia, Aftab, Natalia, Nirajan, Taha, Musaed, Motasem, Akousha, and Christiana, you are my dearest and most valuable friends and colleagues. Sharing this crazy PhD journey with best colleagues like you is something unforgettable. You were always there to help me through the variety of emotions, feelings and experiences that only come with a PhD. You understood the immaturity and susceptibility that comes with this process and I am so obliged to have made it through with you all smoothly walking by my side.

Last but not the least, to all the water scholars and mentors, whom work truly guided me to finish this PhD. I would like to specifically thank Mark Zeitoun, Jeron Warner, Naho Mirumachi, Ursula Daxecker, Margreet Zwarteveen, Pieter van der Zaag and my own promotor Joyeeta Gupta. You have all shaped me and my research in ways I have only begun to understand.

SUMMARY

Problem Definition

Although, transboundary and multi-level water resources are often seen as a source of conflict among countries and regions, recent studies suggest that they can also be a catalyst for cooperation. Such institutionalised cooperation can benefit riparian parties by promoting greater socio-economic development. Despite the existence of institutions at transboundary levels, some basins are better managed than others. This may be because, as hydro hegemony scholars argue, asymmetric power relations between peoples and countries shape institutions in different ways in different basins. However, if power has such a dominant role, the question is why are there so many institutions in other parts of the world (such as between Canada and the USA and among European countries) that do work? Neo-institutionalists argue that power constellations vary from issue to issue, there is often space to develop fair and effective institutions and that institutions, once developed, are also able to shape and curb power.

Ineffective institutions and poor governance practices are creating various challenges for water resource management in the Global South which can be further exacerbated with worldwide changes in social, political, environmental and economic systems. These challenges can be divided into two broad categories: (1) flow-related (i.e., issues of water quality, quantity and related ecosystem services); and (2) administration-related (i.e., issues of power, jurisdiction, coordination, principles and instruments at multiple levels of governance). The scholarly literature suggests that an analytical transboundary water governance framework is essential to address the challenges of water politicisation and securitisation, quality degradation and quantity reduction. Yet, there are four key gaps in scholarly literature regarding transboundary water governance. First, this literature rarely integrates (a) a multi-level approach, (b) an institutional approach (c) an inclusive development approach, or (d) accounts for the uses of different types of water and their varied ecosystem services.

Question

Hence, this thesis responds to the main research question: How can regional hydro-politics and institutions be transformed at multiple levels of governance through inclusive development objectives and incorporate the relationships with non-water sectors in addressing issues of water quality, quantity and climate change'? The following sub-questions are also addressed: In order to answer this question, four sub-questions were developed: 1) How can the concept of biodiversity and ESS be incorporated in a framework to analyse the effectiveness of institutions, and the role of power in governing transboundary water resources? 2) Which principles and instruments address the causes/drivers of freshwater problems in transboundary river basins at multiple geographic levels? 3) How does legal pluralism affect transboundary water cooperation? 4) How do power politics and institutions influence water governance in transboundary river basins at multiple geographic levels? These questions are explored with special reference to the Kabul River that flows through Afghanistan and Pakistan. I chose the Kabul river basin because: (a) it is unique as both countries are contributing water, they are both simultaneously up and down-stream and both countries are in conflict; (b) Afghanistan would now like to develop the river basin unilaterally and to increase its access to the waters for its socio-economic development leading to conflict with Pakistan; Pakistan argues that it was the first user in time and thus has historical rights and because this basin is already the most densely populated basin of Afghanistan and Pakistan (with a population density of 93 persons/km^2 accommodating an estimated 7 million people); (c) its vulnerability to climate change as this river is predominantly fed by glaciers and snow-melt unlike other rivers which are mainly recharged by rainfall; (c) its lack of formal institutionalism coupled with very limited interest from scholars because of the long conflict in the basin area, and (d) since the tribal customs are difficult to understand for foreigners, my personal affinity with the region, affiliation with the tribal customs, and full command over local languages makes it a suitable case study.

Chapter 2 presents the methodology and analytical framework. It argues that addressing transboundary water issues can be improved by using a multi-level governance (MLG) framework. I choose to use type I MLG which treats each administrative level independently. Since power politics influences water institutions and vice versa, I briefly explain the fundamentals of power theory and institutional theory and their relationship. Since water governance approaches have been developed concurrently at different levels of governance indicating plural water laws, I assess whether the rules are consistent or contradictory in a given jurisdiction. Finally, I have chosen to discuss inclusive development as a goal of water governance in contrast to sustainable development, because I would like to emphasise the social/relational (focusing on how power can be shifted to local people through, inter alia, the adoption of principles and instruments) and ecological (aiming at improving the ecosystem services of water) components, rather than how economic growth per se can be enhanced. Additionally, since water law has a number of procedural principles that are difficult to classify as social or ecological, I have clustered them under political principles.

In terms of methodology, I reviewed the literature on hydro-hegemony (from 2004 – 2018), and the relationship between power and institutions (from 1970s – 2018); I reviewed five key international policy and legal documents relevant to this thesis at the global level, 15 at the transboundary level (3 colonial and 12 post-colonial); 22 at the national and local level in Afghanistan (2 pre-colonial, 1 colonial, and 19 post-colonial), and 25 in Pakistan (2 pre-colonial, 4 colonial, and 19 post-colonial) concerning water governance in the case study region of Afghanistan and Pakistan. I conducted a comparative multi-layered case study of Afghanistan and Pakistan on the Kabul River Basin where a total of 70 interviews (30 from Afghanistan and 40 from Pakistan) and two Focus Group Discussions (FGDs) were conducted.

To guide the structure of this research, I have adapted Oran Young's institutional analysis model (also adopted by the Institutional Dimensions of Global Environmental Change) to accommodate the key concepts of my thesis. In this framework, it is first important to understand the context and the driving forces which lead to water governance challenges in the Kabul River Basin. I also look at how power has shaped the existing transboundary, national and local institutions in the basin. I then identify the key principles and instruments that aim to change the behaviour of humans with respect to water use and pollution. Then I analyse whether these instruments change human behaviour, given the context and drivers in such a way as to ensure social and ecological inclusion and alter relational issues (i.e. inclusive development). Finally, based on an assessment of which principles and instruments work and which do not (in terms of addressing contextual challenges and the drivers; and in mobilising changed behaviour in actors), I make recommendations regarding how the institutional approach can be improved and discuss whether this can change the win-set which might influence the underlying power politics and therefore lead to the development of mutually satisfactory conclusions.

Chapter 3 elaborates on the how realist and institutionalist approaches deal with multilevel transboundary water governance. The literature on realism (e.g. hydro hegemony) focuses generally on the role of state power in international cooperation while institutionalism (i.e., hydro institutionalism) focuses generally on the role of institutions in international cooperation. Power theories focus on how power influences multilevel transboundary water governance institutions by: including or excluding actors; including or excluding various issues and thereby shaping the agenda or preferences of different actors involved in the political/negotiation process; and when there is power asymmetry between actors in relation to (i) geography; (ii) material; (iii) bargaining; and (iv) ideational power. Such power can lead to fragile and unsustainable water institutions. Institutionalist theories argue that cooperation, and ultimately peace, can nevertheless emerge through the development of norms which may act as an intermediating factor between power structures of the international system. It argues that cooperation is even possible in the anarchic system in the form of cooperation in trade, human rights and collective security, among other issues. Such cooperation often emerges as different issues have different power constellations.

Hydro Hegemony (HH) and Hydro Institutionalism (HI) (the term I give to an institutionalist approach to water) are two different approaches to water conflict and cooperation in transboundary river basins. They may not essentially be incorporated into a distinct informative framework, but they balance each other and may be productively engaged to answer diverse aspects of the main research question and sub-questions. In reality, there is a complex interplay of power and institutions where power influences transboundary water outcomes when there is asymmetric power relationships among riparians; however, formal and informal institutions in turn limit the role of power politics in shared freshwater resources (a hybrid approach). Furthermore, there might be (a) unresolved historical issues in non-water related sectors (e.g. border issues), and (b) lack of scientific and societal information that may hamper institution building in the water sector. Hence, this thesis assesses the role of power in excluding and including actors and issues; and assesses how existing institutions can be improved based on solving unresolved historical issues and providing additional scientific and societal information which could perhaps change the perceptions of those negotiating. Furthermore, while most of these theories focus only on state-state relations, I explicitly look at the different levels of governance.

Chapter 4 discusses the state-of-the-art knowledge of freshwater systems within the context of a basin. I identify first the key direct drivers (i.e., agricultural development, animal husbandry, the extractive sector and water use in energy; industry including services and infrastructure; municipal water supply and sanitation services, and demographic shifts including migration, population growth, population density, and urbanisation) and indirect drivers (e.g., political dynamics between states/provinces; culture and ethnic attitudes towards water, non-water-sector policies that affect water, the drive for economic growth; poverty; technological advances including agriculture intensification; international trade; climate variability and change; earthquakes) of the problem of flow (quality, quantity, and related ecosystem services of water) in river basins. The drivers can be natural (e.g. tectonic movements, climate and weather variability) and anthropogenic (e.g. climate change).

Since a key addition of this thesis is including the ecosystem services of water, this chapter elaborates on the different ecosystem services (ESS) of freshwater. These include the supporting (e.g. the role of water in erosion control and nutrient recycling), regulating (e.g. climate regulation, hydrological regulation), provisioning (e.g. drinking water, water for food) and cultural (e.g. spiritual, recreation) services of water, the different kinds of freshwater (i.e. rainbow-water, blue surface-water, blue groundwater, green-water, grey-water, black-water and white frozen water or snow) have different ecosystem services with differing contributions to human well-being (i.e. freedom of choice through security, good life, health and good social relations) at different levels of governance. A key hypothesis of this chapter is that if the scope of transboundary river basin discussions includes other water issues in the basin, a better understanding of the ecosystem services of water, and how non-water issues can cause problems in the water sector, then this can enable better collaboration between riparians.

Chapter 5 examines five key institutions at the global level (i.e. customary international water law; the 1992 UNECE Water Law which has been opened up for global participation; the 1997 Watercourses Convention; the Human Right to Water and Sanitation, and the Sustainable Development Goals of Agenda 2030). The Sustainable Development Goals call for incorporating the socio-ecological characteristics in the pursuit of development. The chapter identifies the following principles and instruments from the above documents that are relevant for this thesis: (i) political (e.g., the principles on sovereignty, information sharing, obligation to cooperate, dispute resolution, notifying about planned measures and emergencies); (ii) social-relational (including the human right to water and sanitation, public participation, awareness, education, and access to information, priority of use, rights of women, youth and indigenous people, capacity building, ensuring equitable and reasonable use, intergenerational equity, alleviation of poverty); and (iii) ecological (i.e., EIA, monitoring, prevention of pollution, polluters pay, preservation and protection of ecosystems as well as protection of water recharge and discharge zones). Furthermore, it identifies the obligation to form a river basin organisation and use dispute resolution mechanisms for problem solving.

Chapter 6 describes and analyses relations between Afghanistan and Pakistan in the KRB in order to answer the question of how power and institutions influence international relations between the two countries and how they obstruct or contribute to achieve inclusive and sustainable development. It does so by i) discussing the overall political context, ii) identifying key ESS including biodiversity, iii) recognising direct and indirect drivers of freshwater challenges, and iv) providing a detailed overview of the transboundary level institutions by analysing recent and historic practices including the pre-colonial agreements.

This chapter draws four conclusions. First, there is no formal regulatory framework to equitably share the water resources in the KRB between Afghanistan and Pakistan, and hence each country does what it wants in such an anarchic system. Although Pakistan is utilising a major portion of the water resources in the KRB, it is worried about the future access to water as Afghanistan is planning new dams and irrigation infrastructure. The Pakistani authorities are very much concerned about the unilateral completion of these mega projects as it would significantly affect the agrarian economy of the Khyber Pakhtunkhwa (KP) province of Pakistan. Second, acknowledging the variety of BESS at transboundary level within the Kabul River Basin can provide the basis for similar problem framing as well as highlight the benefits for local livelihoods. Similarly, regulating services of the Kabul River include connected and improved groundwater recharge by removing nutrients and pollutants, climate regulation through carbon sequestration in the floodplains and surrounding forests.

Third, acknowledging and identifying similarities in the key direct and indirect drivers can be a first step towards addressing transboundary water issues in the KRB. Key direct drivers include: (a) agriculture development; (b) industry; and (c) demographic shifts. The main indirect drivers of the freshwater problems in the KRB at the transboundary level are: (a) political dynamics between states; (b) culture and ethnic elements; (c) non-water-related policies; (d) economy; (e) poverty; (f) technological; (g) international trade; and (h) natural change and variability in weather. In particular, political dynamics between states is an important transboundary level driver where Afghanistan does not accept the Durand Line as an internationally recognised border. The issue of the Durand Line also needs to be urgently resolved and this requires the establishment of a fact-finding mission by including concerned authorities and international legal experts. The political context of extremism, Taliban proxies and Pakhtunistan inhibits collaboration and is affected by water related issues between the two neighbours. Fourth, the absence of formal institutions has enabled power politics to prevail between the two countries. Although Afghanistan has a geographic advantage, it is less influential due to the lack of capacity and resources to exploit water resources whereas Pakistan possesses higher material, bargaining and ideational power consuming about 90% of water from the KRB and is less motivated to change the status quo. Due to these power asymmetries, Pakistan has adopted a security oriented foreign policy agenda towards Afghanistan.

Chapter 7 describes and analyses multilevel freshwater governance in Afghanistan. It does so by looking at (1) how different characteristics and drivers of freshwater problems are taken into account at multiple geographic levels in Afghanistan; (2) how freshwater governance frameworks have evolved at multiple geographic levels in Afghanistan; (3) which governance instruments address the drivers of freshwater problems at multiple geographic levels in Afghanistan; (4) how legal pluralism can be observed at multiple geographic levels in Afghanistan; and (5) how power and institutions influence freshwater governance frameworks at multiple geographic levels in Afghanistan. Through these sub-questions, this chapter draws four conclusions.

First, widespread poverty, weak institutional and human capacity due to long-lasting conflicts and instability, and lack of knowledge and capacity to manage water resources have contributed to water challenges in Afghanistan. Additionally, Due to Afghanistan's land-locked status and important geographic location, it has had historical disputes with its neighbours over the water flow from its rivers which are Afghanistan's primary source of water. Moreover, the long-standing civil war and the Soviet incursion has intended that much of the water-related institutional knowledge was lost, water monitoring equipment was destroyed, and the capabilities of water scientists deteriorated. Second, since a large segment of the population in Afghanistan directly and indirectly depends on a variety of biodiversity and ecosystem services (BESS) provided by the Kabul River, deprivation of these dynamic services due to water flows variability can have adverse significances. The contemporary classification divide Afghanistan into 15 minor ecological zones of which only two are comparatively constant and undamaged. The species arrangement of all ecological zones has been considerably concentrated through overgrazing, wood collection for fuel, and manipulation of pastures, forests, and water resources. Approximately 70-80% of Afghanistan's population are dependent on agriculture, animal husbandry and artisanal mining for their livings. The country needs to connect these resources to boost the country from the bottom of the Human Development Index.

Third, it is vital to identify and understand key direct and indirect drivers of water issues for evidence based policies in a changing geopolitical context and given climatic scenarios for sustainable water use in the country. Key direct drivers of water issues in Afghanistan are: (a) agriculture and industrial development; (b) demographic shifts; and (c) increasing demand for clean drinking water and improved sanitation. Indirect drivers are made up of political drivers addressing the State, security and infrastructure; (b) social drivers including poverty; (c) economic drivers; and (d) cultural drivers including, wasteful behaviour towards water consumption and pollution; and (e) natural changes due to climate variability and change such as flooding and glacial lake outbursts and droughts. Fourth, due to a pluralistic legal framework and weak organisational capacity, Afghanistan is unable to address the freshwater drivers. Existing institutions are unable to achieve equitable access and allocation, and hinder the protection and conservation of ecosystem services (ESS) at sub-national and national level. Moreover, donor organisations are able to define and push their agenda in Afghanistan by prioritising their strategic and security interests. Water resources are formally governed by the Constitution, Civil Code, and Water Laws while informally through the local customs and principles of Sharia. Together, there is an overlapping of local customs, Sharia and modern rules of freshwater use in Afghanistan creating a legal pluralistic form of governance where: (a) at local level some principles of Sharia (included in civil codes) contradict local customs, and (b) freshwater governance is disaggregated across multiple texts and often contradict non-water-related policies.

Chapter 8 describes and analyses multilevel freshwater governance in Pakistan. It does so by looking at (1) how the various characteristics including ESS and drivers of freshwater problems are taken into account at multiple geographic levels in Pakistan; (2) how governance frameworks have evolved at multiple geographic levels in Pakistan; (3) which governance instruments address the drivers of freshwater problems at multiple geographic levels in Pakistan; (4) how legal pluralism can be observed at multiple geographic levels in Pakistan; and (5) how power and institutions influence freshwater governance frameworks at multiple geographic levels in Pakistan. Through these sub-questions, this chapter draws five conclusions.

First, water sharing among provinces in Pakistan is a highly politicised issue and is often employed as an instrument to promote the political interests of different national and local actors. The devolution of power from the federal government to the provinces after the 18th Constitutional Amendment has complicated the governance of natural resources including cases where small provinces (e.g. KP and Balochistan) blame large provinces (Punjab and Sindh) for water theft. This has created mistrust among the provinces despite the 1991 Interprovincial Water Apportionment Accord among the provinces. Second, since many people directly and indirectly depend on freshwater related BESS, the severity of water scarcity, poor water governance and climatic changes and variability can directly affect food security, livelihoods and the economy of the country. Pakistan has recently hit the _water scarce mark' (with current annual per capita water availability of 1017 m3) and according to some estimates Pakistan could _run dry' by 2025.

Third, identification of key direct and indirect drivers of freshwater challenges is the first step towards addressing water challenges in the country. Key direct drivers include: a) increasing water demand for a range of practices, and b) water and sanitation needs of the growing population (2% growth rate in 2018) and unsustainable rapid urbanisation. Key indirect drivers of the water problems are: political dynamics at the level of the state, the transboundary level; negligence about the domestic water crisis including climatic and environmental changes and legacies of colonial laws; and demographic, socioeconomic changes, and the impacts of climate change in the Kabul-Indus Basin. Fourth, the institutional architecture is characterised by legal pluralism because of the continued application of colonial laws, local customs and Sharia, as well as three levels of often different water governance frameworks (national, provincial and local). This has reduced the effectiveness of water policies and their implementation in Pakistan. Water laws and other policies created during the colonial administration were mainly to control and regulate water unilaterally and do not serve the needs of current day Pakistan. In addition, the security establishment is a key player in shaping foreign policy development especially with India and Afghanistan, both of which have transboundary water sharing mechanisms with Pakistan. Fifth, the analysis of the inclusion and distribution of principles and instruments over three major eras and across all water governance frameworks reveals weak linkages between instruments and the dimensions of inclusive and sustainable development. Economic principles and normative environmental principles are present because Pakistan has ratified various multilateral environmental and water related agreements. However, in terms of addressing environmental issues there is hardly any effective instrument to protect and preserve ecosystems by endorsing penalties or developing strict regulations or adopting subsides.

Chapter 9 has integrated elements of freshwater governance from the different geographic levels of the KRB in order to answer the question of how power and institutions influence multilevel freshwater governance in the KRB facilitate the achievement of inclusive and sustainable development. It has done so by looking at (1) how various characteristics including biodiversity, ESS and drivers of freshwater problems are taken into account at multiple levels of governance in the KRB; (2) how freshwater governance frameworks have evolved at multiple levels of governance in the KRB; (3) which governance instruments address the drivers of freshwater problems at multiple levels of governance in the KRB; (4) how legal pluralism can be observed at multiple levels of governance in the KRB; (5) how power and institutions influence water sharing at multiple level of governance in the KRB; and (6) how the current designs of the KRB multilevel institutional architecture can become consistent with the key global institutions to achieve inclusive and sustainable development. Through answering these sub-questions, the chapter draws four conclusions.

First, due to four decades of conflict in the KRB, the ideological-based insurgencies have seriously influenced the foreign policies of Afghanistan and Pakistan. These long-standing border disputes restrict both countries in initiating dialogues and solving various bilateral issues including transboundary water issues and water issues are seen through the lens of territorial sovereignty. Pakistan, being a hydro-hegemon in this case can use its powerful position to initiate dialogue for transboundary water cooperation, also by involving international players. Second, since both Afghanistan and Pakistan are signatories of many international environmental conventions and treaties (e.g. SDGs, CBD, Ramsar, HRWS), the BESS based approaches can provide an enabling environment and common ground for cost-effective transboundary cooperation including water. Hydro-energy (provisioning service) is governed at different levels in Afghanistan (federal) and Pakistan (provincial & local) and can have negative consequences for transboundary level interaction since interests and administrative issues at different levels can undermine transboundary water cooperation. Thus, new knowledge and evidence by applying the valuation of ESS can also inform the policy narrative of transboundary water cooperation by highlighting win-win scenarios.

Third, highlighting the anthropogenic and natural drivers and linking them to similar issues in both countries can result in common problem framing at the transboundary level where solutions can be discussed at a similar understanding of issues, and ultimately feed into policy making processes. Moreover, other large regional projects (e.g. CPEC, TAPI) can potentially create an opportunity for powerful actors and donor countries to play their role in bringing stability and cooperation in the KRB. Fourth, as no formal regulatory framework exists at transboundary level in the KRB, there are no goals on social and ecological inclusion. Pakistan's water goals, principles and instruments are mostly based on local priorities while Afghanistan's are heavily influenced by the donors and have some common elements with the global instruments. In this scenario, the UN Watercourses Convention can offer support by addressing legal weaknesses, providing guidance for policy coherence and for sharing water, facilitating the work of bilateral and multilateral institutions in fostering transboundary cooperation by establishing a level playing field among riparian countries so that social and environmental aspects are incorporated into the development and management of transboundary water resources.

The thesis asked how regional power politics can be changed. I believe that a better understanding of the role of water in ecosystems, livelihoods and in the economy can enable change in regional power politics by changing the interests of each party. The thesis ends with seven conclusions and related recommendations. First, current cooperation is frozen as both countries use sovereignty approaches to water. This is also because both countries define their relationship in terms of security and strategic issues and ignore the water related issues. As a consequence, data and information on the river is also securitised and secret. However, water collaboration could provide gains to both countries. Pakistan, as the regional hegemon, could invest in and promote win-win collaboration with Afghanistan (see recommendations below). Establishing a river basin organisation is critical as a first step. Second, the contested Durand line prevents water collaboration. However, the Pakhtun people living on both sides of the Durand line have similar water related customs. Discussing water sharing in line with the Watercourses Convention may be counter-productive as the border itself is disputed. However, one could perhaps address the water problems without waiting for the Durand line problem to be solved. This would require using Pakhtun customs prevalent on both sides of the river to develop common water strategies in the contested border areas. Pakistan needs to take the initiative here as it has thus far ignored these customs. Third, although there are differences with respect to biodiversity and the ecosystem services of the river (see Table 9.1 and Annex K), there are more similarities in recognising the huge social and economic value of protecting these services. In particular, if the water level falls too much, salt water intrusion can destroy agricultural land in coastal Pakistan. This thesis has made a first attempt to list these services and argues that further work on recognising the social and economic value of biodiversity and ecosystem services could enable a collaborative approach that can be more cost-effective for both. Fourth, both countries have similar drivers (see Table 9.3) and face the problem that non-water related policies dictate water use and pollution. The role of China as an investor in trade routes to Pakistan and Afghanistan can also be a major driver of water use and pollution. However, pooling knowledge and resources to address common drivers can be cost-effective. Developing an agricultural, industrial and trade policy that takes water limits into account is critical for the long-term sustainability of development policy. This can also help to develop a common stand with respect to Chinese investment. Fifth, Pakistan and Afghanistan have common elements in their domestic policies (see Tables 9.4 and 9.5) that could be scaled up in an international agreement. However, before doing this they need to address some of the internal contradictions. These are with respect to the differences between Islamic law and its focus on social and ecological dimensions and customary law and post-colonial law; the human right to water and sanitation versus the cost-recovery principle; and the differences in power as water governance is centralised in Afghanistan and is devolved to state level in Pakistan making, for instance, collaboration on dams is difficult. Furthermore, policies in both countries tend to be incomplete and there need to be appropriate policy mixes to ensure that policies can be implemented. Sixth, the research reveals that resource limits seriously hamper the operationalisation and implementation of policy, its monitoring and enforcement. At the same time, instruments that limit the potential for corruption are often deliberately sabotaged by political actors. This leads to a short-term focus on economic growth and dependence on aid agencies. I argue that a socially and ecologically inclusive system has a higher chance of being sustainable in the long-term. A focus on short-term economic growth is more likely to lead to the externalisation of social and ecological impacts with long-term impact on security and livelihoods. Seeking out collaborative, locally developed, cost-effective solutions is critical for enhancing livelihoods and wellbeing. Finally, this thesis has tried to gather and collate the primary and secondary evidence to make a prima facie case for the need to prioritise multi-level, transboundary water issues in the region, especially given the impending water crises and the fact that the Kabul is a glacier fed river extremely vulnerable to the impacts of climate change. However, there is an absence of collective and integrated knowledge based on data and experiences as well as dialogue at different levels of cross-border governance to be able to craft and refine each of the above recommendations in more detail. The need for mobilising cross-border awareness in schools, colleges, universities and institutions as well as discussion between civil society and governments is essential to address the long-standing water-sharing issues. Thus a freshwater governance framework can best contribute to inclusive and sustainable development when constructed on concrete basis.

SAMENVATTING

Probleem Definitie

Hoewel grensoverschrijdende en onderling op meerdere niveaus gedeelde waterbronnen vaak worden gezien als een bron van conflicten tussen landen en regio's, suggereren recente studies dat ze ook een katalysator kunnen zijn voor samenwerking tussen oeverstaten. Een dergelijke geïnstitutionaliseerde samenwerking kan de oeverstaten ten goede komen doordat het een grotere sociaaleconomische ontwikkeling kan bevorderen. Ondanks het bestaan van instituties op grensoverschrijdend niveau, worden sommige stroomgebieden beter beheerd dan andere. Dit kan zijn omdat, zoals vanuit het perspectief van de hydro-hegemonie betoogd wordt, asymmetrische machtsverhoudingen tussen volkeren en landen de instituties op verschillende wijzen vormgeven in de verschillende stroomgebieden. Als macht echter zo een dominante rol heeft, is de vraag waarom er zoveel instituties zijn in andere delen van de wereld (zoals tussen Canada en de VS en tussen Europese landen) die wel werken? Neo-institutionalisten betogen dat machtsconstellaties variëren van kwestie tot kwestie; er is vaak wel degelijk ruimte om eerlijke en effectieve instituties te ontwikkelen die - eenmaal ontwikkeld - ook macht kunnen vormen en beteugelen.

Ineffectieve instituties en slechte bestuurspraktijken creëren verschillende uitdagingen voor het waterbeheer in het Zuiden. Dit kan nog worden verergerd door wereldwijde veranderingen in sociale, politieke, ecologische en economische systemen. Deze uitdagingen kunnen worden onderverdeeld in twee brede categorieën: (1) stromingsgerelateerde (d.w.z. betreffende kwesties van waterkwaliteit, waterkwantiteit en gerelateerde ecosysteemdiensten); en (2) bestuurs- en beheersgerelateerde (d.w.z. betreffende kwesties van macht, jurisdictie, coördinatie, principes en instrumenten op meerdere bestuursniveaus). De wetenschappelijke literatuur geeft aan dat een analytisch kader voor grensoverschrijdend waterbestuur essentieel is om de uitdagingen van verlies van waterkwaliteit en kwantiteit, de politisering en pacificatie (dat wil zeggen iets tot veiligheidsvraagstuk maken ten einde buitengewone maatregelen te kunnen nemen) van het bestuur van water het hoofd te kunnen bieden. Toch zijn er vier belangrijke hiaten in de wetenschappelijke literatuur over grensoverschrijdend waterbestuur. Ten eerste integreert deze literatuur zelden (a) een aanpak op meerdere niveaus, (b) een institutionele benadering (c) een inclusieve ontwikkelingsbenadering, of (d) het meewegen van de implicaties van het gebruik van verschillende soorten water en de daarbij behorende ecosysteemdiensten.

Vraag

Vandaar Derhalve probeert dit proefschrift de volgende hoofdonderzoeksvragen te beantwoorden:

Hoe kunnen regionale waterpolitiek en waterinstituties worden omgevormd door het ontwikkelen van inclusieve ontwikkelingsdoelstellingen, en hoe kunnen deze instituties relaties opbouwen met andere sectoren dan de watersector bij het aanpakken van kwesties als waterkwaliteit, waterkwantiteit en klimaatsverandering?

De volgende sub-vragen worden behandeld:

1. Hoe kunnen de concepten van biodiversiteit en ESS opgenomen worden in een kader dat beoogt de effectiviteit van instituties en de rol van macht in het bestuur van grensoverschrijdende waterbronnen te analyseren?
2. Welke principes en instrumenten richten zich op de oorzaken/*drivers* van zoetwaterproblemen in grensoverschrijdende rivieren op meerdere geografische niveaus?
3. Hoe beïnvloedt juridisch pluralisme in meervoudig gelaagd waterbestuur de grensoverschrijdende samenwerking?
4. Hoe beïnvloeden machtspolitiek en instituties waterbestuur in grensoverschrijdende rivieren op meerdere geografische niveaus?

Deze vragen worden onderzocht aan de hand van de situatie van de Kabul rivier, die door Afghanistan en Pakistan stroomt. Ik koos voor het Kabul stroomgebied omdat:

a) Het uniek is, want beide landen leveren water en liggen zowel boven- als benedenstrooms en in beide landen is sprake van een conflictsituatie; b) Afghanistan nu eenzijdig de rivier wil ontwikkelen en het aandeel in watergebruik wil vergroten voor haar sociaaleconomische ontwikkeling, wat tot conflict leidt met Pakistan; Pakistan werpt tegen, dat het de eerste gebruiker van de rivier was en dus historische rechten heeft c) dit stroomgebied is bovendien al het dichtst bevolkte stroomgebied van Afghanistan en Pakistan met een bevolkingsdichtheid van 93 personen per km^2. Het huisvest naar schatting 7,185,000 personen. Het is bovendien kwetsbaar voor klimaatverandering, omdat deze rivier gevoed wordt door gletsjers en sneeuw. Dit in tegenstelling tot andere rivieren, die voornamelijk door regen van water worden voorzien; d) Er zijn weinig formele instituties. Bovendien is er van academische zijde weinig belangstelling voor dit stroomgebied ten gevolge van het langdurig karakter van het conflict. De gewoonten van de stammen in het gebied zijn moeilijk te begrijpen voor buitenstaanders. Mijn persoonlijke affiniteit met het gebied, affiliatie met de stamgewoonten en een volledige beheersing van de lokale talen maken het voor mij tot een geschikte case studie.

Hoofdstuk 2 presenteert de methodologie en het analytisch kader. Dit hoofdstuk beargumenteert dat de analyse van grensoverschrijdende water kwesties verbeterd kan worden door gebruik te maken van het zogenoemde multi-level governance (MLG) analyse kader. Ik koos voor een kader dat iedere bestuurslaag onafhankelijk behandelt, een zogenoemd _type 1 MLG analyse kader'. Omdat machtspolitiek water instituties beïnvloedt - en vice versa -leg ik kort de basisbeginselen van de machtspolitiek en van de institutionele theorie uit, en hun onderlinge relatie. Omdat verschillende benaderingen van waterbestuur tegelijk ontwikkeld zijn, en op verschillende lagen van bestuur een veelvoud van wetten voort hebben gebracht, onderzoek ik of de regels consistent of tegenstrijdig zijn in een gegeven jurisdictie. Ten slotte, heb ik ervoor gekozen om inclusieve ontwikkeling als een doel voor water bestuur te contrasteren met duurzame ontwikkeling. Dit doe ik omdat ik de sociaal/relationele componenten wil benadrukken (met een focus op hoe macht kan worden overgeheveld naar de lokale bevolking door, onder andere, het hanteren van bepaalde principes en instrumenten). Daarnaast wil ik ecologische componenten benadrukken, om de ecosysteemdiensten van water te verbeteren, in tegenstelling tot te focussen op hoe economische groei kan worden versterkt. En omdat waterrecht een aantal procedurele principes heeft, die moeilijk zijn te classificeren als sociaal of ecologisch, heb ik deze geclusterd onder politieke principes.

In termen van methodologie, heb ik de literatuur over hydro-hegemonie (2004-2018) bestudeerd, en de relatie tussen macht en instituties (1970 - 2018). Ik bestudeerde vijf belangrijke internationale beleids- en juridische documenten relevant voor dit proefschrift op mondiaal niveau met betrekking tot water bestuur in de casus regio van Afghanistan en Pakistan, 15 op grensoverschrijdend niveau (3 koloniale en 12 postkoloniale); 22 op nationaal en lokaal niveau in Afghanistan (2 pre-koloniale, 1 koloniaal, en 19 post-koloniale), en 25 in Pakistan (2 pre-koloniale, 4 koloniale, en 19 post-koloniale). Ik voerde een vergelijkende meervoudig gelaagde case studie uit van Afghanistan en Pakistan en het grensoverschrijdende Kabul stroomgebied uit, waar in totaal 70 interviews (30 in Afghanistan en 40 in Pakistan) en twee Focus Group Discussies (FGDs) werden uitgevoerd.

Ik maak gebruik van een aangepaste versie Oran Young's institutionele analyse model, zoals dit ook in de Institutional Dimensions of Global Environmental Change is gehanteerd. In dit kader is het allereerst van belang om de context en de drijvende krachten (drivers) te begrijpen, die leiden tot uitdagingen voor waterbestuur in het Kabul stroomgebied. Ik analyseer hoe macht de bestaande grensoverschrijdende, nationale en lokale instituties in het stroomgebied gevormd heeft. Vervolgens heb ik de belangrijkste principes en instrumenten geïdentificeerd die gericht zijn op het veranderen van het gedrag van mensen met betrekking tot het gebruik van water (waterkwantiteit) en tot vervuiling (waterkwaliteit). Daarna analyseer ik of deze instrumenten – gegeven de context en de drivers - het menselijk gedrag op zo een manier veranderen dat bijgedragen wordt aan sociale en ecologische inclusie, en ook de relationele aspecten veranderen (m.a.w. is er sprake van inclusieve onwikkeling?). Ten slotte ontwikkel ik aanbevelingen op basis van de beoordeling van welke principes en instrumenten werken en welke niet. Ik bespreek hoe de institutionele aanpak kan worden verbeterd en of deze verbeterde aanpak de onderliggende machtspolitiek zouden kunnen beïnvloeden en daamee tot de ontwikkeling van wederzijds bevredigende uitkomsten zou kunnen leiden.

Hoofdstuk 3 gaat in op de manier waarop realistische en institutionele benaderingen omgaan met grensoverschrijdend waterbestuur op meerdere niveaus. De literatuur over realisme (bijvoorbeeld hydro-hegemonie) richt zich over het algemeen op de rol van de macht van de staat in internationale samenwerking, terwijl institutionalisme (d.w.z. hydro-institutionalisme) zich over het algemeen richt op de rol van instituties in internationale samenwerking. Machtstheorieën analyseren hoe macht grensoverschrijdende instituties van waterbestuur beïnvloedt door: actoren toe te voegen of uit te sluiten; het opnemen of uitsluiten van verschillende kwesties en daarmee het vormgeven van de agenda of voorkeuren van verschillende actoren die betrokken zijn bij het politieke / onderhandelingsproces. Ze analyseren of er machtssymmetrie is tussen actoren in relatie tot (i) geografie; (ii) materiele factoren; (iii) onderhandelen; en (iv) op de kracht van ideeën. Onder bepaalde omstandigheden kan macht resulteren in fragiele en niet duurzame instituties voor waterbestuur. Institutionele theorieën betogen dat samenwerking, en uiteindelijk zelfs vrede, toch kan ontstaan door de ontwikkeling van normen die kunnen fungeren als een intermediaire factor tussen de machtsstructuren van het (inter)nationale systeem. Vanuit institutionele theorie wordt gesteld dat samenwerking zelfs mogelijk is in een anarchistisch systeem, en wel in de vorm van samenwerking op bijvoorbeeld het gebied van handel, mensenrechten en collectieve veiligheid. Een dergelijke samenwerking kan ontstaan omdat verschillende kwesties verschillende machtsconstellaties hebben.

Hydro Hegemonie (HH) en Hydro Institutionalisme (HI) (de term die ik geef aan een institutionele benadering van water) zijn twee afzonderlijke theoretische benaderingen van grensoverschrijdende waterconflicten en samenwerking. Ze hoeven niet noodzakelijkerwijs te worden geïntegreerd in één enkel interpretatiekader, maar ze vullen elkaar aan en kunnen goed worden gebruikt om verschillende aspecten van de onderzoeksvraag en deelvragen te beantwoorden. In werkelijkheid is er een complex samenspel van macht en instituties, waar macht de uitkomst van grensoverschrijdend waterbestuur beïnvloedt wanneer er asymmetrische machtsverhoudingen zijn tussen aangrenzende landen en gebruikers. Formele en informele instituties beperken op hun beurt echter weer de rol van machtspolitiek in het bestuur van water (een hybride benadering). Verder kunnen er (a) onopgeloste historische problemen zijn in niet-water gerelateerde sectoren (bijvoorbeeld grensaangelegenheden) en (b) een gebrek aan wetenschappelijke en maatschappelijke informatie die institutionele opbouw in de watersector kan belemmeren. Daarom behandelt dit proefschrift de rol die macht speelt bij het betrekken van of uitsluiten van actoren en het al dan niet agenderen van bepaalde problemen. Het proefschrift bespreekt hoe bestaande instituties kunnen worden verbeterd. Het probeert aanvullende wetenschappelijke en maatschappelijke inzichten te bieden die wellicht ook de percepties van hen betrokken in de onderhandelingen rondom grensoverschrijdend waterbestuur zouden kunnen veranderen. Waar de meeste van deze theorieën zich alleen richten op het niveau van de natiestaat, kijk ik expliciet naar alle verschillende niveaus.

Hoofdstuk 4 bespreekt de state-of-the-art kennis van zoetwatersystemen in de context van een stroomgebied. Ik identificeer eerst de belangrijkste directe en indirecte drivers van invloed op de kwaliteit, kwantiteit en daarmee samenhangede ecosysteemdiensten van de waterlopen. Onder de directe drivers behandel ik de ontwikkeling van de landbouw, veeteelt, de mijnbouwsector en het watergebruik voor energievoorziening, de industrie en dienstensector, gemeentelijke drinkwater- en sanitaire voorzieningen, en demografische verschuivingen ten gevolge van migratie, bevolkingsgroei, bevolkingsdichtheid, en verstedelijking. Onder indirecte drivers vallen de politieke dynamiek tussen staten / provincies, culturele en etnische verhoudingen tot water; beleid uit andere sectoren dan de watersector dat van invloed is op water, de drang naar economische groei, armoede, technologische vooruitgang (waaronder intensivering van de landbouw), internationale handel; klimaatvariabiliteit en verandering; aardbevingen. De drivers kunnen natuurlijk factoren zijn (bijvoorbeeld tektonische bewegingen, klimaat- en weervariabiliteit) en antropogeen (bijvoorbeeld klimaatverandering).

Omdat een belangrijke toevoeging van dit proefschrift de ecosysteemdiensten van water omvat, gaat dit hoofdstuk in op de verschillende ecosysteemdiensten (ESS) van zoet water. Deze omvatten de ondersteuning (bijvoorbeeld de rol van water bij erosiebestrijding en recycling van voedingsstoffen), regulering (bijv. klimaatregulering, hydrologische regulering), bevoorrading (bijv. drinkwater, water voor voedsel) en culturele (bv. geestelijke, recreatieve) diensten van water. De verschillende soorten zoetwater (d.w.z. atmosferisch water, blauw oppervlaktewater, blauw grondwater, groen water, grijs water, zwart water en wit bevroren water of sneeuw) vervullen verschillende ecosysteemdiensten met verschillende bijdragen tot het welzijn van de mens (d.w.z. keuzevrijheid door veiligheid, goed leven, gezondheid en goede sociale relaties) op verschillende bestuursniveaus. Een belangrijke hypothese van dit hoofdstuk is dat als in discussies over grensoverschrijdende stroomgebieden ook andere waterkwesties in het stroomgebied zouden worden meegenomen, dit een betere samenwerking tussen actoren in aangrenzend landen mogelijk zou maken. Op dezelfde wijze zou een beter begrip van de ecosysteemdiensten van water, en begrip van de wijze waarop niet-water specifieke kwesties problemen kunnen veroorzaken in de watersector, deze samenwerking kunnen verbeteren.

Hoofdstuk 5 onderzoekt vijf belangrijke instituties op mondiaal niveau (dat wil zeggen de gebruikelijke internationale wetgeving op het gebied van water, het Verdrag inzake de bescherming en het gebruik van grensoverschrijdende waterlopen en internationale meren van 1992 (oorspronkelijk EU wetgeving, nu opengesteld voor mondiale participatie); de UN Watercourses Conventie uit 1997, het mensenrecht op water en sanitatie en de Duurzame Ontwikkelingsdoelen (SDGS) van Agenda 2030). De Duurzame Ontwikkelingsdoelen vragen om integratie van de sociale en ecologische aspecten in het streven naar ontwikkeling. Dit hoofdstuk identificeert de volgende beginselen en instrumenten uit de bovenstaande documenten die relevant zijn voor dit proefschrift: (i) politieke (bijv. de principes over soevereiniteit, informatie-uitwisseling, verplichting tot samenwerking, vreedzame oplossing van geschillen, kennisgeving van noodsituaties en kennisgeving van geplande maatregelen); (ii) sociaal-relationele (inclusief het mensenrecht op water en sanitaire voorzieningen, publieke participatie, bewustmaking van het publiek en onderwijs, toegang van het publiek tot informatie, prioriteit van gebruik, rechten van vrouwen, jongeren en inheemse volkeren, capaciteitsopbouw, billijk en redelijk gebruik, intergenerationele gelijkheid, armoedebestrijding); en (iii) ecologische (d.w.z. beoordeling van het milieueffect, monitoring, voorkoming van vervuiling, bescherming en behoud van ecosystemen, betaling van vervuilers en bescherming van waterwingebieden en lozingszones). Verder identificeert het de verplichting om een organisatie voor het bestuur van het stroomgebied in het leven te roepen en om geschillenbeslechtings-mechanismen te gebruiken voor het oplossen van problemen.

Hoofdstuk 6 beschrijft en analyseert de relaties tussen Afghanistan en Pakistan in de KRB (het Kabul stroomgebied) om de vraag te beantwoorden hoe macht en instituties de internationale betrekkingen tussen de twee landen beïnvloeden en hoe ze een inclusieve en duurzame ontwikkeling belemmeren of hieraan bijdragen. Het doet dit door i) de algemene politieke context te bespreken, ii) essentiële ESS inclusief biodiversiteit te identificeren, iii) directe en indirecte drivers in de zoetwaterhuishouding te bespreken, en iv) een gedetailleerd overzicht te geven van de ontwikkeling van de instituties op grensoverschrijdend niveau door zowel historische als recente bestuurspraktijken te analyseren (inclusief de pre-koloniale overeenkomsten).

Dit hoofdstuk trekt vier conclusies. Ten eerste is er geen formeel regelgevend kader om de waterbronnen in de KRB op billijke wijze te delen tussen Afghanistan en Pakistan, en daarom doet elk land wat het wil binnen dit anarchistisch systeem. Hoewel Pakistan een groot deel van de waterbronnen in de KRB gebruikt, maakt het zich zorgen over de toekomstige toegang tot water, aangezien Afghanistan nieuwe dammen en irrigatie-infrastructuur wil ontwikkelen. Het Pakistaanse establishment is van mening dat deze projecten de economie en infrastructuur van de provincie Khyber Pakhtunkhwa (KP) van Pakistan ernstig zouden schade. Ten tweede kan de erkenning van de variëteit van Biodiversiteit en EcoSysteemServices (BESS) op grensoverschrijdend niveau van de Kabul rivier de basis vormen voor een gelijkluidende probleemdefinitie, en dit kan ook de voordelen van BESS voor lokaal levensonderhoud benadrukken. Op dezelfde manier omvatten de regulerende diensten van de rivier de Kabul een aaneengesloten en verbeterde aanvulling van het grondwater door het verwijderen van nutriënten en verontreinigende stoffen, de klimaatregulering door middel van koolstofsekwestratie in de uiterwaarden en de omliggende bossen.

Ten derde kan het erkennen en identificeren van overeenkomsten in de belangrijkste directe en indirecte drivers een eerste stap zijn in de aanpak van grensoverschrijdende waterkwesties in de KRB. Belangrijke directe drivers zijn: (a) ontwikkeling van de landbouw; (b) industrie; en (c) demografische verschuivingen. De belangrijkste indirecte oorzaken van de zoetwaterproblemen in de KRB op grensoverschrijdend niveau zijn: (a) politieke dynamiek tussen staten; (b) cultuur en etnische elementen; (c) niet-watergerelateerd beleid; (d) economie; (e) armoede; (f) technologisch; (g) internationale handel; en (h) klimaatvariabiliteit en klimaatverandering. Met name de politieke dynamiek tussen staten is een belangrijke grensoverschrijdende driver, daar Afghanistan de Durand-lijn niet als een internationaal erkende grens accepteert. De kwestie van de Durand-lijn vraagt dringend om een oplossing. Hiervoor is een fact finding missie nodig, waarin zowel de betrokken autoriteiten als en internationale juristen zijn opgenomen. De politieke context van extremisme, zetbazen van de Taliban- en Pakhtunistan hindert de samenwerking en wordt beïnvloed door aan water gerelateerde kwesties tussen de twee buren. Ten vierde heeft de afwezigheid van formele instituties ervoor gezorgd dat machtspolitiek de overhand heeft gekregen tussen de twee landen. Hoewel Afghanistan een geografisch voordeel heeft, heeft het minder invloed ten gevolge van het gebrek aan capaciteit en middelen om waterbronnen te exploiteren. Pakistan beschikt over meer materieel, onderhandelings- en ideëel vermogen. Daar Pakistan ongeveer 90% van het water van het KRB verbruikt is het minder gemotiveerd om de status quo te wijzigen. Vanwege deze machtsasymmetrieën heeft Pakistan een op veiligheid gerichte agenda voor het buitenlandse beleid ten aanzien van Afghanistan aangenomen.

Hoofdstuk 7 beschrijft en analyseert zoetwaterbeheer op meerdere niveaus in Afghanistan. Het doet dit door te kijken naar (1) hoe verschillende kenmerken en drivers van zoetwaterproblemen uitwerken op de verschillende geografische niveaus in Afghanistan; (2) hoe kaders voor zoetwaterbeheer zich op meerdere geografische niveaus in Afghanistan hebben ontwikkeld; (3) welke instrumenten de drivers van zoetwaterproblemen op meerdere geografische niveaus in Afghanistan aanpakken; (4) hoe juridisch pluralisme op meerdere geografische niveaus in Afghanistan kan worden waargenomen; en (5) hoe macht en instituties de kaders voor het bestuur van zoetwater beïnvloeden. Dit leidt tot vier conclusies.

Ten eerste hebben wijdverspreide armoede, zwakke institutionele en menselijke capaciteit als gevolg van langdurige conflicten en instabiliteit en een gebrek aan kennis en capaciteit om watervoorraden te beheren, bijgedragen aan de uitdagingen ten aanzien van water in Afghanistan. Omdat Afghanistan een binnenstaat is, en geografisch gesproken een belangrijke locatie is, kenmerkt haar historie zich door geschillen met de buurlanden over het water uit de bergrivieren, de belangrijkste waterbron van Afghanistan. Bovendien heeft de burgeroorlog en de Sovjet invasie ertoe geleid dat veel van de institutionele kennis met betrekking tot watervoorraden verloren is gegaan. De meeste apparatuur voor watermonitoring van het land werd vernietigd en de capaciteiten van waterwetenschappers stagneerden. Ten tweede, omdat een groot deel van de bevolking in Afghanistan direct en indirect afhankelijk is van een verscheidenheid aan BESS van Kabul River, kan degradatie van deze vitale diensten als gevolg van variabiliteit van de watertoevoer negatieve gevolgen hebben. Een recente classificatie verdeelt Afghanistan in 15 kleinere eco-regio's waarvan slechts twee relatief stabiel en intact zijn. De soorten rijkdom in alle eco-regio's is aanzienlijk verminderd door een combinatie van overbegrazing, het verzamelen van brandhout en de exploitatie van weidegebieden. Naar schatting 70-80% van de bevolking van het land is afhankelijk van landbouw, veeteelt en ambachtelijke mijnbouw om in hun levensonderhoud te kunnen voorzien. Het land moet een beroep doen op deze mogelijkheden om zijn positie onder aan de ranglijst van de Human Development Index te verbeteren.

Ten derde is het van essentieel belang om de belangrijkste directe en indirecte drivers van waterkwesties te identificeren en te begrijpen. Dit om in de context van veranderend geopolitiek en klimaatscenario's op wetenschappelijke inzichten gebaseerd beleid te kunnen formuleren voor duurzaam watergebruik in het land. De belangrijkste directe oorzaken van de waterproblemen in Afghanistan zijn: (a) landbouw en industriële ontwikkeling; (b) demografische verschuivingen; (c) toenemende vraag naar schoon drinkwater en verbeterde sanitaire voorzieningen; en (d) veranderingen ten gevolge van klimaatvariabiliteit en klimaatverandering (zoals overstromingen, ontstaan van gletsjermeren en perioden van droogte). Indirecte drivers zijn o.a. de politieke drivers op het gebied van de staat, de veiligheid en infrastructuur; (b) sociale factoren, waaronder armoede; (c) economische factoren; en (d) culturele drijfveren, waaronder verspillend waterverbruik en vervuiling. Ten vierde kan Afghanistan, vanwege een pluralistisch juridisch kader en een zwakke organisatorische capaciteit, de zoetwaterbestuurders niet aanspreken. Bestaande instituties zijn niet in staat gelijke toegang tot en toedeling van water te bewerkstelligen, en belemmeren de bescherming en het behoud van ecosysteemdiensten (ESS) op sub-nationaal en nationaal niveau. Bovendien kunnen donororganisaties hun eigen agenda in Afghanistan definiëren en pushen door prioriteit te geven aan hun strategische en veiligheidsbelangen. Water valt formeel onder de grondwet, het burgerlijk wetboek en de waterwetten, terwijl het informeel door lokale gebruiken en principes van de sharia worden bestuurd. Daarmee ontstaat een overlap van lokale gebruiken, sharia en moderne regels voor zoetwatergebruik in Afghanistan. Dit creëert een juridisch pluralistische vorm van bestuur, waarbij: (a) op lokaal niveau sommige principes van de sharia (vervat in het burgerlijk recht) in tegenspraak zijn met de lokale gebruiken, en (b) regels voor zoetwaterbestuur zijn opgesplitst in meerdere teksten, die vaak in tegenspraak zijn met niet-water gerelateerd beleid.

Hoofdstuk 8 beschrijft en analyseert zoetwaterbestuur op meerdere niveaus in Pakistan. Het doet dit door te kijken naar (1) hoe rekening wordt gehouden met de verschillende kenmerken, waaronder ESS en oorzaken van zoetwaterproblemen, op meerdere geografische niveaus; (2) hoe bestuurlijke kaders zijn geëvolueerd op meerdere geografische niveaus in Pakistan; (3) welke bestuursinstrumenten de drivers van zoetwaterproblemen op meerdere geografische niveaus in Pakistan aanpakken; (4) hoe juridisch pluralisme op meerdere geografische niveaus in Pakistan kan worden waargenomen; en (5) hoe macht en instituties de kaders voor zoetwater bestuur beïnvloeden op meerdere geografische niveaus. Dit hoofdstuk trekt vijf conclusies.

Ten eerste is het delen van water tussen provincies in Pakistan een sterk gepolitiseerde kwestie en wordt het vaak gebruikt als een instrument om de politieke belangen van verschillende nationale en lokale actoren te bevorderen. De decentralisatie van macht van de federale overheid naar de provincies na het 18e constitutionele amendement heeft het beheer van natuurlijke hulpbronnen gecompliceerd, waaronder gevallen waarin kleine provincies (bijvoorbeeld KP en Balochistan) grote provincies (Punjab en Sindh) de schuld geven van waterdiefstal. Dit heeft geleid tot wantrouwen bij de provincies, ondanks het Interprovinciale Water Akkoord uit 1991 tussen de provincies. Ten tweede, omdat een groot deel van de bevolking direct en indirect afhankelijk is van zoetwater gerelateerde BESS, kunnen de ernst van waterschaarste, slecht waterbeheer en klimaatvariabiliteit en klimaatverandering direct van invloed zijn op de voedselzekerheid, de levensstandaard en de economie van het land. Een groot deel van de bevolking vertrouwt direct en indirect op de BESS die rechtstreeks afhankelijk is van zoetwatervoorziening. Pakistan heeft onlangs het 'waterschaarste niveau' (met de huidige jaarlijkse waterbeschikbaarheid per inwoner van 1017 m3) bereikt en volgens sommige schattingen zou Pakistan in 2025 'droog kunnen vallen'.

Ten derde is het identificeren van de belangrijkste directe en indirecte drivers voor zoetwatergerelateerde uitdagingen de eerste stap naar het aanpakken van deze uitdagingen. Belangrijke directe drivers zijn: a) de toenemende vraag naar water voor een reeks van activiteiten en b) water- en sanitatiebehoeften van de groeiende bevolking (groeipercentage van 2% in 2018) en niet-duurzame en snelle verstedelijking. Belangrijke indirecte oorzaken van de waterproblemen zijn: politieke dynamiek op het niveau van de staat en tussen de buurlanden; veronachtzaming van de binnenlandse watercrisis, (resultante van klimatologische - en milieuveranderingen en de erfenis van de koloniale wetten); en demografische, sociaaleconomische veranderingen en de gevolgen van klimaatverandering in het stroomgebied van Kaboel-Indus. Ten vierde wordt de institutionele architectuur gekenmerkt door juridisch pluralisme ten gevolge van de voortdurende toepassing van koloniale wetten, lokale gewoonten en sharia, evenals drie niveaus met vaak verschillende kaders voor waterbeheer (nationaal, provinciaal en lokaal). Dit heeft in Pakistan de effectiviteit van het waterbeleid en de implementatie hiervan verminderd. Waterwetten en andere beleidsmaatregelen die tijdens het koloniale bestuur zijn ingesteld, dienden voornamelijk om water eenzijdig te beheren en te reguleren en zijn slecht toepasbaar op de behoeften van het huidige Pakistan. Bovendien is de veiligheidselite een belangrijke speler bij het vormgeven van de ontwikkeling van het buitenlands beleid, vooral ten aanzien van India en Afghanistan, die beide mechanismen hebben voor het delen van het water met Pakistan. Ten vijfde blijkt uit de analyse van de mate waarin principes en instrumenten voor het waterbestuur voorkomen (zowel in drie verschillende tijdvakken, als in de kaders voor waterbestuur op verschillende niveaus), dat er maar weinig verband is tussen deze instrumenten en inclusieve en duurzame ontwikkelingsdoelstellingen. Economische principes en normatieve milieubeginselen zijn aanwezig omdat Pakistan verschillende multilaterale milieuovereenkomsten heeft geratificeerd. Wat de aanpak van milieukwesties betreft, is er echter nauwelijks een doeltreffend instrument om ecosystemen te beschermen en te behouden of om strikte regelgeving te ontwikkelen, door middel van toepassing van sancties of het toekennen van subsidies.

Hoofdstuk 9 heeft de elementen van zoetwaterbestuur uit de verschillende geografische niveaus van het KRB geïntegreerd om een antwoord te bieden op de vraag hoe macht en instituties zoetwaterbestuur op meerdere niveaus in het KRB kunnen beïnvloeden om inclusieve en duurzame ontwikkeling te bevorderen. Dit is gedaan door te kijken naar (1) hoe verschillende kenmerken, waaronder biodiversiteit, ESS en drivers van zoetwaterproblemen, in aanmerking worden genomen op meerdere bestuursniveaus in de KRB; (2) hoe de kaders voor zoetwaterbestuur zijn geëvolueerd op meerdere bestuursniveaus in de KRB; (3) welke beheersinstrumenten de drivers van zoetwaterproblemen op meerdere bestuursniveaus in het KRB aanpakken; (4) hoe juridisch pluralisme kan worden waargenomen op meerdere bestuursniveaus in het KRB; (5) hoe macht en instituties het delen van water op meerdere bestuursniveaus in het KRB beïnvloeden; en (6) hoe de huidige ontwerpen van het KRB institutionele architectuur op meerdere niveaus consistent kan worden met de belangrijkste mondiale instituties om inclusieve en duurzame ontwikkeling te bereiken. Dit hoofdstuk trekt vier conclusies.

Ten eerste hebben, vanwege vier decennia van conflicten in het KRB, de op ideologie gebaseerde opstanden het buitenlandse beleid van Afghanistan en Pakistan ernstig beïnvloed. Deze langdurige grensconflicten beperken beide landen bij het initiëren van dialogen en bij het oplossen van verschillende bilaterale kwesties, waaronder grensoverschrijdende waterproblemen. Waterproblemen worden gezien door de lens van territoriale soevereiniteit. Pakistan, dat in dit geval een hydro-hegemoon is, kan zijn krachtige positie gebruiken om de dialoog over grensoverschrijdende samenwerking op het gebied van water op te starten, ook door internationale spelers te betrekken. Ten tweede, aangezien zowel Afghanistan als Pakistan vele internationale milieuconventies en -verdragen hebben ondertekend (bijvoorbeeld SDG's, CBD, Ramsar, HRWS), kunnen de op BESS gebaseerde benaderingen zorgen voor een gunstig gespreksklimaat en een gemeenschappelijke basis voor kosteneffectieve grensoverschrijdende samenwerking, waaronder op het gebied van water. Het opwekken van waterkracht wordt in Afghanistan op een ander bestuursniveau beheerd dan in Pakistan (respectievelijk nationaal versus provinciaal en lokaal). Dit kan negatieve gevolgen hebben voor grensoverschrijdende interactie, omdat belangen en administratieve kwesties op verschillende niveaus grensoverschrijdende samenwerking op het gebied van water kunnen ondermijnen. Nieuwe wetenschappelijke kennis voor het aantonen van de meerwaarde van een grensoverschrijdende benadering van BESS kan dus ook de beleidsargumentatie van grensoverschrijdende samenwerking in het waterbestuur informeren, door de win-winscenario's te benadrukken.

Ten derde kan het benadrukken van de overeenkomsten tussen de antropogene en natuurlijke factoren in de beide landen, en het koppelen van vraagstukken die op vergelijkbare wijze spelen in beide landen, resulteren in een gemeenschappelijke probleemstelling op grensoverschrijdend niveau.

Dan kunnen oplossingen worden besproken op basis van een gedeeld inzicht in kwesties die voor de beide landen op vergelijkbare wijze spelen. Dit inzicht kan uiteindelijk worden gebruikt voor beleidsvormingsprocessen. Bovendien kunnen andere grote regionale projecten (bijvoorbeeld CPEC, TAPI) potentiële machthebbers en donorlanden een kans bieden om hun rol te spelen bij het brengen van stabiliteit en samenwerking in het KRB. Ten vierde zijn er geen doelstellingen op het gebied van sociale en ecologische inclusie, omdat er geen formeel regelgevend kader bestaat op grensoverschrijdend niveau in het KRB. De doelstellingen, principes en instrumenten met betrekking tot water in Pakistan zijn meestal gebaseerd op lokale prioriteiten, terwijl Afghanistan sterk beïnvloed wordt door de donoren en daardoor enkele elementen gemeenschappelijk heeft met de mondiale instrumenten. In dit scenario kan het UNWC steun bieden door juridische tekortkomingen aan te pakken, richtlijnen te bieden voor beleidscoherentie. Het UNWC kan het werk van bilaterale en multilaterale instellingen vergemakkelijken door grensoverschrijdende gelijkwaardige samenwerking te bevorderen door een gelijk speelveld te creëren tussen oeverstaten en sociale en milieuoverwegingen te integreren in het beheer en de ontwikkeling van internationale waterlopen.

Dit proefschrift stelde de vraag hoe regionale machtspolitiek kan worden veranderd. Ik geloof dat een beter begrip van de rol van water in ecosystemen, het voorzien in het levensonderhoud en in de economie verandering in de regionale machtspolitiek mogelijk kan maken door de belangen van elke partij te veranderen. Het proefschrift eindigt met zeven conclusies en bijbehorende aanbevelingen. Ten eerste is de huidige samenwerking bevroren omdat beide landen gebruikmaken van soevereiniteitsbenaderingen voor water. Dit komt ook omdat beide landen hun relatie definiëren in termen van veiligheid en strategische kwesties en de water gerelateerde kwesties negeren. Als gevolg hiervan worden gegevens en informatie over de rivier ook geheim gehouden. Samenwerking op het gebied van water kan echter voordelen opleveren voor beide landen. Pakistan zou, als regionale hegemon, kunnen investeren in een win-win-samenwerking met Afghanistan (zie onderstaande aanbevelingen). Het opzetten van een organisatie voor het beheer van het stroomgebied is als eerste stap van cruciaal belang. Ten tweede verhindert de betwiste Durand-lijn samenwerking op het gebied van water. De Pakhtun-bevolking die aan beide zijden van de Durand-lijn woont, heeft echter vergelijkbare water gerelateerde gebruiken. Het bespreken van het verdelen van water in overeenstemming met de Watercourses Conventie kan contraproductief zijn, aangezien de grens zelf wordt betwist. Men zou echter de waterproblemen kunnen aanpakken zonder te wachten tot het Durand-lijn probleem is opgelost. Dit vereist het gebruik van Pakhtun-gewoonten die aan weerszijden van de rivier voorkomen om gemeenschappelijke waterstrategieën te ontwikkelen in de betwiste grensgebieden. Pakistan moet hier het initiatief nemen omdat het tot dusverre deze gewoonten heeft genegeerd. Ten derde, hoewel er verschillen zijn met betrekking tot de biodiversiteit en de ecosysteemdiensten van de rivier (zie tabel 9.1 en 9.2), zijn er overeenkomsten in het erkennen van de enorme sociale en economische waarde van bescherming van deze diensten. In het bijzonder, als het waterniveau te veel daalt, kan verzilting landbouwgrond vernietigen in het kustgebied van Pakistan. Dit proefschrift heeft een eerste poging gedaan om deze diensten te benoemen en betoogt dat verder onderzoek naar het erkennen van de sociale en economische waarde van biodiversiteit en ecosysteemdiensten een gezamenlijke aanpak mogelijk zou kunnen maken die voor beide kosten effectiever kan zijn. Ten vierde hebben beide landen vergelijkbare drivers (zie tabel 9.3) en staan zij voor het probleem dat niet-water gerelateerd beleid het gebruik van water en vervuiling bepaalt. De rol van China als investeerder in handelsroutes naar Pakistan en Afghanistan kan ook een belangrijke motor zijn voor watergebruik en vervuiling. Het bundelen van kennis en middelen om gemeenschappelijke drivers aan te pakken kan kosteneffectief zijn. Het ontwikkelen van een landbouw-, industrie- en handelsbeleid dat rekening houdt met waterbeperkingen is van cruciaal belang voor de duurzaamheid van het ontwikkelingsbeleid op lange termijn. Dit kan ook helpen om een gemeenschappelijke standaard te ontwikkelen met betrekking tot Chinese investeringen. Ten vijfde hebben Pakistan en Afghanistan gemeenschappelijke elementen in hun binnenlands beleid (zie tabellen 9.4 en 9.5) die zouden kunnen worden opgeschaald in een internationale overeenkomst. Voordat ze dit doen, moeten ze echter een aantal interne contradicties aanpakken. Deze hebben betrekking op de verschillen tussen islamitische wetgeving en de focus op sociale en ecologische dimensies en gewoonterecht en postkoloniale wetgeving, het mensenrecht op water en sanitaire voorzieningen versus het principe van het terugverdienen van de kosten. Daarnaast moet rekening worden gehouden met verschillen in macht, omdat waterbestuur gecentraliseerd is in Afghanistan, en gedelegeerd is naar 'provinciaal niveau' in Pakistan, waardoor bijvoorbeeld samenwerking op het gebied van dammen moeilijk is. Bovendien is het beleid in beide landen vaak onvolledig en moet het beleid op adequate wijze gemixt worden om ervoor te zorgen dat beleid kan worden uitgevoerd. Ten zesde toont het onderzoek aan dat de beperkte beschikbaarheid van water het operationaliseren en implementatie van beleid, en de monitoring en handhaving hiervan ernstig bemoeilijken. Tegelijkertijd worden instrumenten die de mogelijkheid tot corruptie beperken vaak opzettelijk gesaboteerd door politieke actoren. Dit leidt tot een korte termijn focus op economische groei en afhankelijkheid van hulporganisaties. Ik beargumenteer dat een sociaal en ecologisch inclusief systeem een betere kans heeft om op lange termijn duurzaam te zijn. Een focus op economische groei op korte termijn zal eerder leiden tot het negeren van sociale en ecologische gevolgen met een langetermijneffect op de veiligheid en de levensstandaard. Het zoeken naar op samenwerking gebaseerde, lokaal ontwikkelde, kosteneffectieve oplossingen is van cruciaal belang voor het verbeteren van de levensstandaard en het welzijn. Tot slot heeft dit proefschrift getracht het primaire en secundaire bewijsmateriaal te verzamelen om een prima facie casus te maken voor de noodzaak om prioriteit te geven aan meervoudige, grensoverschrijdende waterproblemen in de regio, vooral gezien de dreigende watercrises en het feit dat de Kabul een gletsjer gevoede rivier is, en daardoor bijzonder kwetsbaar is voor de gevolgen van klimaatverandering. Op dit moment ontbreekt gedeelde en geïntegreerde kennis, en ook de dialoog op de verschillende niveaus van bestuur in het grensoverschrijdend stroomgebied. Deze zijn echter wel nodig om elk van de bovenstaande aanbevelingen in meer detail te kunnen uitvoeren en verfijnen. Grensoverschrijdende kennisontwikkeling op scholen, universiteiten en instellingen voor levenslang leren en dialoog tussen het maatschappelijk middenveld en de regeringen mobiliseren, is van cruciaal belang om de lange termijn problemen van het delen van water aan te pakken. Want een bestuurlijk kader voor zoetwater bestuur kan het beste bijdragen aan inclusieve en duurzame ontwikkeling wanneer het berust op een solide basis.

LIST OF ACRONYMS

ADB	Asian Development Bank
AFD	French Development Agency / Agence Française de Développement
AREU	The Afghanistan Research and Evaluation Unit
CABI	Centre for Agriculture and Biosciences International
CDKN	Climate and Development Knowledge Network
CDM	Clean Development Mechanism
CPEC	China-Pakistan-Economic-Corridor
DFID	Department for International Development
EIA	Environmental Impact Assessment
EPA	Environmental Protection Agency
ESS	Ecosystem Services
FATA/NA	Federally Administered Tribal Areas/Northern Areas
FO	Farmers Organization
GB	Gilgit Baltistan
GDP	Gross Domestic Product
GEF	Global Environment Facility
GHGs	Green House Gases
GIRoA	Government of the Islamic Republic of Afghanistan
GLOFs	Glacier Lake Outburst Floods
GoP	Government of Pakistan
HDR	Human Development Report
IBRS	Indus Basin River System
IFAD	International Fund for Agriculture
IMF	International Monetary Fund
IPCC	Intergovernmental Panel for Climate Change
IRSA	Indus Rivers System Authority
IUCN	International Union for Conservation of Nature
IWC	Indus Water Commission
IWT	Indus Waters Treaty
KP	Khyber Pakhtunkhwa
LDCs	Least Developed Countries
LEAD	Leadership for Environment and Development
M&E	Monitoring and Evaluation
MAF	Million Acre Feet
MDGs	Millennium Development Goals
MoF	The Ministry of Finance
MoFA	The Ministry of Foreign Affairs
NWFP	North West Frontier Province
OECD	Organization for Economic Co-operation and Development
OFWM	On-Farm Water Management
PNRDP	Pakistan Network of Rivers, Dams and People
PPAF	Pakistan Poverty Alleviation Fund

SDC	Swiss Agency for Development and Co-operation
SDPI	Sustainable Development Policy Institute
SEI	Stockholm Environment Institute
SPO	Strengthening Participatory Organization
TAPI	Turkmenistan–Afghanistan–Pakistan–India
UNDP	United Nations Development Programme
UNEP	United Nations Environment Programme
UNESCO	United Nations Educational, Scientific, and Cultural Organization
UNFCCC	United Nations Framework Convention on Climate Change
WAPDA	Water and Power Development Authority WB World Bank
WASH	Water, Sanitation and Hygiene
WFP	World Food Program
WUA	Water User Association
WWDR	World Water Development Report
WWF	World Wide Fund for Nature

TABLE OF CONTENTS

LIST OF FIGURES

LIST OF TABLES

1
INTRODUCTION

1.1   INTRODUCTION

This thesis aims to understand how transboundary water governance can be improved with special reference to the Kabul River between Afghanistan and Pakistan. In this chapter, the problem definition will first be discussed including the real-life problem (see 1.2) as well as the gaps in the scientific literature (see 1.3). The real-life problem takes into account the flow and administration-related issues in transboundary river basins. It explains the consequences of reduction in freshwater flow or degradation of freshwater quality in Transboundary Rivers. The literature review has provided insights into various relevant approaches such as power, institutions, multilevel governance and inclusive development as well as its implication for this thesis. The gaps in knowledge have highlighted the areas that need further research leading to the main research question and sub-questions (see 1.4) with special reference to the Kabul River. Third, the focus and limits (see 1.5) of the research are discussed. Fourth, section 1.6 presents the structure of the thesis.

1.2   REAL LIFE ISSUES

Currently about 276 transboundary river basins (Barraqué 2011a; Bakker and Duncan 2017) support the production of approximately 60% of global food (Sadeqinazhad et al. 2018), supply domestic water needs of approximately 40% (Munia et al. 2016) and accommodate more than 70% of the global population (Earle and Neal 2017). With the increase in population rate, fast urbanization and commercial agriculture practices the pressure over transboundary water resources will surge which may ultimately influence both the quality and quantity of freshwater (Cosgrove and Loucks 2015a; Munia et al. 2016). These issues may lead to different intensities of conflicts among states and societies (Kasymov 2011; Munia et al. 2016). Overall, water issues can be differentiated in terms of: (1) flow-related i.e., issues of water quality (see 1.2.1), quantity (see 1.2.2) and climate change (see 1.2.3); and (2) administration-related i.e., issues of power, jurisdiction, coordination, principles and instruments at multiple levels of governance. In the sub-sections below, I explain these challenges in detail.

1.2.1   Water Quality Issues

Water is an essential resource provided by nature which is used in our lives for drinking, cleaning, bathing and other developmental purposes (Bibi et al. 2016). Clean and safe drinking water is vital for human health worldwide. However, as a universal solvent, freshwater can also transmit infections (WHO/UNICEF 2015). According to a report (WHO/UNICEF 2015), water quality in several developing countries does not meet the WHO quality standards, and this poor water quality is responsible for 80% of diseases. Furthermore, unhygienic and poor quality of water is also responsible for more than three percent of deaths (Pawari and Gawande 2015).

The release of domestic, industrial and radioactive waste, dumping of waste directly into rivers, and leakage from water tanks are some of the most important causes of water pollution (Islam and Tanaka 2004; Halder and Islam 2015). Discarded heavy metals and industrial waste in lakes, streams and rivers have proven to be harmful to humans and animals (Halder and Islam 2015). Harmful industrial and domestic waste are the key elements that contribute to causes severe poisoning, reproductive failure, and immune-suppression (Islam and Tanaka 2004). Transferrable diseases like typhoid and cholera as well as other diseases like diarrhoea, gastroenteritis, skin and kidney problems are mainly due to polluted water (Khan et al. 2013). Additionally plants and animals are also affected by poor water quality which can, in turn, affect human health (Haseena et al. 2017). Water pollutants can kill seaweeds, mollusks, seabirds, fish, crustaceans and other marine organisms that are the main sources of human food (Owa 2014). The excessive and increasing use of pesticides (e.g DDT) in agriculture is harmful for human health (ibid).

Table 1.1 shows the impacts of contaminated water on human health and the environment (see also Hunter et al. 2010). Water-borne diseases are transmitting from human to human (Halder and Islam 2015) carried by pathogens (Ashbolt 2004) that exist in specific regions or are common around the globe (Kamble 2014). Extreme weather conditions leads to heavy rainfall and floods in both underdeveloped and developed countries (McMichael, Woodruff, and Hales 2006). Contaminated water is being used in some developing countries to grow vegetables and other food items which is consumed by 10% of the global population (Corcoran 2010). Using contaminated water can lead to various chronic diseases including respiratory, cardiovascular, cancer, diarrhoea, and neurological (Ullah et al. 2014). Furthermore, Nitrogen-containing chemicals in water is one of the main reason for blue baby syndrome and cancer (Krishnan and Indu 2006). Mortality rates due to cancer is comparatively higher in rural areas than in urban areas because inhabitants of rural areas have less access to treated water (Angoua et al. 2018). Poor people face an increased risk of illness due to lack of access to improved sanitation, hygiene and supply of clean drinking water (Jabeen et al. 2011). Pregnant women are particularly vulnerable to negative impacts of contaminated water (Collier 2007).

Various factors such as dissolved oxygen, nutrients, turbidity, and water temperature contribute towards the growth of animals and plants (Kamal et al. 2007). Additionally, the chemical oxygen demand (COD) and the biological oxygen demand (BOD) specify the pollution level of a given water body (Shrestha and Kazama 2007). Each element has a specific and important role to play in the aquatic ecosystem. These ecosystems have a significant impact on the fisheries which are the main source of food for thousands of people living in coastal areas. Surface freshwater resources have a distinct role in keeping the environmental balance in the estuaries and at the mouth of the rivers (Postel and Richter 2012). Monitoring of surface water resources is essential for flood forecasting and aquatic resource management (Jacobs 2002). Contaminated water not only affect the quality of crops but also harmful for aquatic life (Khan and Ghouri 2011). The above water quality issues can affect relations between upstream and downstream countries.

Table 1.1: Risks to human health & the environment due to poor water quality

Wastewater Constituents	Parameters	Risks
Solids **Dissolved inorganic substances**	TSS, TDS, VSS, EC, Na, Ca, Mg, Cl, and B	Increasing soil osmotic pressure, blockage of irrigation system and increase of sludge deposits for salinity, Phytotoxicity and soil absorptivity and configuration
Other organic materials **Biodegradable organic Materials Nutrients**	Detergents, phenols, fat, oil, grease, pesticides, solvents, and cyanide, BOD, COD, N, P, ammonium	Poisonous effects, visual problems, bioaccumulation in the food chain, degradation in dissolved oxygen in receiving water body, decrease in fish production and increase in fish mortality, lack of oxygen, toxic effects.
Metals	Hg, Pb, Cd, Cr, Cu, Ni	Poisonous effect, bio-accumulation, can make wastewater inappropriate for irrigation, conceivable health effects
Hydrogen ion concentrations Microorganisms	Pathogenic bacteria, pH, virus and worm eggs	Possible adverse impact on plant growth due to acidity or alkalinity, cause diseases

Source: Jabeen et al. 2011; M. Khan and Ghouri 2011; Juneja and Chaudhary 2013; S. Khan et al. 2013; Owa 2014; WHO/UNICEF 2015; Pawari and Gawande 2015; Halder and Islam 2015; Bibi et al. 2016; Haseena et al. 2017

1.2.2   Impacts of Water Variability on Society & the Environment

Flow variability includes flooding as the result of higher glacial and/or snow melt rates and higher rainfall; droughts as the result of reduced contribution from glacial and/or snow melt and rainfall, low rainfall and high evaporation. Such variability can have significant consequences on social and environmental systems.

Flooding impacts include the destruction of crops, deterioration of health conditions due to waterborne diseases, loss of livestock and human lives, and damage to property (Armah et al. 2010). Additionally, communication and some economic activities can be disturbed, and people may be forcefully displaced (Corvalan et al. 2005). Floods can have long term psychological impacts on victims and their families (Doris et al. 2018). The loss of family members has a major impact on children. Forced displacements of human populations, loss of assets and disruption of social affairs and business due to floods can add to ongoing trauma (Cardoso et al. 2008). Floods in agricultural regions can damage crops and may also lead to the loss of livestock (Onifade et al. 2014). Increasing salinity in the soil, damage to the crops due to rain, and delays in harvesting are worsened by transportation problems due to flooded roads and impaired infrastructure (ibid). This may lead to reduced production in the agricultural sector and higher prices of food items due to supply shortages (Armah et al. 2010). However, on the other hand floods can have some long-term benefits for the agricultural sector in the arid and semi-arid zones as it may recharge underground aquifers.

They also enhance soil fertility due to mineral and silt deposition in the soil (Dokhani and Ramezani 2017). A much larger portion of the population is affected when floods damage the roads, railways tracks and shipping ports. These damages can have a significant impacts on the GDP of both developed and developing economies (Cardoso et al. 2008).

The social impacts of low rainfall and droughts include public well-being, health, disputes among water users, concentrated value of life and disparities in the dissemination of impacts, and disaster relief. Many of the economic and environmental impacts also have social components. For instance, population migration to areas where there are better food and water supplies puts pressure on those areas (Cosgrove and Loucks 2015b). Migration usually occurs to urban areas within the stressed area or to areas outside of drought zones. Migration can even be international to neighbouring countries (Selby et al. 2017). Migrants rarely go back to their homes even after the drought passes which results in lack of valuable human resources in rural areas. Drought migrants put enormous stress on the cities' social infrastructure, leading to an social disorder and rise in poverty level (Wilhite et al. 2007). These impacts of water variability affect relations between transboundary countries.

❖ *Environmental Impacts of Flow Variability*

Floods play a distinct role in sustaining key functions of the ecosystem and biodiversity in various natural systems (Schindler et al., 2016). They connect the river with the surrounding land, rechaarge the groundwater aquifers, fill the lakes and streams with freshwater and strengthen connection between the water habitats. Flooding is one of the main sources for sediment and nutrients transport into the sea and around the landscape (Onifade et al. 2014). Floods enable reproduction, migration and dispersal for many species (ibid). Natural systems such as mangroves provides adequate resistance to the effects of all kinds of floods except the largest (Liao 2012; Onifade et al. 2014). Increase in fish production, replenishing groundwater resources and preserving the recreational environment are some of the environmental benefits of flooding (Cosgrove and Loucks 2015a; Onifade et al. 2014). Floods can further degrade ecosystems (Holman et al. 2003) by transporting sediments and nutrients, increasing soil erosion, disturbing vegetation surrounding the rivers, enrichment of channel size, dams, and floodplains (Pressey and Middleton 2009).

Similarly, low rainfall and drought have various environmental impacts including damage to plant and animal species, habitats of wildlife, degradation of air and water quality, forest loss and forest fires, degraded landscape quality, biodiversity loss and soil erosion (Acevedo-Whitehouse and Duffus 2009; Wilhite et al. 2007). Some of the impacts of drought are short-term where conditions are normalised soon after the drought while some persist for long-term and may even become permanent (Meir et al. 2018). For example, due to the degradation of lakes, wetlands, and vegetation some of the wildlife can be affected where some species recover soon from the temporary deviation while some takes longer. Deteriorating landscape quality such as increased soil erosion can lead to an enduring loss of biological productivity (see Table 1.2). Increasing understanding of the social and ecological damages of climate change can exacerbate tensions between riparian states.

Table 1.2: Social-ecological impacts of climate variability and change as well as water flow variability

	Flood	Drought	Low rainfall
Social Impact	- Affects agriculture, fisheries, industries, energy, water quality, infrastructure, navigation & trade - Water supply more than demand - Affects human lives, health and wellbeing, damage to property - Affects biodiversity; - Catalyses cooperation & conflict at multiple levels, increases migration	- Affects agriculture, fisheries, industries, energy (e.g. hydropower), water quality, navigation & trade - Water demand more than supply - Affects human lives, livelihoods, health & development - Affects wildlife, habitats, plant species; - Catalyses conflicts between water users	- Reduces water levels in reservoirs and affects irrigation - Affects rainfed agriculture - Affects livelihoods and health - Affects plants, wildlife and ecosystems - Catalyses conflicts between water users
Ecological impact	***Positive Impacts*** - Positive impacts on aquatic biodiversity; Sustaining biodiversity and key ecosystem functions, transporting sediment and nutrients - Linking the river with the surrounding land and connectivity between aquatic habitats - Recharging groundwater systems & filling wetlands ***Negative Impacts*** - Removes vegetation, erosion & transfer of sediment & nutrients, disperses weeds, degrades freshwater habitats, hill-slopes, rivers & floodplains and wetlands - Increases channel size, dams, levee bank & catchment - Changes in freshwater flow - Releases pollutants	- Threatens freshwater & marine ecosystem, air quality, water quality - Forest & range fires - Degradation of soil and landscape quality - Loss of biodiversity and biological productivity	- Increases water temperatures - Increased wildfire risk - Stresses aquatic life

Sources: Wolf, Yoffe, and Giordano 2003; Agardy et al. 2005; MEA 2005; WRI 2005; Thornton 2006; UNESCO 2006; UNDP 2006; Luzi 2007; Warner 2008; Sanchez and Roberts 2014; UN-WWAP 2015

1.2.3   Climate Change Related Challenges

Global climate change is leading to higher temperatures causing more evaporation, glacier melt, changes in the flow of rivers, changes in rainfall patterns and sea level rise. Climate change can have positive as well as negative impacts (Bates, Kundzewicz and Wu 2008). However, in the long-term positive impacts might be outnumberd by the negative impacts which can lead to the destabilization of entire climatic system. Even for those who argue in terms of winners and losers, the harmful impacts of climate change on the global freshwater system far exceed the benefits globally (Bates, Kundzewicz, and Wu 2008). The 4th Assessment Report of the Intergovernmental Panel on Climate Change (IPCC) have highlited both the positive as well as negative impacts of human-induced climate change (Bates et al. 2008). This was also summarised in the recent 5th Assessment Report of the IPCC (Team et al. 2014). According to different estimates of the IPCC, approximately 20% of the world's population will be subject to increased flooding by 2080 while people living in severely stressed river basins could rise from 1.4 - 1.6 billion people in 1995 to 4.3 - 6.9 billion by 2050 (Larigauderie and Mooney 2010). Different climate change models for the 21st century forecast a rise in rainfall in high altitude regions and some parts of the tropics while low precipitation may occur in mid-latitude and sub-tropical regions (Bates, Kundzewicz, and Wu 2008). In addition, these models also predict a change in seasonality of precipitation in many regions i.e. more rainfall in winter and low rainfall in summer which may increase the risk of drought and flood (Jiménez-Cisneros et al. 2014).

1.3   THEORETICAL GAPS IN TRANSBOUNDARY WATER GOVERNANCE LITERATURE

Surface and ground water (or their catchment areas) that crosses international borders are referred to as transboundary waters (Elhance 1999: 3) while their management and regulation is referred to as transboundary water governance (Orme et al. 2015). The dependence on transboundary waters by the upstream and downstream states make their governance inherently political and contested (Duratovic 2016; Zeitoun 2015; Zeitoun and Warner 2006). Although transboundary water is often seen as a source of conflict, exisiting studies argue that it can be a basis of cooperation among countries (see Biswas 2011; Wouters 2013; Bhaduri et al. 2011; Morris and de Loë 2016; Jager 2016; Neal et al. 2016; Guo et al. 2016; Paula Hanasz 2017), which will serve as a source for lagrer socio-economic development and will hence benefit all the parties (Catafago 2005). There is a documented evidence of Senegal River Basin, where the participating countries have not only seen increase in cooperation, leading to overcoming the existing agircultural challenges but have also achieved growth in socio-economic development (Kipping 2009; Vick 2006).

Freshwater resources are finite and subject to political and national boundaries and jurisdictional issues (Griffiths and Lambert 2013). Therefore, its distribution, ownership and uses are also highly contested (Kehl 2017) and sometimes become a source of conflict which are driven by factors of both temporary and permanent water scarcity (UN Water 2007), differences in international goals (Cosgrove and Loucks 2015a), complex social and historical contexts (Levy and Sidel 2011;

Mollinga 2003), misinterpretation of data (Damkjaer and Taylor 2017; Döll et al. 2016), asymmetric power relations between riparian states (Warner and Zawahri 2012; Zeitoun and Warner 2006), lack of cooperation among States and disputes over particular projects (Hanasz 2017; Levy and Sidel 2011). Moreover, the uncertain impacts of climate change make transboundary water governance more complicated and may enhance competition among water scarce states and societies (Cosgrove and Loucks 2015a; Mgquba and Majozi 2018). The scarcity, depletion, reduced supply, increased demand, as well as unequal distribution of freshwater resources may affect the very existence of states and may translate into aggressive tendencies, power and dominance by many countries (Brochmann and Gleditsch 2012; Damkjaer and Taylor 2017). Transboundary water governance is a continuing political process which involves a multitude of actors, institutions and drivers (Luzi 2007; Paula Hanasz 2017) which is connected to multiple levels of governance (Gupta and Pahl-Wostl 2013a; Warner and Zeitoun 2008). Therefore, the occurrences of conflict and cooperation at one level may significantly affect the situations for conflict and cooperation at another level (Warner and Zeitoun 2008). The politics of scaling also influences the choice of deciding how and at what level water should be governed (Gupta and Pahl-Wostl 2013a).

Transboundary water governance literature rarely takes these multilevel complex linkages into account and mainly focuses on power relations at the transboundary level among sovereign states. Taking into account the importance of multilevel water governance institutions can potentially influence power politics not only at the formal level of interstate cooperation, but also among actors at multiple level of governance. For win-win outcomes, institutional frameworks can potentially reduce the influence of power politics over freshwater-related policies and decisions. Since 805 AD more than 3,000 agreements (concerning navigational, boundary delineation, or fisheries related matters) have been recorded (in Atlas of international freshwater agreements) in contrast to few conflictual events (Wolf 2002). These agreements clearly indicate that transboundary river basins are sources of cooperation rather than conflict. However, conflict still exists in some transboundary river basins despite governance institutions. This shows that institutions in these basins are struggling to address current and potential challenges. This may be due to imbalanced power relationships among states and societies which influence the performance of governance institutions in different river basins but may also be triggered when societies pass ecological tipping points.

According to Hydro Hegemony (HH) scholarship, conflict exists alongside cooperation but with different intensities, and the role of power shapes and influences water governance institutions with pre-determined outcomes (Zeitoun and Warner 2006). Institutional performance varies from region to region, country to country and from one geographic level to another (Leeds 2009; Rodríguez-Pose 2013). Neo-institutionalists believe that power patterns also vary from issue to issue and that there is always an opportunity to create equitable and efficient transboundary water governance institutions (Jacobs 2012; Warner et al. 2017; Hassenforder and Barone 2018). Once effective institutions are created, they can potentially shape and curb power (Brady et al. 2016). Transboundary water challenges in the Global South are rapidly increasing due to rapid economic growth externalising the impacts on the environment and relatively poor water institutions and governance practices and with worldwide changes in social, political, environmental and economic systems (Friend and Thinphanga 2018; Saleth and

Dinar 2000). The scholarly literature suggests that an analytical transboundary water governance framework is necessary to address the challenges of water quality degradation and quantity reduction as well as water politicisation and securitisation.

Nevertheless, there are four key gaps in scholarly literature regarding the administration of transboundary river basins. First, transboundary water governance (TWG) literature rarely combines the role of institutions in dealing with HH at multiple geographic levels. Second, TWG literature scarcely links international relations (IR) scholarship with multilevel governance scholarship promoted by European Union scholars; I argue that an understanding of basin politics requires examining both flow and administration related issues at multiple , levels of governance. This is because inequitable water distribution and biased decisions made at one geographic level may influence the social and environmental outcomes at another geographic level. Third, TWG literature insufficiently highlights the role of water outside the basin (e.g. rainwater, snow and green-water), the ecosystem services of water, and non-water related issues and actors in transboundary water research and policies. There are various non-water related actors, factors and driving forces that influence water-related policies and decision-making. Their role requires analysis in sustainable and effective freshwater resources governance, especially in the context of the Anthropocene – where the ‚great acceleration‘ in demand for resources will have to make us reconsider how we use the shrinking ecospace that we have (Gupta 2016a). Fourth, TWG literature scarcely focuses on including the inclusive development approach of international development studies which prioritises inequality and focuses on socio-relational and ecological aspects. This thesis addresses these gaps through a conceptual framework linking the HH scholarship of the transboundary freshwater governance literature with institutional approaches and with theories on multilevel governance. In doing so it addresses issues of power, jurisdiction, coordination, principles and instruments and promotes inclusive and sustainable development.

1.4   MAIN RESEARCH QUESTION AND SUB-QUESTIONS

The four key research gaps that are discussed above lead to the formulation of one overarching question and four sub-questions.

1.4.1   Main Research Question

How can regional hydro-politics and institutions be transformed at multiple levels of governance through inclusive development objectives and incorporate the relationships with non-water sectors in addressing issues of water quality, quantity and climate change?

1.4.2   Sub-research Questions

The following sub-questions which are linked to the above knowledge gaps will be further explored with special reference to the Kabul River that flows through Afghanistan and Pakistan. These research questions are briefly elaborated below.

a. **How do power politics and institutions influence water governance in transboundary river basins at multiple geographic levels?**

This question positions the foundation of this thesis. Through the literature review on the evolution of multilevel transboundary water governance, it explores the intrinsic relationship between power politics and multilevel freshwater governance institutions in transboundary river basins. The literature discusses that sharing transboundary water resources are new issues in international relations, and if harmony is to be established within imbalanced power relations, efforts have to be made to resolve historical and existing conflict and promote genuine cooperation over water at multiple geographic levels. In addition, geography, exploitation potential and the structural power of states in their relations inter se occupies significance in international politics where the geo-strategic position of one country becomes important for the other.

b. **How can the concept of biodiversity and ESS be incorporated in a framework to analyse the effectiveness of institutions, and the role of power, in governing transboundary water resources?**

This question helps analyse how the concepts of biodiversity and ecosystem services can be taken into account to assess relevant institutions and the role of power in order to understand transboundary water governance and enhance cooperation. It explains that the inclusion of social and ecological aspects within the existing institutional frameworks can inform the transboundary water cooperation narrative for human wellbeing. Additionally, it also assesses the consequences of ecosystem change as there is a dynamic interaction between ecosystems and human wellbeing including the drivers of these changes for better governance, understanding (unequal) distribution of benefits from ESS across different population groups, and managing risks and opportunities and quantifying the ESS for better communication with policymakers.

c. **Which principles & instruments address the causes/drivers of freshwater problems in transboundary river basins at multiple geographic levels?**

This question is responded by using the results from the previous question regarding the evolution of transboundary water governance institutions to examine the patterns of principle and instruments inclusion and their role in achieving inclusive and sustainable development. This question allows the analysis of the different leading features of each governance framework. By analysing the content of numerous policies, laws and acts, I show which principles and instruments are included and excluded within a particular framework, how these principles and instruments address the challenges of freshwater systems, and how they relate to the goals of inclusive and sustainable development.

d. **How does legal pluralism affect transboundary water cooperation?**

This question highlights the impacts of plural legal systems in a transboundary river basin. It confers that different actors can create diverse rules which can have conflicting inferences, and costs for the same populations and the resources upon which they rely. This concept is fundamentally employed to understand the dynamics of pluralism by examining: (a) the disintegration of the legal system; and (b) the concurrence of multiple rules in a single jurisdiction.

1.5   FOCUS AND LIMITS

1.5.1   International Development Studies Focus

The focus of this thesis moves from a broad framework of International Development Studies (IDS), angled through a lens of sustainable development, explored through inclusive development and then finally focuses on an institutional analysis within this established context. Emerging and evolving as a field of study seeking to bring desirable change to society with a traditional focus on economic growth (Valters 2014), IDS addresses the multi-scalar and institutional contexts in which poverty as well as improved human well-being occur. Sustainable development approaches have enriched traditional IDS and have arguably become the dominant development discourse in the arena of international politics (Jacobs 2002). Recognising three important _pillars' – social, ecological and economic, sustainable development aims to provide solutions to global ecological and development problems (Hopwood et al. 2005; Roberts 2007) for current and future generations in both the developing and developed countries (Lewis and Wiser 2005; Montaldo 2013).

The focus on inclusive development draws on sustainable development and sharpens it further with an emphasis on social inclusiveness (i.e., to enhance the well-being of people, communities and states) (Gough and McGregor 2007); ecological inclusiveness (e.g., to know that natural resources are finite and ecosystems need to be maintained); and political geography (for example, to protect socio-ecological goals and address power politics) (Mosse 2010) that can lead to the demonstration of relational inclusiveness (Bos and Gupta 2016; Rauniyar and Kanbur 2010). It refers to, but is not limited to, development and social wellbeing coupled with equal opportunities (Gupta et al. 2015; Vella 2014), and argues for transparency and accountability and cooperation between a variety of participants including government, private sector, and civil society. Importantly, it focuses less on growth and more on shared wellbeing (Bos and Gupta 2016; Gupta et al. 2015). Finally, it recognises that the well-being of all people, particularly the poor, is closely related to the protection and conservation of ecosystem services (MEA 2005; Hayat and Gupta 2016). By analysing the various layers and aspects of the institutional setting in which transboundary water governance takes place, this thesis aims to show what is possible in terms of generating change toward improved inclusive development.

1.6   LIMITS OF THE THESIS

There are some limitations to this thesis. As a scholar of International Development Studies, I have pursued a multi-disciplinary approach which is in line with the spirit of the field. However, multi-disciplinarity can also have its challenges. For instance, there is an important International Relations component to shaping this dissertation, but my own background and expertise is limited. I have tried to draw on and appropriately use the most important approaches and concepts from IR for this thesis, but I recognise that it may be limited in analytical depth. Likewise, it is in the nature of a multi-scalar analysis that broad strokes are drawn in order to make links between scales and levels visible. However, this requires another trade-off in terms of the depth that can be reached in describing and analysing the specifics of the case study. The attempt to analyse ecosystem services was limited by the limited knowledge in the field and access to needed resources. In terms of the real-world, continuing conflicts and tensions in the case study area created a range of limits to the eventual claims and conclusions that this thesis makes. Specifically, with the withdrawal of US and NATO forces from Afghanistan, the (so-called) Taliban, ISIS, or other extremists and terrorists' groups have the potential to become active in such a way that could lead to a new proxy war over shared water resources between the two countries. This new proxy war can potentially deteriorate the already tense relationships between the governments and peoples of both countries. The movement of people across the border, bilateral trade, and visa policies and have been interrupted already. The Government of Pakistan had decided to send back the three million registered as well as all other unregistered Afghan refugees (since the 1979 Soviet–Afghan War and US invasion of Afghanistan in 2001) by December 2018. As this has been taking place during the time of writing, I have not only had limited access to relevant data, but there is uncertainty in the region which may impact the predictive quality of this thesis.

1.7   STRUCTURE OF THE THESIS

Following this introduction, Chapter 2 furnishes the research methodology and the theoretical framework. In Chapter 3, I elaborate key realist and institutional approaches of transboundary water governance to establish the theoretical foundation of this thesis. Chapter 4 explores how addressing transboundary water issues can give us a better understanding of the causes of water challenges in the context of updated knowledge about the role of the ecosystem services of water. Chapter 5 examines the five key global institutions related to water and how they relate to inclusive development. Chapter 6 links the theoretical analysis of transboundary water issues with the empirical data on the Kabul River Basin (KRB). Chapters 7 and 8 describe and analyse multilevel freshwater governance in Afghanistan and Pakistan, respectively. Chapter 9 combines the transboundary, national, provincial, and local levels of the KRB case and compares each level with the key elements of the UN Watercourses Convention on equitable and efficient water sharing. Finally, Chapter 10 concludes the thesis by answering the research question and reflecting on the implications of the findings.

2

METHODOLOGY AND ANALYTICAL FRAMEWORK

2.1   INTRODUCTION

This chapter presents the methodology and analytical framework in terms of six steps. I explain my method to review the literature (see 2.2), to assess the content of policies (see 2.3), my choice of case study (see 2.4), the key concepts used in the thesis (see 2.5), and how these are integrated in the institutional analysis approach (see 2.6); methodological limits (see 2.7) and my ethical approach (see 2.8).

2.2   LITERATURE REVIEW

The first step of my methodology was to undertake a literature review. The literature review for this thesis demonstrates the current state of knowledge about the research focus. The literature review informed the gaps in the scientific literature (see 1.3) and research questions (see 1.4), and the key concepts presented in this chapter. I systematically reviewed the literature on hydro hegemony (HH) and hydro institutionalism (HI); on ecosystem services (ESS); and in relation to the case study area. For the literature review, particularly on power, institutions and ecosystem services I covered the period from 1970 – 2017, while on HH and HI, I considered the period from 2004-2018. For the case study area, I covered the literature since 1860.

The publications I reviewed include: peer-reviewed journal articles, books and grey literature including official reports and communications, Memorandums of Understanding (MoUs) at multiple levels of governance, as well as the informal literature which I sourced through different archives in Pakistan. I searched for all these resources in various databanks such as Elsevier, Web of Science, JStor, Wiley, Nature, Science Direct and EBSCO. For literature on transboundary freshwater resources, I searched catalogues of specific journals such as Water, Third-World Quarterly, Water Policy, Global Environmental Change, Water Alternatives, Regional Environmental Change, Water Resources Development, International Environmental Agreements, Water International, International Environmental Governance, Environmental Policy and others. I also actively followed water-related articles in leading English newspapers of Afghanistan (The Afghanistan Times, Daily Outlook Afghanistan, etc.) and Pakistan (The DAWN, The NEWS, The Express Tribune, etc.) as well as international media sources such as the BBC and the Wall Street Journal. For the case study, I also reviewed the scholarly literature in the local languages.

2.3   CONTENT ANALYSIS

The second step was to conduct a qualitative content analysis. The content analysis method focuses on understanding the informal and formal laws and policies relevant for my multilevel case study. I analysed the content of laws and policies to identify: (1) the drivers of freshwater problems; (2) the pattern of inclusion for governance principles and instruments; (3) the way these principles and instruments address the drivers of freshwater problems; (4) the conditions of legal pluralism; and (5) the contribution of principles and instruments in achieving inclusive and sustainable development. Content analysis is a very useful method to determine major trends and perceive any changes in such trends or in the meaning of concepts as they are used in different texts (McLellan and Porter 2007).

In my thesis, the content analysis showed how governance principles have emerged from various levels of governance and how they are included in various governance frameworks through instruments (such as regulatory, economic, management, suasive, etc.). To conduct the content analysis, I followed four key steps: (a) defined my units of analysis based on the literature review which resulted in (b) two types of drivers (direct and indirect); (c) four categories of principles (political, social-relational, ecological and economic); and (d) four categories of instruments (regulatory, economic, management, and suasive).

In this regard, I reviewed five key institutions (i.e. customary international water law; the 1992 UNECE Water Law which has been opened up for global participation; the 1997 UN Watercourses Convention; the Human Right to Water and Sanitation, and the Sustainable Development Goals of Agenda 2030) at the global level; 14 at the transboundary level (3 colonial and 11 post-colonial); 14 at the national and local level in Afghanistan (2 precolonial, 1 colonial and 11 post-colonial), and 24 in Pakistan (2 precolonial, 3 colonial and 19 post-colonial) concerning water governance in the case study region of Afghanistan and Pakistan.

2.4   CHOICE OF CASE STUDY

I chose a unique and multi-faceted case study to understand a complex transboundary problem in its real-life context. Although, there is a general belief that a single case study has no external legitimacy, Easton (2010) and Yin (2013) argue that the results can be generalised from a single case study and can truly contribute to theory development. The case study method is suitable for steering research because it is based on several sources of evidence and has the advantage of using specific theoretical suggestions to monitor data collection and analysis (Baxter and Jack 2008; Stoecker 1991).

2.4.1   Importance of the Case Study Method

The case study method is appropriate for examining the 'how' of research questions; when the research covers existing phenomenon, such as dealing with the issues of water quality and quantity or the problem related to power politics in transboundary river basins, which is located within a real-life context; where the researcher has less control over the events or people involved; or the boundaries are vague between the phenomenon and context (Yin 2013). The real-life problem of existing and future water challenges make the case study method appropriate for analysing the role of power politics in transboundary water sharing. The case study method can provide practical evidence of how power influences institutions and how institutions in turn influence power in Transboundary Rivers.

2.4.2   Choice of a Case Study

I chose the case of the Kabul River Basin (KRB) for the following reasons (McMurray and Tarlock 2003; Hanasz 2011; Renner 2013; Vick 2014; Azam 2015; Akhtar and Iqbal 2017; Kakakhel 2018):

Uniqueness: (a) Both countries contribute water to the KRB and are simultaneously up- and down-stream; (b) Afghanistan would now like to develop the KRB unilaterally and to increase its access to the waters for its socio-economic development leading to conflict with Pakistan which argues that as first user it has historical rights, and because the KRB is one of the densely populated basins of Afghanistan and Pakistan (with a population density of 93 persons/km^2 accommodating an estimated 7 million); (c) The Kabul River is extremely susceptible to the impacts of climate change as this river is primarily fed by glaciers and snow-melt unlike other rivers that are mainly recharged by rainfall; (d) The Kabul River lacks formal river basin institutions; (e) There is relative lack of interest from scholars because of the long conflict in the basin.

Affinity: A major portion (80%) of the Kabul River length is located in Afghanistan while the major deposit of water flow is from Pakistan. The Kabul River in both countries is mainly located in the tribal areas. The Pakhtun inhabitants of the KRB across the border follow a non-written ethical code called _Pakhtunwali' (Barfield 2003; Ali 2013; Mehsud 2015), which is a system of law and governance that is preserved and still in use in the tribal areas (Banting 2003). This ethical code and other customs in the KRB are not easily understandable for outsiders (Renner 2013). Having a tribal background (as I belong to the Yousafzai Tribe), and the Pashto language as my mother tongue, I have familiarity with local culture, social practices and following Pakhtun ethical code of conducts (Pakhtunwali), which gives me an added advantage to conduct the KRB case study.

Generalizability: The KRB case provides an opportunity to replicate the case study in other transboundary river basins where power asymmetries exist between the countries, political institutions are weak, where there is lack of, or no, coordination at all among the different levels, and where the riparian countries suffer from the problems of instability and insecurity.

Having selected the KRB case, I continued to conduct a desk study to create the existing state of knowledge on transboundary water governance between Afghanistan and Pakistan. As the KRB is an underdeveloped river basin, a very limited amount of scientific literature is available. Most of the available literature is grey-literature (published by IUCN, IWMI, the World Bank, NATO, and USAID etc.) or in the local languages, and includes project reports by various water and non-water related ministries. The available academic literature on the KRB mostly focuses on quality related issues in the Kabul River; however, in the last few years the focus has moved to include climate change, flow regimes and governance-related challenges mainly by international research think-tanks. The scholarly literature focusing on the performance of institutions and their contribution to inclusive development is yet to be taken into account. It was therefore important to collect primary data and address the research questions in the context of the case study.

To obtain primary data for my case study, I conducted 70 interviews (30 in Afghanistan, 40 in Pakistan) and two focus group discussions. Details of the interviewees' (excluding code & names), professional background, organisation and country are provided in Annex C. Due to security issues it was not recommended to visit Afghanistan during the fieldwork, so most of the interviews about freshwater   governance   in   Afghanistan   were   conducted   during   international conferences/seminars/workshops at a neutral location. Some of the interviews were also conducted through Skype and with Afghanis residing and working in Pakistan. Afghan foreign missions and students studying in Pakistan or in the Netherlands were also a great help where they connected me with relevant people. These interviews covered international, national, and local levels of freshwater governance in the KRB. Through a purposive samling technique, a total of ten interviewees were chosen from both Afghanistan and Pakistan.   Additionally, snowball methods were applied to identify the other interviewees. I conducted the interviews in three phases (i.e., in 2015, 2016 and 2017). These interviews lasted between 30 to 90 minutes each, covering questions concerning institutions, power, principles and instruments, ESS, water quality degradation and quantity reduction, current and potential threats to freshwater system and its remedies, and issues related to inclusive and sustainable development. Key approaches in the thesis were applied to examine the data derived from the interviews and were integrated with the outcomes of the content analysis and literature review. This helped in understanding the theoretical gaps and in generalizing results which were valuable in the development of the conclusions and recommendations of my thesis.

2.5   THE CONCEPTUAL APPROACH

This section discusses the following concepts and approaches that form the building blocks of the theoretical framework (see Chapter 3): (a) multilevel governance (b) power (c) institutions (d) legal pluralism (e) inclusive and sustainable development. Chapter 3 discusses in more detail theories of relevance to transboundary water governance.

2.5.1   Multi-Level Governance (MLG)

Water is governed at multiple geographic levels making it vital to analyse the academic literature on Multi Level Governance (MLG) approach. The academic literature on bilateral cooperation and global governance can guide states to develop international treaties (Keohane 1982). Theories in the social sciences are fragile in terms of their competence to realize the multi-layered and contextual dynamics of such governance systems (Harrison 2012; Ostrom 2007). Scientists have investigated that globalisation extends political authority to local and international institutions (Pattberg and Stripple 2008; Andonova et al. 2009) and to non-state actors in international governance (Keck and Sikkink 1999). In this thesis, I have drawn on the MLG approach to understand the interaction among institutions (both formal and informal) and their performance at multiple geographic levels. MLG can be defined as _an arrangement for making binding decisions that comprise a variety of politically independent but otherwise mutually dependent actors – private and public – at different levels of territorial aggregation in more-or-less continuous negotiation […]' (Schmitter and Kim 2005). _Multilevel' refers to the connectedness between different political and geographic levels (national, sub-national, supranational), while _governance' indicates the interdependence between public authorities and non-governmental actors at various geographic levels (Bache and Flinders 2004: 3).

The MLG approach points to mutual changes in political mobilisation, decision-making and the restructuring of political structures. In particular it specifies: (a) the involvement of sub-national authorities in policy formulation; (b) the mobilisation of social actors at regional and state levels; and (c) the creation and institutionalisation of intergovernmental agreements (Piattoni 2010). In general, MLG literature specifies that political power moves at the same time in two directions: up to the international level of governance and down to local communities (Pierre and Peters 2000: 1), but gradually local communities gain more political power within the state and in an international atmosphere (Eckerberg and Joas 2004). There are two types of MLG: Type-I usually has features that are revealed in federalist thinking, involving authority and clear accountability in non-overlapping jurisdictions (Marks and Hooghe 2003). Type-II denotes the recognition of overlapping jurisdictions and a multifaceted fluid patchwork of numerous intersecting authorities that address policy problems (Bache and Flinders 2004; Hooghe et al. 2002).

There are three different dimensions of MLG. The first one is called _centre vs periphery' (e.g., decentralised systems of governance); the second is _domestic vs international' (i.e., structured modes of international cooperation); and the third is _state vs society' (e.g. by involving of NGOs and civil society organisations in authoritative decision-making and policy implementation) (Piattoni 2009). The relevant actors' vertical interdependencies at various levels of governance happens where advanced levels of government are anxious with results at a lesser level when there are mutual duties (Tasan-Kok and Vranken 2011). Compared to government-led arrangements, MLG is based on more inclusive, comprehensive, and horizontally interacted relations among political, socio-cultural, and business elites where confidence among the participants is high, despite conflicts and oppositional agendas (Swyngedouw 2005). Furthermore, MLG recognises the significance of incorporating scientific technical information with the indiginious knowledge of society (Jasanoff et al. 2004).

The conceptual lens in political sciences can be broaden with the MLG approach (Kohler-Koch and Rittberger 2006: 38) which can also highlight the importance of different levels of governance within modern forms of governance (Awesti 2007: 5). It encourages a re-evaluation of the traditional conflict between international and local policies (Bache and Flinders 2004: 94). The approach highlights the value of endless, non-hierarchical and interacted relationships between diverse levels of government (Bernard 2002; Peters and Pierre 2004).

Although the general characteristics may be the same, the literature presents the challenges for different forms of MLG: for example, there are no strong and visible lines of responsibility and expressive democracy (Peters and Pierre 2004: 85) and the development of systems that adhere to the rules and work in our legal systems or to understand the meaning of _meta-governance' (Daniell 2012; Glasbergen 2011). Additionally, the local implementation of federal regulations in this area may be inadequate and regulated from top to bottom and often not locally compatible (Burby and May 1998). The MLG approach is primarily descriptive and does not contain any binding guidelines (Panara 2015).

Explaining the tensions of MLG approaches is applicable in this thesis as the sovereignty of state and government centralisation become outdated in addressing existing and potential transboundary water challenges. Therefore, highlighting direct linkages between multiple levels of governance is the central focus in my thesis (Type-1 MLG i.e. transboundary, national, provincial, and local). The structure of transboundary water governance institutions has similar weaknesses as MLG. Those identified in the literature are: lack of cohesion and harmonisation among sub-systems and various policiesas well as the difficulties of securing compliance and authorisation (Enderlein et al. 2010). Therefore, the applicability of Type-1 MLG allows examining the freshwater governance framework at multiple geographic levels. This contributes to input legitimacy (political equality) and output legitimacy (policy implementation) i.e., equality in reality as well as on paper.

2.5.2   Power

Water is so important for all aspects of society that power is wielded to maintain control over water. This section explains how power is relevant for the analysis of transboundary river basins. International Relations (IR) scholars from the realist and neo-realist schools focus on power as key to explaining outcomes of international cooperation (Schmidt 2007). Empirically, power has been measured by the size of population, strong economy and military capabilities (Nye 2004: 178), but more complex interpretations include the idea that ‗power is the capability to influence the behaviour of others to get an anticipated outcome‘ (Nye 2004: 181). The drive for survival and the actualisation of self-interest are the primary factors influencing the behaviour of actors involved (Kagan 2003). Power has two types i.e., hard power and soft power where hard power is associated with military capabilities and economic strength, while soft power includes diplomatic channels and intimidation in relations between states or between states and international organisations (Wilson 2008). States sometimes have difficulty in controlling and using soft power, but this does not reduce the importance of this form. (Vedrine and Moisi 2001). If a country develops its institutions and strictly follows rules that encourage other countries to channel or modify their activities in the way they prefer, it will not need expensive carrots and sticks (Nye 2004).

While violence, coercion and oppression are terms of hard power, they can also be implemented, monitored and measured through intervention-based measures. Daoudy (2009) and Turton et al. (2006) established these concepts further and point to two broad forms of power: ‗puissance‘ and ‗pouvoir‘ where puissance is potential power or physical power while pouvoir is actualised power. Puissance is the most visible form of power such as military capabilities, strong economy and knowledge supremacy (Strange 1987), as well as dominant geographic position and support from strong international allies (Zeitoun and Warner 2006). The Pouvoir kind of power talks about the rules for controlling the game (Lukes and Haglund 2005). Some scholars (Lukes and Haglund 2005; Strange 1994) recognize a third kind of power employed through the knowledge structure i.e., to implant ideas into the weak‘s minds where the weak genuinely belief that the assessment of the strong is true and right (Strange 1994: 176). The concept of hydro-hegemony (HH) is basically developed by combining the above mentioned three forms of power. It is eventually rooted in realism / neo-realism, which inadequately explains how cooperation in an anarchic system arises at several geographical levels (Warner and Zeitoun 2008). Power and HH theories are discussed in detail in Chapter 3.

19

2.5.3   Institutions

Through history, societies have institutionalised customary rules into formal rules to address water issues. Hence, it is critical to understand institutions. Normally, institutions are norms and rules which are identified under firm situations by a social group's members which are either sovereign or controlled by a peripheral power (Rutherford 2001 in Raadgever and Mostert 2005). These rules are created by actors, which hamper and support their behaviour (Héritier 2007). Institutions and regimes are identical and very related ideas as both appoint the procedures that shape behaviour of humans by weakening the confusion of the numerous number of possible but clear set of instructions (Raadgever and Mostert 2005). In the scholarly literature, the word _institution' has now largely replaced _regime'. Institutions can be constitutive, regulative and procedural and affect the system, the actors in the system and their activities (Duffield 2007).

In neo-institutionalism (see 3.3.3), the word _institution' denotes the informal and formal rules that govern the behaviour of actors (Douglass 1990). According to Helmke and Levitsky (2004: 37) _formal institutions are openly codified, in the sense that they are established and communicated through channels that are widely accepted as official', while _informal institutions are socially shared rules, usually unwritten, that are formed, communicated and executed outside of formal legitimate channels'. Although, informal institutions are unwritten, they may be more influential than formal ones (Douglass 1990). In the past, the literature on governance largely bypassed and ignored the contribution of local customs (Sokile and Van Koppen 2004) but this has changed dramatically (Delgado and Zwarteveen 2008; Zwarteveen and Boelens 2014). This is because, in reality, the role of these informal mechanisms cannot be overstated, especially in transboundary river basins where no formal water sharing mechanism or treaty exists (e.g., Kabul River Basin). Unlike formal institutions, the informal ones are not intentionally formed at one moment; they rather evolve through centuries of continuous communication in response to the existing situations (Saleth and Dinar 2004a). The strength of institutions varies at different geographical levels, for example, at the lowest institutional level, it is argued that informal arrangements take precedence over formal ones (Sokile and Van Koppen 2004). Institutional and neo-institutional approaches as well as hydro-intuitionalism are discussed in more detail in Chapter 3.

2.5.4   Legal Pluralism

Since water institutions have developed at different levels of governance, they are not always coherent. Legal pluralism is a situation whereby diverse rules are applicable to identical legal jurisdiction (Bavinck and Gupta 2014: 1). Diverse actors can set different rules that can have contradictory implications and costs for the same population and the resources on which they depend. Empirically, this approach was employed to define the interface between colonial laws and customary practices (Benda-Beckmann 2001). However, it is now used to understand the dynamics of pluralism (Nobles and Schiff 2012) by observing: (a) the disintegration of the legitimate system (Koskenniemi and Leino 2002); and (b) the occurrence of various rules in a single system (Tamanaha 2011). Plural legal system can happen in both ways i.e., _horizontally' (when diverse rules are applied at the same geographical level) and _vertically' (when different rules are shaped at different geographic levels) (Conti and Gupta 2014). Thus, the concept of legal pluralism is interpreted differently in different disciplines: law and anthropology use the same concept but with diverse methodological approaches. In this thesis, I focus on legal pluralism in freshwater governance.

At the international and regional levels, the ability of communities of practice has improved diversity in global law (Cullet 2013). Furthermore, the disparity between customary and formal water use principles at the national and sub-national levels may not only hinder access to water and its equitable allocation (Boelens 2009; Gupta and Lebel 2010) but can also lead to legal pluralism.

In this thesis, I apply the legal pluralism typology presented in Table 2.1. This shows that when there are different rules applicable to the same jurisdiction, it is critical to address them if there are clear contradictions between the rules. When there is indifference, i.e. when there are different rules, only one set applies in practice, there is only need to intervene if the rules in practice are problematic. Where there is mutual support, there is no need to intervene. Where there is accommodation, participatory approaches attempt to align both sets of rules.

Table2.1: Adaptation of the typology of legal pluralism to water governance

Quality/Intensity	Weak relations	Strong relations
Contrary	Type 1: **Indifference** Goals/principles/instruments included in water law frameworks lack operational relationship, although they theoretically apply to same situations.	Type 2: **Competition** There is contradiction between goals/principles/instruments included in water law frameworks, which apply in the same jurisdiction.
Affirmative	Type 3: **Accommodation** Recognition of goals/principles/instruments included in water legal frameworks but are not formally integrated in one single law or code, although they might be in practice.	Type 4: **Mutual support** Goals/principles/instruments included in water law frameworks support each other as a result of explicit arrangement or provision made in the legal frameworks.

Source: Bavinck and Gupta 2014

2.5.5 Inclusive Development (ID)

As stated in 0, while sustainable development aims at addressing social, ecological and economic aspects, the application of sustainable development approaches in some parts of the world has tended to emphasise the economic aspects over and above the social and ecological aspects. Hence, I have used the inclusive development (ID) approach which emphasises the social and ecological aspects. Adopting an ID approach in transboundary water governance requires ensuring equitable and sustainable distribution of water resources among all stakeholders including the poor, less powerful, vulnerable, and marginalised in order to protect and enhance their dignity and engage them actively in the development processes. Equality, efficiency and sustainability dimensions are essential to improve freshwater governance. These dimensions are multi-faceted and require renegotiating the political context. Since the ID approach calls for social, ecological and relational inclusiveness, I therefore adopt this approach as a guiding norm for my thesis. However, where necessary I also look at economic _growth' related issues (Gupta et al. 2015; Gupta and Pouw 2017; Gupta and Vegelin 2016; Rauniyar and Kanbur 2010).

There are three main components of ID: first, social inclusiveness which promotes overall well-being for the poorest, vulnerable and most marginalised through capacity building (Fritz et al. 2009) and institutionalising equitable principles, participatory approaches including the use of local knowledge, and non-discrimination (Lawson 2010). Second, relational inclusiveness tackles the direct and underlying drivers of inequitable development (Lawson 2010) through addressing the underlying power dimensions. Third, ecological inclusiveness supports development within the carrying capacity of the earth and enhanced human resilience (Gupta et al. 2015; Gupta and Vegelin 2016). Although the neo-classical notion of efficacy proposes that well-organized allocation of natural resources would result in sustainable development, a number of case studies indicate that uncertainties of time preference, the strong focus on economic growth and the substitution of natural capital, using technology and innovation can result in exclusion in development (Basu and Shankar 2015). I adapt the three components of ID (i.e., social, ecological and relational) in my analytical framework to explore how trade-offs between them can be minimised and to test the performance of freshwater governance frameworks at multiple levels of governance in Chapters 6, 7 and 8 of my thesis.

2.6   INTEGRATED INSTITUTIONAL ANALYSIS

The above concepts are now integrated within an institutional analysis model. A conceptual framework is a useful tool that identifies the researcher's world view of the research topic and defines his/her assumptions and preconceptions (Lacey 2010). I will now incorporate the concepts of multi-level governance, power, institutions, legal pluralism, and inclusive development in my institutional analysis model (see Figure 2.1). This model is inspired by the Young et al. (2005) framework.

This framework, requires an analysis of the non-institutional drivers and context of a problem, the institutions that deal with the problem, the instruments within the institutions, the actors on which the instruments are levied, analysis of the effectiveness of the instruments in changing the behaviour of the actors given the drivers in terms of impacts in relation to certain goals, and then based on the analysis of the contextual effectiveness of the instruments – a proposal to redesign them. This framework has been modified to accommodate multi-level governance by examining the drivers, institutions and instruments at multiple levels of governance (see Table 2.2). The relationship between instruments has been analysed using the legal pluralism approach. The impacts are assessed against the goals of socio-relational, and ecological inclusiveness where the ecosystem services of water have also been incorporated. Power is analysed in the context, as well as in relation to how it influences the formation of rules. In adapting the framework, I have been inspired by the work of other PhD colleagues – Margot Hurlbert (2016), Kirstin Conti (2017), Pedi Obani (2018) and others.

Table 2.2: Adapted Young's framework Vs original framework

Elements	Definition	Adaptation in this thesis
Institutions	The set of rules, decision-making procedures, norms (principles), and programmes that describe social practices	At *multiple levels of governance*, taking into account how *power* influences these institutions
Instruments	Isolation of specific principles and tools, or incentives and disincentives used to change behaviour	The application of *legal pluralism* Categorised as: - *Regulatory* - *Economic* - *Suasive* - *Management*
Actors	Organisations, communities, persons, or states participating in or affected by instruments	Same
Drivers	Causes of biophysical phenomenon or anthropogenic behaviours that affect actors' behaviour (e.g. changes in land cover, climate change, urbanisation, and economic growth)	Direct and indirect drivers, context including *power*
Performance of instruments	Resource's biophysical condition (i.e. abundant, clean and safe vs scarce and polluted)	How instruments impact on socio-relational and ecological inclusiveness (*ID*)
Redesign	Improving the performance of current institutions and principles to address ongoing actor stress of actors and drivers	Same

Source: Building further on Gupta, Van Der Grijp, and Kuik 2013

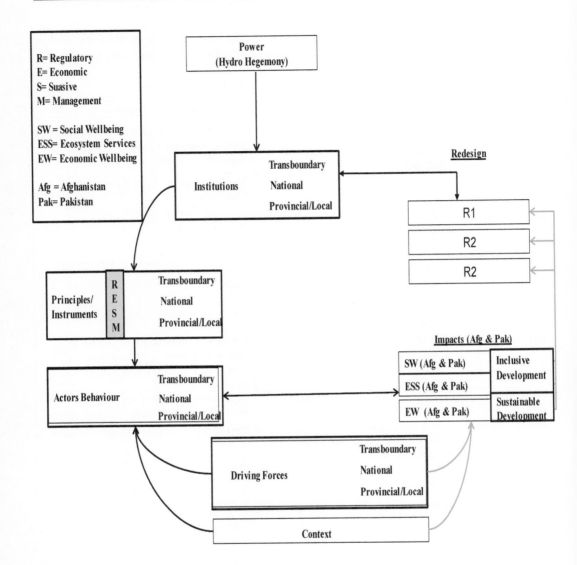

Figure 2.1: Schematic of conceptual framework

2.6.1   Defining Key Terms in the Institutional Analysis Framework

Before moving further, it is important to define and explain a few of the other terms used here. As seen above, institutions refer to the rules, decision-making procedures, norms (principles), and programmes that describe social practices. Principles are drawn from institutions and are the formal and informal legal norms with general application to all similar situations (Alexy 2000). Finnemore and Sikkink (1998: 891) define norm as the _standard of appropriate behaviour for actors with a given identity'. Hence, principles reflect the general guideline against which behaviour can be tested. There are different meanings of principles at the international level including: (a) a basis of international law, (b) an essential norm of international law that needs to be taken into consideration in the relations between the states, and (c) _a measure of the changing rules of international law' (De Sadeleer, 2002: p.237). I conceptualise principles as rules with broad implications in my investigation because these are enclosed in several sources of law and policies and which are invoked through different instruments of governance. I have therefore put particular emphasis on the principles reflected in the instruments of water policy and have consciously framed the structure of institutions at multiple geographic levels.

To continue, instruments are used by the state and other authorities to guarantee sustenance and outcome or prevent social change (Majoor and Schwartz 2015; Vedung and Van der Doelen 1998). Although literature from instrumentalist and functionalist approaches sees instruments as technical, pragmatic, rational and neutral, political sociologists emphasize that instruments are neither rational nor merely technical, but rather disclose the balance of power (Kassim and Le Galès 2010), notions of social control, and the interactions between the citizens and government (Majoor and Schwartz 2015). Table 2.2 shows that instruments are tools used in specific laws, policies or customs governing a transboundary river and include the rules, principles or measures used to affect behavioural change. Instruments can be categorised based on: (a) the mechanisms or resources which the instruments apply to, (b) the purpose of the instruments, (c) the mode of application, or (d) the impact of the instruments (Rivera 2004).

In my thesis, instruments are arranged according to who uses them (at multiple levels of governance), for whom (which actors), and are grouped as (i) regulatory (including different governance rules and binding policies), (b) economic, (c) suasive, and (d) management. It is essential to note that economic instruments can be part of regulatory instruments and that the distinction between these different instruments is very difficult to make. Regulatory instruments encourage or exclude certain activities and comprise of permits and licencing measures, sanctions and environmental impact assessments. Economic instruments promote attitude and behavioral changes through different market signals or financial incentives or disincentives and include: (i) property rights, (ii) taxes, (iii) tradable quotas, (iv) duties, and (v) grants and subsidies (Rivera 2004). Suasive instruments effect behaviour through significant tools including (i) awareness campaigns, education and training, (ii) award mechanisms and (iii) disclosure requirements (Vedung and Van der Doelen 1998). Management instruments usually include processes of self-regulation and voluntary management by the actors (Rivera 2004).

The pros and cons of these four categories of instrumnts are further disntingueshed in the available literature (Majoor and Schwartz 2015) which further reflects the authors' capacity and capability to apply a mix of instruments (Howlett 2000; McLellan and Porter 2007). I examine a range of water governance instruments at multiple levels of governance on the basis of this categorization. These instruments are adopted by both states and private/non-State actors. I believe the instruments to be valuable and hence further examine the core principles in each of the instruments and referr them to the three components of inclusive development.

2.7   KEY LIMITATIONS OF THE THESIS

There were a few methodological limits in this thesis. First, some of the pre-colonial and colonial laws and policies (such _Riwajat-i-Aabpashi' or called _Customs of Irrigation') were utilsed for the content analysis at the local level in Pakistan. As most of these laws and policies were in native Pashto and Persian languages, I had to trust informal interpretations on some occasions. However, to ensure reliability, I compared the outcomes across some legal databases. The hard sources of laws and policies were mainly examined at the international level, while soft policies and laws were analysed where there was evidence of customary international water law. Furthermore, the national laws and constitutions of countries were examined at the national level as they are solely responsible for the legal basis for policy direction at the national and sub-national levels.

Second, the number of stakeholders that were interviewed for the case study were constrained by limited time, financial resources and bureaucratic bottlenecks. Second, various important stakeholders that were interviewed for the case study had various constraints incuding time and financial resources as well as bureaucratic constriction. It was also particularly difficult to interview some of the stakeholders from key ministries (such as defence, foreign and interior) as well as from the military research and information wing who are indirectly involved in water-related decision-making in both Afghanistan and Pakistan. They were simply unwilling to participate in the study despite several requests. It was also difficult to conduct interviews with influential local people (identified by fellow villagers) who questioned the main objectives of the research and were scared of being recognized and prosecuted by the irrigation officials. Conducting interviews with such people could have helped in framing various water governance insights for future research. This underlines the limitations of water governance in transboundary river basins at the local, provincial, and international levels.

Third, flow data provided by the Government of Pakistan is not sufficient to accurately characterise how data limitation influences the norms and principles locally, provincially, nationally, and regionally. Additional features of governance such as accountability, adaptiveness, non-water related actors, participation, and politics are not addressed comprehensively for the same reason. Fourth, there is a lack of research studies i in the underdeveloped Kabul River Basin (KRB). There is very limited relevant published work and that is mostly carried out by international donors (i.e., Asian Development Bank, Department for International Development, the World Bank, etc.) and research think tanks (IUCN, WWF, IWMI, and SIWI, etc.).

Finally, apart from the flow and utilisation data, Afghan Government officials working in water-related organisations are strictly prohibited to discuss their views on shared-water resources particularly with researchers and officials from Iran and Pakistan. Therefore, most of the findings in the accessible official policy documents were cross-verified with water experts from INGOs, donor organisations and research think tanks working in Afghanistan. In Pakistan, the big hurdle was to get access to the official agreements and master plan of USD 62 billion worth of China-Pakistan Economic Corridor (CPEC) projects (see 8.2.1); which according to some leaked reports to the international media and leading Pakistani newspapers (such as The DAWN, The NEWS, The Express Tribune) are related to China's geo-strategic influence and the development of commercial agri-businesses by having access to fresh- and marine-water of Pakistan.

Additional limitations were including the views of indigenous Pakhtun women at the local level in the KRB due to cultural barriers where it is almost impossible to interview a Pakhtun woman. I could try to include the voices of Pukhtun women by considering the gender aspects in my analysis but it would have been inadequate as had to rely on a female researcher. The changing geo-political and geo-strategic situation in the region has significantly affected the mobilization across the border forcommon people in both Afghanistan and Pakistan. It is very unfortunate that the KRB has, for four decades, hosted the conflict in the region.

2.8   RESEARCH ETHICS

I am affiliated with both IHE Delft, Institute for Water Education and the University of Amsterdam for my PhD while my research was funded by NUFFIC's Netherlands Fellowship Programme (NFP). Additionally, after utilising the NFP scholarship, I secured partial funding from the Higher Education Commission (HEC) of Pakistan. I have presented parts of my research at international conferences, workshops, and seminars during my PhD. Additionally, parts of my thesis findings were published in peer-reviewed articles. I have always included details of both affiliations and funding sources in my oral presentations and articles.

In my research design, I followed the ethical rules of the AISSR[1] at the University of Amsterdam. These rules provide specific guidelines for conducting human research, including prior permission for conducting interviews, ensuring confidentiality, and anonymity. Moreover, at the start of each interview, I always introduced myself as a PhD researcher associated with IHE Delft and the University of Amsterdam. I presented an official letter from the affiliated institutes to the respondents and informed them about their roles and purpose of the research.

I fixed appointments for conducting interviews with the representatives of business communities and officials from NGOs/INGOs/donor agencies by submitting a letter of introduction to the heads of those organisations. During the interviews, I avoided voice recording devices due to the sensitive nature of some of the information shared by the interviewees. Although all the respondents allowed

---

[1] AISSR Ethical Procedure and Questions: http://aissr.uva.nl/research/ethics-and-integrity/ethics-and-integrity.html

to note-taking during the interviews, I still assured them about the confidentiality and anonymity of the process. I have also anonymised the names of individuals, communities and households in order to protect their identities (see Annex C)[2].

---

[2] AISSR ethical consideration are not technical and too general and obvious to be referred

3

APPROACHES TO TRANSBOUNDARY WATER GOVERNANCE

3.1 INRODUCTION

The previous chapter discussed general aspects of power and institutional theory. This chapter goes into more detail about (neo-) realist and institutionalist theories in international relations before zooming into their application to water. Hence, this chapter addresses the question: How does power shape institutions and how do institutions limit the role of power in transboundary water governance at multiple geographic levels? To answer this question, this chapter will address subsidiary questions: How do realist and institutionalist perspectives differ in international relations as well as in transboundary water governance? How does power influence freshwater governance institutions at multiple geographic levels? How does a combined approach of water governance and institutions help understand the influence of power at multiple geographic levels?

To answer these questions, this chapter first discusses key approaches in International Relations (IR) i.e., realism and neo-realism (see 3.2) as well as institutionalism and neo-institutionalism (see 3.3). I then discuss concepts of hydro hegemony (see 3.4) and water governance and institutions (see 3.5) which are particularly relevant to transboundary water resources. Third, the concepts of power and institutions are presented as a hybrid approach to see how power shapes institutions and how institutions shape power (see 3.6) before inferences are drawn (see 3.7).

3.2 INTERNATIONAL RELATIONS (IR) APPROACHES IN TRANSBOUNDARY GOVERNANCE

3.2.1 Introduction

Two of the important and relevant theories of international relations (IR) that will be discussed in this thesis are realism and institutionalism. These two theories lead to the debate on interaction between power and institutions in transboundary river basins. These theories are of particular importance for this thesis because the impact of both is noticeable on the literature and politics concerning environmental security, conflict and cooperative management.

3.2.2 Realist and Neo-Realist Approaches in Transboundary Governance

The theory of realism emphasises that all sovereign states are driven by their national interests (Keithly 2013; Meagher 2017). For some states the attainment of more resources or land might be the key goal (Mearsheimer 2007), while others may want to extend their own political or economic systems to other areas (Meagher 2017) and some states may merely wish to be left alone (Mearsheimer 2007). For realists, survival is key and this can be achieved by maximising power (Russell 2000; Sheldon 2003). All nation-states must rely on their own resources to safeguard their national interests and implement treaties with other states or maintain a desirable national and international order (Goddard and Nexon 2015). Pease (2012) presents four main assumptions of realism: first, the political state is the strongest actor to contest with other states in the global political system (Mearsheimer 2001); second, the administrative structure at global level is anarchic because no universal power exists that can impose rules over the countries (Lechner 2017; Ozkan and Cetin 2016); third, rational actors in the global political system strive to enrich their own interests (Mearsheimer 2009); fourth, countries compete to gain power (Antunes and Camisão 2017).

The theory of realism has traditionally been widely accepted in the field because: first, realists argue that it is a realistic and practical approach (Bortolotti and Antrobus 2015; Horton 2017). Second, it is better able to explain outcomes of international relations (Booth 2011; Voinea 2013). Third, it draws on clear assumptions about human nature and behaviour which allow for predictions about conflict and cooperation among states (Mearsheimer 2007; Brown 2009). Realism has four main branches: (1) Classical realism which considers states as the only actors and gives importance to military power and maximisation of power (Mearsheimer 2009); (2) Neo-realism where states seek to survive within an anarchic system (Waltz 2010); (3) Defensive realism in which states are considered as security maximisers (Person 2017; Toft 2005a); and (4) Offensive realism where states are considered as power maximisers (Person 2017).

Neo-realism is the most relevant in this thesis. Neo-realism is defined by its ordering principle ‗anarchy' and by the distribution of competencies (Gorissen 2016; Humphreys 2007; Waltz 2010). The principle of anarchic organisation of the international structure is dispersed which means that there is no recognized central authority and that all sovereign states are formally equal in this system (Powell 1994; Andreatta and Koenig-Archibugi 2010). These states act for self-help and pursue their own interests and do not subordinate their interest to the interests of other states (Mearsheimer 2014). For states, the dynamic force of survival is central in influencing their behaviour and in turn states acuire more offensive capabilities to enhance their relative power (Ngan 2016; Pashakhanlou 2009; Toft 2005a). Because states can never be sure what the future of other states will look like, there is a lack of trust between states, which means they havd to protect themselves against relative power losses. This is labelled as the security dilemma which is the result of trust deficiency in uncertain situations (Mearsheimer 2014). The wish and comparative capability of each state to enhance its relative power is mutually exclusive, leading to a power balance which shapes international relations (Toft 2005b).

Another relevant theory for this thesis is Hegemonic Stability Theory (HST) which states that the international system remains more stable when a single country is the dominant world power (or hegemon) (Gilpin and Palan 1987; Goldstein 2005: 81-82). A hegemon deploys its ‗preponderance of power' when it exercises leadership either through diplomacy, coercion or persuasion (Goldstein 2005: 81). The fundamental notion behind HST is that the strength of the global system in terms of politics, international law, etc., relies on the hegemons to develop and implement the rules of the system (Liu and Ming-Te 2011; Silvia and Stanaitis 2013). For a country to rise to the level of hegemon, it must have some qualities (Adams-Jack 2015). First, it must forge political stability, military and economic strength and strong national power to make new international laws and organisation or strengthen the existing laws and policies in terms of implementation (Toft 2005a); second, a growing economy (Silvia and Stanaitis 2013); and third, a superior geographic position, technological innovation, ideology, superior resources, and other factors (Yilmaz 2010; Liu and

Ming-Te 2011). According to Mearsheimer (2001), the hegemon creates, shapes and maintains the system by coercion as long as it is in its interest. Neo-liberals approach, the neo-realists in this regard argue that the hegemon wants to uphold its dominant position without incurring the implementation costs, thus endows influence to those institutions that support their own interests (Dirzauskaite and Ilinca 2017; Kromah 2009). This happens through institutions which favour the hegemon, but also offer protection and stability to the rest of the world (Ikenberry 2005; Ikenberry 2008; Dirzauskaite and Ilinca 2017). With the decline of the hegemon, the institutions do not automatically disintegrate because they were created in a way that benefited all stakeholders (Keohane 2005; Koremenos et al. 2001) instead, they acquire a life of their own. The hegemon bears the costs because it is good for all actors and thus creates stability in the system (Ikenberry 2014). However, it is also possible that the decline of a hegemonic power may lead to the possibility of a collective power, or the disintegration of the regime (Snidal 1985; Morris 2015). This theory could be downscaled to regional level – something I will try to come back to in the last chapter.

3.2.3 Institutionalist and Neo-Institutionalist Approaches in Transboundary Governance

Neoliberal institutionalists contend that countries act in their own interests, yet hold a positive view on cooperation. Keohane argues that cooperation among the states can create tension as it is not an easy task, but states also know that there could be potential benefits from cooperation (Keohane 1988). Snidal (1991) believes that relative gains are likely to have minimal impact on collaboration when the absolute benefits of collaboration are substantial (Snidal 1991cited in Keohane and Martin 1995). Institutions offer a coordination mechanism which allows states to take advantage of the potential benefits of cooperation. This _built-in focus' increases the opportunity for collective outcomes (Keohane and Martin 1995: 45). Institutionalists argue that states through cooperation seek to maximise their absolute gains and care less about the relative gains of other states (Rees 2010). In this approach, cooperation and peace can be achieved by norms, regimes, and institutions that mediate between the international power arrangements and the negotiations that take place within them (Jägerskog 2001 cited in Rees 2010: 12).

The three important forms of institutions include regimes, constitutions, and social institutions. A regime is denoted by the process of giving something the character of an institution while constitutions are spoken or written law and the techniques for making laws themselves (Badie et al. 2011). The interactions between citizens within a society is regulated by social institutions which leads to the creation of a social order (Tuomela 2007). Similarly, constitutions legalize the relationships amongst the state, its citizens, and their representatives, and thereby create a civil mandate (Murphy 2007; Weingast 1993). Regimes often exist on the international stage, where international agreements and treaties are designed (Keohane 1982; Young 2014). Markets are ealso considered as economic institutions (Reis 2012; Weingast 1993). Some earlier theories state that institutions can influence individuals in two ways: they can encourage individuals within institutions to enhance benefits (regulatory institutions, also referred to as rational choice institutionalism), or to act out of duty (normative institutions) (Valli 2015). Generally, the empirical literature on institutions includes historical institutionalism (Capoccia 2016; Steinmo 2008; Tilly 1984), neo-institutionalism (Tsakatika 2004; Lecours 2005; Ball and Craig 2010; Bates and Block 2013) as well as modern transaction cost theories of institutions (Nolan and Trew 2015; North 1990). In this section I will explain institutionalist and neo-institutionalist approaches along with their strengths and weaknesses.

Institutionalism takes into account the processes that affect the formation of rules and norms as authorised guidelines for social behaviour and the actual outcomes of such processes (Puffer et al. 2016; Scott 2008a). The theory on institutions fundamentally describes how these processes are formed, implemented, altered (Scott 2005; Suddaby et al. 2013), and discarded (Katsikas et al. 2016) over time and space and how they become increasingly resilient (Scott 2008b).

Institutions include normative, cognitive, regulatory elements and related activities as well as means which give strength and importance to social life (Palthe 2014a). Institutions can be communicated through various types of carriers, such as symbolic systems, relational systems, routines, and objects (Palthe 2014b). Usually, institutions function at different levels of governance, from global to local and interpersonal relationships (Guirdham 2011). Institutions provide stability and are also subject to incremental and irregular change processes (Tokaranyaset 2013). It is argued that ‒institutional theory is policy-making that indicates the formal and legitimate characteristics of government structures" (Kraft and Furlong 2012: p.81). Furthermore it can be claimed that _institutional theory is a broadly recognised theoretical posture that points out rational myths and legitimacy‘ (Scott 2008a: p.183).

Moving from this, neo-institutionalism advocates that _institutions matter‘ since they form various administrative methods and apply a sovereign influence on political results (Steinmo, Thelen, and Longstreth 1992: 7). Neo-institutionalism offers a way of seeing institutions outside the outmoded view, explaining why and how institutions evolve in a particular way within a given context (Leicht and Jenkins 2009; Lewis 2015). Neo-institutionalism suggests that institutions operate in an open environment in the presence of other institutions, labelled as the institutional environment (Caravella 2011; Vijge 2013). Therefore, institutions are a critical variable in policy analysis while structuring the contribution of social, economic and political forces and can affect policy outcomes (Bulmer, 1998, p.369). Wider contextutal factors shape institutions where the main objective is to survive and gain legitimacy (Deephouse and Suchman 2008; Glover et al. 2014). To achieve this, they should be economically successful as well as legitimised within the institutional world (Powell and Staton 2009; Glover et al. 2014). According to Hay (2002), neo-institutionalism stresses on the arbitrating role of the institutional setting in which political developments take place. By considering institutions more as social structures and taking into account the effect of institutions on individual preferences and actions, neo-institutionalism has stimulated away from its institutional origins and has become a descriptive discipline within politics (Lecours 2005). Neo-institutionalism should not be reflected as a coherent and integrated theoretical perception, but to a certain extent as involving various conflicting choices (Awesti, 2007, p.8–9). Institutional approach correlate MLG by offering a situation of inadequate impact of stakeholders in policy making. Its relation to power is explored below (see 3.6).

3.3   Hydro Hegemony (HH)

Scholars working on transboundary water have developed the theory of Hydro Hegemony during the last two decades. It was first formally coined in 2004 by the _London Water Research Group (LWRG)' based at School of Oriental and African Studies (SOAS) London, King's College London; and the London School of Economics and Political Science (Zeitoun and Warner 2006). These scholars examine power asymmetries by taking into account the varying intensities of conflict and geographical position of the riparian States and between riparians states. The evolution of hydro hegemony is summarised in Table 3.1.

Table 3.1: Evolution of hydro hegemony (HH) concept

Year	Context	Objectives	Contribution to the Theory
2004-2006	Conceptualisation & theorisation of HH	Politicising water-based relations using IR theories	Analysing power, position & exploitation potential in transboundary rivers
2007	HH & water resource sustainability	Ensuring equitable & efficient use of water resources	Political stability, socio-economic growth; counter-HH strategies
2008	HH & International Water Law (IWL)	Linking water resources with international law to counter HH	Integration of international legal principles in negotiation
2009	Transboundary water governance institutions	Role of the relative distribution of power for resolving water disputes	Contribution of institutions in conflict management
2010	HH & critical hydro-politics	Effective implementation of transboundary water management	Enhancing stakeholder participation in freshwater negotiations
2011	The use of soft power in transboundary rivers	Evaluating the impacts of soft power	Soft power influences choices & treaties to manage, delay, or perpetuate conflict
2012	Bilateral management through common basin governance	Initiating discussions concerning peace & development	Ensuring equity, human rights & social justice
2013	Transboundary water interactions and the UN Watercourses Convention	Applying transboundary level rules & conventions for improved water resources management	Management & governance through political processes
2014	HH theory to inform virtual water trade	Exposing relationships between access to water, global trade, and power	Virtual water shaped by material, bargaining, & ideational powers
2015	Power shift & HH	Political & economic changes challenge the regional balance of power	Donor's investment leads to power changes & new kind of relationships
2016- & beyond?	Countering multilevel HH	Multilevel institutional analysis of transboundary rivers	Inclusive and sustainable development approaches to depoliticise transboundary rivers

HH scholars draw from the realist/neo-realist accounts of power (e.g. Mearsheimer 2001; Lukes and Haglund 2005, see 2.1.2) and hegemony (e.g. (Gilpin 2005; Lustick 2002); institutionalist/neo-institutionalist theories of regimes (Keohane 1982), knowledge and discourse (Hajer 1995), water conflict (e.g. Wolf 2004), and water conflict intensity (e.g. Yoffe, Wolf, and Giordano 2003), and has roots in sovereignty theory in law. In this context, some scholars (Zeitoun and Warner 2006;

Zeitoun and Allan 2008; Woodhouse and Zeitoun 2008; Daoudy 2009) have revealed that different forms of power regularly used in water-based relations include incentives (e.g. treaties), coercion (e.g. pressure), manipulation (e.g. threat of military operations) and perception change control (e.g. through knowledge construction). These forms of power, used in various combinations, allow one riparian state to influence another state and thus control the flow of shared water resources. It has been observed that hegemonic struggle varies from institutionalised rules not requiring enforcement (Finnemore and Sikkink 1998; Lukes and Haglund 2005) to a direct threat (hard power) to a situation where power and consent are only implied (soft power).

HH scholars claim that powerful riparian states can gain maximum benefits from shared rivers; however, if the powerful riparians intend to control the flow or persuade co-riparians to take decisions in line with their own preferences, there may be different water outcomes (Zeitoun and Warner 2006: 439). The outcomes may include, (i) benign outcomes (e.g. when states agree to cooperate in situations of water stress); (ii) neutral restrictive outcomes (e.g. when human rights are not fulfilled); (iii) obstructive oppressive outcomes (e.g. when states intervene to allocate water according to their own preferences); and (iv) contested control outcomes (when changing water-related circumstances may ignite a conflict due to no agreement) (Woodhouse and Zeitoun 2008: 117). HH may be _negative and dominative', in which the hegemon desires to maintain power asymmetries and structural inequalities (Zeitoun and Warner 2006: 439). Sometimes, a powerful riparian state can take a leading position to provide benefits to all riparian states and share the water resources more equitably as has occurred in the Orange River Basin (Turton and Funke 2008). Sometimes, colonial legacies influence power-based outcomes in river basins (Sanchez and Gupta 2011), for example favouring lower riparians Egypt and Sudan on the Nile (Waterbury and Whittington 1998). These treaties are biased and are in favour of the more powerful states (Dinar 2000; Cascão 2008).

Essential elements of hydro hegemony theory are: (a) hydro-interactions (the relationship over shared-water resources among riparians) which range from cooperation to conflict with various intensities in between (see (Furlong 2006; Zeitoun and Mirumachi 2008; Zeitoun and Warner 2006); (b) asymmetric power relations (where one riparian state is powerful and the other one is weak in a transboundary river basin) (Zeitoun and Warner 2006; Warner and Zeitoun 2008); (c) exploitation potential (resource capture strategies of a powerful country to affect water quality or quantity by developing dams and reservoirs over the shared water resources) (see (Allan 2003; Warner 2006); and (d) the geographic position of riparian states in a transboundary river basin (Zeitoun and Warner 2006). The physical geography of a river has a substantial effect on the quantity, quality, and nature of relations among riparian states. The geographic position of a riparian state in a transboundary river basin plays an essential role in determining the foreign policy and negotiating process of riparian states. Furthermore, it also influences the degree of similarity or difference of interests and capabilities among them (Dolatyar and Gray 2000). Different elements of HH and how they influence different situations are demonstrated in Table 3.2.

Table 3.2: Influence of geographic power on material, bargaining and ideational power

Elements of Political Power				Elements of Geographic Power		
	Type	Elements	Upstream Riparian (UR) Position	Downstream Riparian (DR) Position	River length & drainage area
	Material	Economy	UR can potentially divert river waters to enhance trade and aid (Kehl 2017) UR is more likely to attract (more) foreign direct investment (Warner et al. 2014) UR can potentially prevent/increase pollution (Arfanuzzaman and Syed 2018)	DR has access to fertile flood plains; Agricultural production/trade dependence (Kehl 2017) DR may control the port, hence trade dependence (Kehl 2017)	Having a longer river length and higher drainage area can increase water-related power (Arfanuzzaman and Syed 2018; Kehl 2017)
		Military	Negligible, no causal relationship	Negligible, no causal relationship	Negligible, no causal relationship
		Population	Negligible, no causal relationship	Negligible, no causal relationship	Negligible, no causal relationship
		Technology	Negligible, no causal relationship	Negligible, no causal relationship	Negligible, no causal relationship
		Pol. Stability	Negligible, no causal relationship	Negligible, no causal relationship	Negligible, no causal relationship
	Bargaining	Strategic relations with powerful states	UR has a better negotiating position as it can divert the water (Brochmann and Hensel 2011; Kehl 2017; Song and Whittington 2004) UR could potentially coerce DR (Menga and Mirumachi 2016)	DR has weak negotiation power in non-navigable rivers (Brochmann and Hensel 2011; Dinar et al. 2013; LeMarquand 1977)	Negligible, no causal relationship
	Ideational	Power of ideas	UR could use water/climate information to shape ideas and norms (Petersen-Perlman and Fischhendler 2018)	DR could attracting support from powerful international actors including researchers (Abdolvand et al. 2015; Eynon 2016; Hensengerth et al. 2012)	Negligible, no causal relationship

HH Theory has been applied in various transboundary river basins: in the Nile (Link et al. 2010), Tigris and Euphrates (Warner 2006), Upper Jordan Waterscape (Zeitoun et al. 2013), Amu Darya Basin (Wegerich 2008), Orange (Turton and Funke 2008), and Brahmaputra (Lahiri and Sinha 2012) basins. Various scholars have evaluated HH theory in terms of its contribution in water resource literature. Some of the significant strengths and weaknesses of HH theory are presented in Table 3.3 below.

Table 3.3: Strength and weaknesses of hydro hegemony theory

Strengths	Weaknesses
Explains important role of power in transboundary water management and allocation (Cascão 2008) especially when treaties are made in an asymmetric environment (Zeitoun et al. 2013: p.201); Explains hydro-political interactions by reference to power (e.g. see (Frey and Naff 1985)	Ignores the role of institutions (Selby 2007); Assumes state sovereignty as an uncontested attribute in international relations (Zeitoun and Warner 2006)
Illustrates how power is exercised (Zeitoun et al. 2013: p.201)	Unable to explain how and why cooperation emerges (Selby 2013); Views inter-state interactions as zero-sum power games, (Lopes 2012)
Highlights and inform the law about the injustices (Woodhouse and Zeitoun 2008)	Can never address issues of justice (Zeitoun 2013)

3.4   WATER GOVERNANCE AND INSTITUTIONS

Water governance and institutions scholarship tend to focus on what needs to happen as well as what is already happening. These include discussions on the role of power explained above. Water governance scholars claim that institutions are formal and informal cooperative processes on water that aim at genuine cooperation, reducing transaction costs and leading to positive-sum interactions (Lopes 2012) using the latest science.

Water governance scholars call for covering all kinds of water i.e., surface, underground and reclaimed or recycled sources, all uses of water and all users of water (Hayat and Gupta 2016). This research analyses both individual and collective actions as well as decision-making for water resource development, allocation and utilisation (Rutten and Mwangi 2014). Due to the formal nature of rules in terms of their administrative arrangement, policy environment, and the legal framework, (Salman and Bradlow 2006), water governance can be conceptualised as interactions between water law, water policy and water administration (Saleth and Dinar 2000) as well as in relation to sociological and anthropological elements of water (Zwarteveen and Boelens 2014). Water governance considers both the formal as well as informal perspectives (Sehring 2009). There are a variety of factors that influence these three formal dimensions of water institutions (Saleth and Dinar 2003). For logical purposes, these elements can be clustered into endogenous elements (e.g., water conflicts, water inefficiency, degradation of water infrastructure, and water scarcity) that are internal to the water sector and exogenous elements (e.g., natural calamities including floods and droughts, international commitments, changing social values, political reforms, technical progress, demographic growth, and economic development) that are outside the strict boundaries of water governance (Saleth and Dinar 2004b; Hashemi et al. 2015). Since both the internal and external elements are linked in one way or the other, and their comparative effects differ by situation, it is quite difficult to separate their distinct roles or to generalise the direction of their effects (Saleth and Dinar 2000 2004a).

Water governance refers to a set of social, political, administrative, and economic procedures that oversee the development, management and regulation of water resources at multiple levels of governance (Rogers and Hall 2003). More specifically, institutions can be termed as procedures that define action sets; offer incentives; regulate outcomes, use, provision, and management; and explain action positions (Saleth and Dinar 2006). Veeman and Polityolo (2003: 322) see institutions as ―social decision systems that provide rules for the use of resources and for the distribution of resultant income or other benefit streams". This indicates that human actions are shaped and stabilised by institutions. However, the shaped human actions could in turn affect future institutions. In this regard, tradition and customary practices also become rules when society adapts them. These informal and formal rules define the social roles of individuals and groups in a particular situation of human relations (O'Riordan and Jordan 1999). Thus, water institutions may be thought of as consensual arrangements of behaviour constituting water conventions, or as rules and entitlements that define individual and group choice sets.

Water governance has taken place at multiple geographic levels (Barraqué 2011; Yang et al. 2014). Growing competition over freshwater resources has resulted in the generation of specialised literature within International Relations, which observe the conditions for cooperation and conflict in terms of water resources. The relationship between political decision-making and environmental stress have been well acknowledged in various academic articles examining the different scales at which water is managed during the 1950s (Toynbee 1946; Wittfogel, 1956).

Broadly, water management under certain rules can be divided into five phases. Initially, freshwater resources were managed collectively by communities through societal norms using simple infrastructures that required less monitoring and slight maintenance (Hellin et al. 2018; Pahl-Wostl et al. 2007). Customary water governance practices dates back to 2500 BC when Umma and Lagash (the two Sumerian city-states) drafted a treaty for ending a water conflict along the Tigris River (Wolf 2002). Subsequently in the second phase, when states emerged, the agrarian state was introduced, for instance, in Ancient Egypt, Mesopotamia and China (Ruddell and Sanchez 2012). In the third phase during the post-industrial revolution, water was tactically used outside irrigation and was stretched to all dynamic activities including the energy sector (Pingali 2012). During the eighteenth to nineteenth centuries, this practice was happening principally in Western Europe with the growth of transportation, the establishment of large-scale and significant agriculture, urbanization, and alteration of human settlements (Mosello 2015). In these two phases, a rich body of water institutions have evolved (Yoffe et al. 2003).

The Food and Agricultural Organisation (FAO) of the United Nations has revealed that from AD 805-1984, more than 3000 freshwater treaties were drafted (Hamner and Wolf 1998). The fourth phase is characterised by the predominance of state regulation and the rise of the private sector. During this period, large hydroelectric and irrigation projects were designed and employed as tools of regional politics for economic development of less developed countries (Mosello 2015). These governance modalities resulted in imbalances between the cost bearer and those who enjoy the benefits. Such imbalances occur when existing laws are not properly implemented by the state to protect the population and environment (Ciervo 2009; Mosello 2015). During this period, the state introduced a large number of rules for water governance. These were often inspired by scholarly works, such as those coming from the Institute of International Law (IIL). In 1911, through the Madrid Declaration, the IIL recommended the formation of joint water commissions to avoid modifications of transboundary rivers and discouraging basin alteration unilaterally. Furthermore, the recommendation in the Madrid Declaration led to the development of the Helsinki Rules of 1966 which introduced two essential water governance principles concerning the obligation of avoiding _significant harm' and the _equitable utilisation' of water resources among co-riparians of a transboundary river basin (Caponera 1985).

The UN asked the International Law Commission (ILC) in 1970 (i.e., four years post Helsinki Rules) to codify the rules for the non-navigational uses of transboundary watercourses. This assignment led to the adoption of the Convention on the Law of the Non-Navigational Uses of International Watercourses (UN Watercourses Convention 1997). This Convention endorsed the principles of _avoiding significant harm' and _equitable and reasonable utilisation'. It also shaped a framework for information and data sharing, conflict resolution, protection and preservation of shared water resources, and the instrument for joint management among the states sharing transboundary watercourses (Wouters 2000). More details regarding the legal institutions of water governance are provided in Chapter 5.

3.5 THE INTERRELATED ROLE OF POWER AND INSTITUTIONS

As discussed above, the role of institutions in changing human behaviour and shaping political outcomes gave birth to _Neo-Institutionalism (NI)' (Crawford and Ostrom 1995; Rutherford 1995). NI was built on the attributes of the _old or Historical-Institutionalism (HI)' (Peters 2000). Hall and Taylor (1996: 940) outline three distinctive characteristics of HI: (1) it emphasises the power asymmetries associated with the operation and development of institutions; (2) it integrates institutional analysis for political outcomes along with the contribution of other factors; and (3) it takes a view on institutional dynamics that underline path dependence and unplanned consequences. However, this can be seen as challenging the views of those who argue that environmental rules reveal preferences of actors who have exerted power during regime creation (Dimitrov 2003). To this, (Young 2004: p.215) argues that even _powerful actors are limited by their understanding of the institutional options available to them, a fact that highlights the role of knowledge in the growth of institutions'.

From the perspective of sociological institutionalism (SI), the influence of institutions is seen in almost every aspect of human life (Schofer et al. 2012). They can be habits and social protocols, cultural templates and frames of meaning (O'Riordan and Jordan 1999). From this viewpoint, institutional practices and techniques are seen to be embraced not on grounds of judiciousness but because of their _embeddedness' in society, culture, and organisational identity (Pai and Sharma 2005). These are aspects which may be significant in the study of local informal and formal institutions in water resources management (Hall and Taylor 1996; Peters 2000). Levi (1990) is of the view that formal rules can empower some groups where underprivileged groups may try for institutional change by retreating from their agreed position from the existing institutional arrangements. Collective action can be one form through which they withdraw their consent but it might also happen when many indivduals take decentralised actions.

Williamson (2000) argues that informal rules can provide the experience within which formal institutions are rooted and classifies four levels of institutions: First, informal institutions or _institutions of embeddedness'; second, _the high-level formal rules' e.g., property rights, laws, and constitutions; third, the _institutions of governance', which manage day-to-day dealings to reduce transaction costs; and fourth, _the prices and quantities' indicated in different agreements. For Bandaragoda (2000), numerous rules related to water are designed to constrain the socially undesirable behaviour in the distribution and use of water. The capability to attain the predictable effect depends largely on the institutional engagements that form the incentives and restrict human action (Alaerts 1997; Barrett et al. 2005).

There are number of institutions that must be examined in an attempt to change or strengthen hydro-institutional arrangements. First, the rules for water diversions from streams, rivers, lakes and even groundwater. Second, those rules or policies that set the conditions for allocating and reallocating water among different users (Easter 2004: p.1):1). The third set of institutions are the laws establishing water use rights (Agyenim 2011). Thus, institutions are dependent as well as independent variables; they shape and are shaped by the strategy of individual actors. When preferences shift, the institutions shift accordingly in order to balance power.

3.6   CONCLUSION

Using a lens of water governance and institutions is worthwhile in enlightening the reasons for conflicts over freshwater and their intensities as well as poor freshwater governance. Similarly, the concept of hydro hegemony explains the role of state power in sharing transboundary water resources with its riparian countries. As becomes evident, power can influence multilevel transboundary water governance institutions by: including or excluding relevant water and non-water related actors, including or excluding various water and non-water related issues and thereby shaping the agenda or preferences of different actors involved in the political/negotiation process, and when there is power asymmetry between actors in relation to (i) geographic, (ii) material, (iii) bargaining, and (iv) ideational power. The resultant institutions which are created by states with visible power asymmetries are fragile, inefficient, ineffective, and unsustainable. Institutionalist theories on the other hand claim that cooperation, and eventually peace, can nonetheless materialise through the development of rules which may act as an arbitrating element between the power arrangements of the global system. It promotes cooperation in trade, human rights and collective security among other issues even if there is anarchy. Such cooperation often emerges as different issues have different power constellations.

Hydro Hegemony (HH), water governance, and institutionalism are different theoretical approaches to transboundary water cooperation and conflict that are difficult to integrate within a single explanatory framework, but which balance each other and can be engaged to describe diverse parts of the research question and sub-questions on which this thesis is based. Whether they can be integrated into a _hydro-institutionalism' approach will be discussed in Chapter 10. In reality, the complex interplay of power and institutions is indeed a hybrid approach where, on one hand power influences transboundary water outcomes when there are asymmetric power relationships among riparians, while on the other hand, informal and formal institutions restraint the role of power politics in transboundary freshwater resources. Furthermore, there might be (a) unresolved historical non-water related issues (e.g. border dispute) and (b) lack of scientific and societal information that may hamper institution building and strenthening in the water sector. Hence, it is necessary to evaluate the role of power in in/excluding actors and issues; and how existing institutions can be improved based on diffusing unresolved historical issues first and then by providing additional scientific and societal information which could perhaps change the perceptions of riparian States in water negotiations.

4

ECOSYSTEM SERVICES AND HUMAN WELL-BEING

4.1   INTRODUCTION

Addressing transboundary water governance entails a thorough understanding of the reasons of water challenges in the context of rationalized knowledge about the vital role of biodiversity and the ecosystem services of water. This chapter builds further on Hayat and Gupta's (2016) article on the ecosystem services of freshwater. It discusses the state-of-the-art knowledge of freshwater systems within the context of a basin. It addresses the question: How can the various drivers of freshwater problems affect the ecosystem services (ESS) of different kinds of freshwater and how does this, in turn, affect human well-being? To answer this question, I first define the natural and anthropogenic, and the direct and indirect drivers (see 4.2) of the problem of flow (quality, quantity and related ecosystem services of water) in river basins. This chapter then discusses freshwater and its different kinds (including rainbow-water, blue surface-water, blue groundwater, green water, grey water, black water and white frozen water or snow) (see 4.3); various types of the ecosystem services of nature (ESS) (see 4.4) with differing contributions to human well-being (i.e. good life, health, good social relations, and freedom of choice through security), and at different levels of governance (in section 4.5). The last section (section 4.6) draws inferences.

4.2   DRIVERS OF FRESHWATER PROBLEM AT MULTIPLE LEVELS

In the scholarly literature, the _driver' symbolises the cause of a problem which can occur in two ways: directly (e.g., affecting the behaviour of an actor to withdraw or pollute freshwater) or indirectly (e.g. to influence the direct drivers and thus affect freshwater problems indirectly). The literature on freshwater resources states that demand for freshwater is surpassing its supply in various parts of the world (Postel, Daily, and Ehrlich 1996). Scarcity of freshwater resources and inadequate access to water supply and sanitation services intimidate health, socio-economic growth, and sovereignity of countries (Palaniappan et al. 2010). Both natural (e.g., winds, storms, precipitation, tectonic movements, climate and weather variability) and anthropogenic (e.g. land cover changes, urbanisation, industrialisation, technological advances and climate change) activities can drive or cause problems. In some situations, these drivers can worsen the existing freshwater problems. Direct and indirect drivers are discussed in the following sub-sections in more detail (see also Table 4.1). For analytical purposes, I have clustered these drivers into direct and indirect, however clearly the drivers in one category are interlinked with drivers in another category.

4.2.1   Direct Drivers

The key direct drivers are agricultural development (e.g. commercial agricultural practices including animal husbandry, the extractive sector and water use in energy), industry (including services and infrastructure), municipal water supply and sanitation services e.g., household uses (drinking water, sanitation and hygiene) and demographic shifts (i.e., migration, population growth, increase in population density and urbanisation).

The core direct driver of freshwater problems is agricultural development (Gupta and Pahl-Wostl 2013b). This comprises commercial agriculture containing animal husbandry at large-scale large-scale (Dore et al. 2012). Agricultural uses including subsistence agriculture are responsible for about 70% of water use world-wide.

Industries, services and infrastructure also use water. Beyond agriculture, extraction of minerals such as petroleum products (Braune and Adams 2013); energy production (Van Weert and van der Gun 2012); production of mineral water (Rodwan 2014) and other industries also use water. The service industry also uses water, as does different infrastructures in society. Although industrial development is more relevant at the national level, it can strongly influence the local level (in terms of reducing or restricting access for marginalised people) and transboundary (increasing demand in transboundary river basins) water management relevant at the transboundary level.

Household uses (e.g., drinking water, sanitation and hygiene) make up a significantly smaller use of freshwater resources. While volumes of these uses are small in comparison, they still constitute important drivers that can contribute to poverty eradication (Moench 2002) while contributing to the cumulative problem of water contamination (Gupta and Conti 2017).

Various other factors that put direct stress on freshwater resources include changes in demography such as population growth, migration, and localised increase in the population density via urbanisation at all geographic levels (Gupta and Pahl-Wostl 2013a). Moreover, it can also enhance the competition for freshwater while too little or too much water can also affect human populations. Furthermore, the increase in population can also influence the quality of freshwater resources (MEA 2005; UN-Water 2012). Table 4.1 below offers the summary of direct drivers and their applicability at multiple geographic levels.

4.2.2   Indirect Drivers

The indirect drivers affecting the governance of transboundary water resources at multiple levels of governance include political dynamics between states occurring at all geographic levels, culture and ethnic elements (e.g., using resources inefficiently, attitudes concerning access and allocation, etc.), non-water-related policies (economic development, land tenure, agriculture and food security, and land use), the drive for economic growth, poverty, technological advances (agriculture intensification), international trade, climate variability and change and other natural causes.

The political dynamics between states can aggravate the direct driver of unequal access or demand, and further intensify the lack of agreement regarding resource management (Zeitoun et al. 2013; Zeitoun and Warner 2006). Both ethnic elements and culture can also serve as indirect drivers of freshwater problems. The attitude of actors towards inefficient use of resources, access and allocation, and public obligation in terms of environmental quality can result in behaviours that result in unnecessary use or pollution (Cullet and Gupta 2009). People's attitudes linked to culture are usually local through to national in character although there can also be a regional dimension.

Non-water related policies including those on growth and poverty eradication, agriculture, industry, infrastructure, land use, land tenure and trade are key drivers of water use and pollution (Braune and Adams 2013; Foster and Garduño 2013; Gupta and Pahl-Wostl 2013a; Moench 2002). Similarly, encouraging economic development for poverty alleviation may also result in land use changes that can contaminate freshwater or decrease its flow (Hoff 2009; Warner et al. 2013). This can also result in economic activities that cause unsustainable use of freshwater resources (Warner, Sebastian, and Empinotti 2013). The pursuit of profit in the economy often leads to the externalisation of the environment. Poor people also make choices that may lead to water problems. Technology is also an indirect driver as it can potentially lead to agricultural intensification and enhance the ability of water users to affect the quality and flow of freshwater resources (Söderbaum and Tortajada 2011).

In terms of economy and trade, market biases and demand for water-intensive products can take place by subsidies or _fee-market' trade rules. These actions can possibly drive freshwater challenges if they escalate demand and production in the agricultural and other related industries, primarily if these demands result from market alterations or marginalisation of environmental costs in pricing (Gupta and Pahl-Wostl 2013a; Söderbaum and Tortajada 2011). Table 4.1 offers an overview of indirect drivers and their applicability at multiple geographic levels. Climate variability and change is an indirect driver which is the most difficult to grasp as it is extremely complex in origin and operates across all geographic levels (Gupta and Pahl-Wostl 2013a). The special effects of climate variability and change on freshwater resources and how it can possibly distress the quality and quantity of freshwater are discussed in Chapter 1 (see 1.2.3). The increased frequency of droughts and floods due to climate variability (Villholth et al. 2013) may at one point reduce water flow in the rivers while at another point the river might overflow due to excessive rains or melting of glaciers. Moreover, it may put some regions at risk due to arsenic and fluoride mobilisation as well as salinisation of freshwater resources (Van Steenbergen 2006). In addition, there are other natural factors such as tectonic movements, where earthquakes lead to changes in freshwater flow and level, and drinkable water resources become contaminated affecting micro-biota therein (Galassi et al. 2014), or hydraulic connections are formed between two diverse resources that were not earlier linked (Malakootian and Nouri 2010).

Table 4.1: Drivers of freshwater challenges at multiple geographic levels

Direct Drivers	Geographic Levels			Key References
	T	N	P/L	
Agriculture development (e.g., commercial agriculture practices including animal husbandry, the extractive sector & water use in energy)	X	X	X	Gupta and Pahl-Wostl 2013; Dore, Lebel, and Molle 2012; Van Weert and van der Gun 2012; Rodwan 2014
Industry (including services and infrastructure)	X	X	X	Van Weert and van der Gun 2012; Rodwan 2014
Municipal water supply and sanitation services e.g., household uses (drinking water, sanitation, and hygiene) and subsistence agriculture	NA	NA	X	Moench 2002; Braune and Adams 2013; Gupta and Conti 2017
Demographic shifts (i.e., migration, population growth, increase in population density, urbanisation, population growth)	X	X	X	Dore, Lebel, and Molle 2012; Gupta and Pahl-Wostl 2013; Assessment 2005; Water 2012a
Indirect Drivers				
Political dynamics between states	X	NA	NA	Zeitoun et al. 2013; Zeitoun and Warner 2006
Culture and ethnic elements (attitudes regarding access and allocation, wasteful use of resources, etc.)	X	X	X	Gupta and Pahl-Wostl 2013; Dore, Lebel, and Molle 2012
Non-water-related policies (agriculture & food security, land use, land tenure, economic development)	X	X	X	Foster and Garduño 2013; Hoff 2009; J. Warner, Sebastian, and Empinotti 2013; Gupta and Pahl-Wostl 2013
Economy (economic growth)	X	X	X	Hoff 2009; Gupta and Pahl-Wostl 2013
Poverty	X	X	X	Moench 2002; Braune and Adams 2013
Technological advances (agriculture intensification)	X	X	X	Söderbaum and Tortajada 2011; Gupta and Pahl-Wostl 2013
International trade (e.g. globalisation' or trade in virtual water)	X	X	X	Hoff 2009; Söderbaum and Tortajada 2011
Natural change and variability in weather, Droughts; Floods; Earthquakes; Landslides, tectonic movement.	X	X	X	Lashkaripour and Hussaini 2008; The World Bank 2009; IUCN 2012

4.3   FRESHWATER AND ITS TYPES

Having discussed the drivers of freshwater problems, I now turn to discuss the topic of freshwater itself. Freshwater is natural water in the ground and on the surface which has low absorptions of total dissolved solids and salts (Penuel et al. 2013). Freshwater does not mean that the water is directly potable; it is _sweet water' in contrast to _salt water' (Hendrickson III 2014). Freshwater habitats can be divided into still water systems including swamps, mires lakes, and ponds; running-water systems such as rivers; and groundwater structures which flow in aquifers and rocks. In addition, there is another type of water system known as green water which connects running-water systems with groundwater and which underlies many larger rivers containing more water than open water networks (Vihervaara et al. 2010).

The precipation from the atmosphere in the form of mist, rain and snow is the primary source of almost all freshwater. Two-thirds of the global freshwater resources are locked up in permanent snow and glaciers while the remaining one-third is groundwater (Shiklomanov 1997). Out of the total water on earth, only 2.5 - 2.75% is freshwater; approximately 0.3% is surface-water (Ostfeld et al. 2012) containing 1.75 - 2% which is locked in glaciers, ice and snow; groundwater and soil moisture is roughly 0.5 - 0.75%; while surface water in rivers, swamps, and lakes is less than 0.01% (Alcamo et al. 2017; Pidwirny 2006). Approximately 87% of the surface freshwater is contained in lakes comprising the African Great Lakes which has 29% of the freshwater, the North American Great Lakes which has about 21%, the Lake Baikal in Russia which stores around 20%, while other lakes which have 14% of the surface freshwater resources. This means that only a tiny portion of freshwater is in rivers. Freshwater provides various ecosystem services to more than six billion people and supports approximately 126,000 freshwater species (Chapagain and Orr 2008).

Freshwater includes green and blue water including groundwater (Falkenmark 2003; Mekonnen and Hoekstra 2011). When these waters are in their vapourised stage (0.04% of total water) (Gleick 1996), they are referred to as rainbow water (Braun and Smirnov 1993); once it is used by humans, crops and industries, it emerges as grey (polluted water which needs water to dilute it) and black water (polluted with human feces and urine) (Andersson 2016). The water balance models of hydrologists have revealed that blue water (blue surface water and blue groundwater) is only one-third of the total global precipitation; this has led water resource scholars to investigate all steps between rainfall and blue water flows (Ukkola and Prentice 2013). With the current trend of agricultural intensification, there will not be enough blue water to sustain the irrigation needed for the agricultural sector (Barlow 2009; Barlow and Clarke 2017). The distinction between these water resources is essential because each kind of water has distinct biological, chemical and physical characteristics, and hence with specific ESS (see 4.4).

4.3.1   Rainbow Water / Atmospheric Moisture

Rainbow-water, also known as atmospheric moisture, is depleted by precipitation and replenished by evaporation (Sigurdsson et al. 2000); 90% of the evaporation is from oceans and water bodies and

10% from terrestrial evapotranspiration (Van Noordwijk et al. 2014). Its volume is about 12,900 km^3 or 0.001% of the total water volume. If all atmospheric water rained down at once, it would only cover the Earth to a depth of about 2.5 centimetres (Sheil 2018). Water exists in three main forms in the atmosphere: (i) gas (as water vapour), (ii) liquid (as rain drops) and (iii) solid (as ice crystals). Atmospheric rivers refer to intense humidity in the atmosphere and include water vapour (Zhu and Newell 1998). Atmospheric rivers are long in length (i.e. several thousand kilometres) while they are only a few hundred kilometres wide, but have the capacity to take more water than the largest river on Earth – the Amazon River (ibid).

4.3.2   Blue water

The concept of blue water, first coined in 1995 (Zhao et al. 2015), has been used by managers and scholars globally (Falkenmark and Rockström 2006; Pittock 2011). In the early days, water consumption data only included the surface blue water extraction by the agricultural sector, industries and municipalities, while not taking into account its other possible uses (Vanham 2012). Blue water includes surface water and groundwater resources (Wang et al. 2013) and is roughly 42,700 km^3 (Shiklomanov 1997; Vaux 2012) stored in underground aquifers, lakes, and rivers (Pittock 2011). Surface blue water requires to be sustainably used to avoid negative impacts on the environment (Hoekstra et al. 2012). In arid and semi-arid zones (e.g., Southern Europe, the Southwest of the United States, North Africa, the Arabian Peninsula, Central and South Asia and parts of Australia), the share of surface blue water is the biggest (Mekonnen and Hoekstra 2011).

4.3.3   Groundwater

Groundwater is a subset of blue water and refers to water in the Earth's crust in all physical states (Winter 1999). It is fed primarily by rainwater: water that does not flow in surface-water streams and is not utilised by plants and trees or evaporates, percolates into the underground aquifers (Sophocleous 2002). Some of the ESS that is associated with groundwater is of high economic value as these services support a range of production and consumption processes (Emerton and Bos 2004). In comparison to blue surface water, blue groundwater has a very long residence time – averaging about 300 years (Robinson and Ward 1990).

4.3.4   Green water

The volume of water that is deposited in the soil after rainfall is known as green water (Vaux 2012). In other words, green water is the rainwater that does not become run-off (Wang et al. 2013) or groundwater. The concept of green water - also coined in 1995 (Zhao et al. 2015) but has received less attention than blue water (Falkenmark and Rockström 2006). Approximately 60% of freshwater flow is green water (Dent 2005), and accounts for approximately 80% of global crop production (Liu et al., 2009). Green water is used by the roots of plants and evaporates through plant transpiration processes (Zang et al. 2015).

4.3.5   Grey water

The wastewater produced from household uses like washing clothes or bathing can be reffered to as grey water or polluted water (Wang et al. 2013). Greywater has comparatively higher potential for reuse due to the small level of contamination (Allen, Christian-Smith, and Palaniappan 2010). The reuse of greywater can potentially compensate the demand for new water supply up to some extent (López-Zavala, Castillo-Vega, and López-Miranda 2016). Furthermore it can meet a wide ranges of economic and social needs (Allen, Christian-Smith, and Palaniappan 2010) and may also possibly reduces the energy and carbon footprint of water services (Griffiths-Sattenspiel and Wilson 2009). This polluted water includes leached nutrients and pesticides from agricultural practices (Tsuzuki et al. 2010) and drain water from hand basins, kitchen sinks and laundries, without any input from toilets (Bergdolt et al. 2012).

4.3.6   Black water

Black water, also called drain water, sewage, brown water and foul water, coined in the 1970s, is the water generated by toilets (Tsuzuki et al. 2010). It is different from industrial wastewater which has been used for making a commercial product. It is drain water containing urine, feces and discharge sewage from flush toilets including anal cleansing water (Bergdolt et al. 2012). It contains a higher amount of organic solids, nutrients and pathogens (ibid) and is distinct from grey water (Oteng-Peprah and Acheampong 2018) by having lesser amounts of detergents and greater concentrations of organic pollutants and nutrients (Santos et al. 2014). It is essential to decompose pathogens that exist in black water on a priority basis before releasing water safely into the environment. However, decomposition can be challenging if it contains large quantities of excess water or pathogens (ibid).

4.3.7   White Frozen water / Glaciers

The water cycle expresses how water travels above, on, and through the Earth (Vörösmarty and Sahagian 2000). However, more water is stored in ice-sheets and glaciers as compared to the water in the water cycle at any point in time on the Earth's surface (Radić and Hock 2014; Siegert 2006). Approximately 90% of the ice mass of Earth is in Antarctica (Rignot et al. 2011). Similarly the Greenland ice cap covers 10% of the total global ice quantity (Hanna and Braithwaite 2003; Nordhaus 2018). The ice cap averages about 5,000 feet (approximately 1524 meters) in thickness, but can be as thick as 14,000 feet (approximately 4,268 meters) (USGS 2018). The National Snow and Ice Data Centre of the United States of America (USA) reveals that seas would rise by about 230 feet (approximately 70 meters) if all glaciers melted today (ibid).

4.4   ECOSYSTEM SERVICES OF DIFFERENT KINDS OF FRESHWATER

4.4.1   Defining Ecosystem Services

The ecosystem services concept was developed in the framework of the Millennium Ecosystem Assessment (MEA), which provided technical information and evaluated the significance of ecosystem change on human well-being (MEA 2005). The MEA framework discusses that a vibrant relation exists between ecosystems and humans and that it is crucial to study the connections among four key components i.e., direct drivers, indirect drivers, ecosystem services, and human well-being (Liu et al. 2007). The MEA framework observed that linkages between ecosystems and humans is perhaps less explored and least well-understood among the above four key components (Yang et al. 2013). However as stated by Abdallah et al. (2008), the linkages between human well-being and the social factors that influence this relationship have been given more attention. To better understand the relationship, Yang et al. (2013) states four reasons for evaluation of human reliance on ecosystem services: first, governance can be enhanced by thoroughly understanding the relationship between poverty and ecosystem services (Shackleton et al. 2008; Suich et al. 2015); second, the equity provisions in governance can be appropriately planned once the unequal distribution of benefits from ecosystem services across diverse population groups can be better assumed (Liu and Ming-Te 2011); third, a proper understanding of unmanaged threats and unexploited opportunities that arise with ecosystem change (e.g., droughts, floods, landlised, and storms) (Yang et al. 2013) facilitates a risk management approach; and fourth, for meaningful communication between policymakers and politicians, the quantitative measurement is essential to better understand the human-nature interactions (Alberti et al. 2011).

Since the publishing of the Millennium Ecosystem Assessment, some authors have felt that the concept of ecosystem services unduly focuses on the quantification of ecosystem services and overlooks the quality of the holistic and comprehensive nature of relations between humans and nature (cf. Larigauderie and Mooney 2010; Perrings et al. 2011). For this purpose Pascual et al. (2017) developed a new concept of Nature's Contributions to Humans. This concept argues that nature's contributions are more holistic and reflect complex relations between humans and nature. However, it continues to draw on the four elements of the ecosystem services.

4.4.2 Freshwater and ecosystem services

Freshwater is a vital resource for the survival of all ecosystems. For example, a key concern for hydrological ecosystems is acquiring minimum stream flow, particularly maintaining and restoring water allocations (Palaniappan et al. 2010). Human utilisation of freshwater for irrigation, domestic use and industrial applications can have adverse impacts on down-stream ecosystems (Gordon et al. 2010) see also Table 1.2 in Chapter 1. This section links different kinds of freshwater to the ESS concept, which shows the contribution of biophysical ecosystem processes to human well-being (MEA 2005; NEA 2011). The concept of ESS is essentially anthropocentric because all the processes and components of ecosystems are studied as services which humans require, demand or benefit either directly or indirectly (Boyd and Banzhaf 2007).

Nature supports the biodiversity of genes, species and ecosystems. In addition to protecting biodiversity, ecosystems provide *supporting* services (see 4.4.3; such as erosion control, climate regulation and oxygen production), *provisioning* services (see 4.4.4; including food, freshwater, fibres, genetic materials and ornamental materials etc. (MEA 2005; Nellemann 2009), *regulating* services (see 4.4.5; comprising environmental benefits and climate regulation, disease regulation, flood management, water treatment and waste management etc.) (MEA 2005; Nellemann 2009) and *cultural* services (see 4.4.6; including material benefits to communities through a variety of spiritual and religious services) (MEA 2005). Table 4.2 summarises how different kinds of water are linked to specific ESS.

4.4.3   Supporting Services

Supporting services of ecosystems underlie the sustainability of all other ESS. The supporting services of freshwater are intrinsic for all water colours. These include (a) supporting the hydrological cycle, recharge, evaporation and transporting of water (including vapour, which is done by wind and rainbow-water), and precipitation; (b) storing water; (c) soil moisture, soil formation, erosion and erosion control; (d) nutrient cycling of e.g. oxygen, carbon, nitrogen, phosphorous, salt concentration and/or dilution (Laruelle 2009); (e) primary production and supporting photosynthesis (Alahuhta et al. 2013; Nellemann 2009); and (f) supporting biodiversity: rainbow water provides a habitat for birds, surface water for fish and other aquatic organisms, and green-water through promoting landscapes for various terrestrial life forms; and ecosystems associated with groundwater are dependent on water quality, discharge fluctuation and the pressure level of the aquifer (Merz et al. 2001).

Supporting services differ from other ESS by their indirect impacts on humans over long periods of time. In other words, changes in the provisioning, regulating and cultural context of ESS have direct impacts and last for short-to-medium periods. Some services, such as erosion control, climate regulation and oxygen production classified as supporting and regulating services, serve both functions (MEA 2005; NEA 2011) see Table 4.2) and have short-to-long-term effects on human well-being.

4.4.4   Provisioning Services

The literature does not discuss in detail the provisioning services of rainbow water, except through its role of providing fresh, clean water directly to the surface of earth. A range of provisioning services are provided by clean blue and green water. Even grey water is seen as a useful, predictable source of water that can be reused. Black water is polluted water and can spread disease, but can be reused after thorough treatment.

In terms of security, the availability of adequate freshwater resources ensures safe and equitable access to all people and communities; where resources are low or controlled through privatisation and/or hegemonic control, the safety of access and the quality of water can be negatively influenced. This has led to, inter alia, the tipping points hypotheses at the global level, water wars hypotheses at the transboundary level (Jarvis and Wolf 2010), and the human water security hypotheses at the national level (Vörösmarty et al. 2010).

In terms of health, the supply of good quality freshwater fosters good health; polluted water can be harmful for both aquatic organisms and the humans who consume it (Dallas and Rivers-Moore 2014) or bathe in it leading to symptoms like fever, skin irritation, eye infections, as well as stomach pain and disorder etc. Surface water contamination is mainly caused by pesticides use which may affect soil fertility and impact humans (Pendleton et al. 2012).

In terms of good social relations, the availability of good quality water is critical for social cohesion and mutual respect while the poor and/or reduced quality of freshwater can create social stress and mutual disregard. The provisioning ESS of freshwater can occur at multiple levels (local, provincial, national, regional, and global levels) and in many different ways; for example, the deterioration of freshwater quality or overflow in lakes and rivers may on the one hand, affect some kinds of employment (e.g. fishing or farming activities), while on the other hand, they may create new types of employment (e.g. flood protection) at multiple geographic levels (cf. Berkes 2004).

4.4.5 Regulating Services

All kinds of water contribute to regulating services, in particular rainbow water, blue surface water, blue groundwater, green water and white ice. Grey water takes water out of the system, alters its state from good to bad, or temporarily removes it locally through the piped water system (Al-Jayyousi 2003); it may have a lower contribution to regulating services (Brown 2007). On the other hand, if grey and black water are returned to the freshwater system untreated, they can lead to the transmission of diseases, disturb other ESS and may reduce benefits for humans (Naidoo and Olaniran 2014).

In terms of safety, regulating services support biodiversity and human life (Dıaz et al. 2005; Wall and Nielsen 2012). However, when the regulating services are disturbed by climate change, this can lead to heatwaves, floods and droughts which probably affect human security (Dallas and Rivers-Moore 2014). In terms of achieving a good quality of life, regulating services are very important (McMichael et al. 2005). However, rise in temperature of water, reduction in the absorption of dissolved oxygen, mobilisation of chemical pollutants (including metals, pesticides and pathogens), or changes in water flows may affect water and food availability and the feeling of well-being (Nellemann 2009). These pollutants can affect human health and may reduce their productivity (Schwarzenbach et al. 2010; WHO 2013). There is a lower causal relationship between regulating services and good social relations, except where climate change is seen as affecting trust between countries and peoples (Hurlbert and Gupta 2015).

4.4.6 Cultural Services

Rainbow, surface and groundwater have positive influences on cultural services and enhancing human well-being. Grey and black water have limited influence on educational services; however, a thorough knowledge about their influence can contribute to human well-being. Cultural services can affect security where freshwater bodies marking state boundaries are affected by the uncertain impacts of climate change, which may affect inter-state relations (López-Hoffman 2010; Sanchez and Roberts 2014; M. Young 2015). Cultural services may affect the quality of life by providing aesthetic and recreational options (Daniel et al. 2012). However, pollution, vector migration resulting from climate change and invasive species reduce water quality for recreation, biodiversity and tourism, and may reduce the value for disadvantaged/indigenous communities (Dallas and Rivers-Moore 2014). This may lead to weak social bonds due to the dispersion of communities, increase in climate refugees and enhanced inequality. The spiritual, recreational and cultural aspects of clean rainbow and blue water can enhance the feeling of well-being of people, facilitate social cohesion and mutual respect, and thereby contribute to the institutionalisation of cooperative mechanisms (Dellapenna and Gupta 2009: eds.).

Table 4.2: Ecosystem services of different kinds of freshwater

Kinds of Freshwater	Supporting Services		
	Supporting hydrological cycle, transporting water (including vapor); climate regulation; water storage; enriching soil & erosion control; nutrient cycling of e.g. oxygen, carbon, nitrogen, phosphorous (Keys et al. 2012); habitat, primary production, photosynthesis (Alahuhta et al. 2013)		
	Provisioning	**Regulating**	**Cultural**
Rainbow water	Huge storage of water on Earth; habitat for birds and insects	Climate regulation, hydrological regulation	Aesthetic (inspiration for art), spiritual (rain Gods/ Gods of thunder), inspiring knowledge)
Blue surface water	Fodder, clean water, food, hydropower, fish and aquatic life, navigation (Alahuhta et al. 2013)	Climate/hydrological regulation, sediments transport (Keys et al. 2012)	Recreational (fishing, boating), spiritual (water for religious activities) (Sarvilinna et al. 2012)
Blue groundwater	Clean water, food production, heated & cooling water for power plants, providing nutrients such as sulphate/nitrate (Carr and Neary 2008); supports land – extraction leads to land subsidence	Water purification, waste treatment, salinity regulation, climate regulation through CO_2 leakage into groundwater (UNEP 2006)	Educational value, water for various religious ceremonies (UNEP 2006)
Green water	Fodder, food, pastureland, herbs and shrubs (Keys et al. 2012)	Evaporation (flowing downwind to later fall as precipitation); aquifer recharge (Keys et al. 2012)	Forests and landscapes for tourism, spiritual needs and education (Keys et al. 2012)
Grey water	vegetable & fodder production, energy production, mining (UNEP 2006), fire-engines, toilet flushing, preserving wetlands	Climate and water regulation, evaporation flowing downwind to later fall as precipitation (UNEP 2006)	Promotes tourism and education (UNEP 2006)
Black water	Fodder, insects & worms as birds' food	Spreads disease unless managed	Educational services regarding its negative effects
Frozen water / glaciers	Habitat for animals; storage of water	Albedo effect	Preserving data for humans, preserving life forms frozen in the past

Source: Hayat and Gupta 2016

4.5 ECOSYSTEM SERVICES AND HUMAN-WELLBEING

This section explains how human well-being can be potentially influenced by the ESS of different kinds of freshwater (MEA 2005; WHO 2013; WHO/UNICEF 2015). In the MEA framework, human well-being represents _security' (safety, security from disasters, safe access to resources); basic material for the _good life' (in terms of livelihoods, shelter, water, and food); _health' (in terms of physical power, feeling fit and having access to a safe and clean environment); and _good social relations' (e.g. respect for each other, strong social bonds, and the capability of helping others) (MEA 2005). All these enhance the freedom of choice and action (see Table 4.3).

Poverty is a multidimensional issue and can be defined as the distinct deficiency of well-being (McGregor and Pouw 2016). The definition, experience and expression of well-being, ill-being or poverty depends on the context, that reveals limited physical, social, and individual elements such as culture, gender, age, geography, and the environment. In all perspectives, ESS are crucial for human well-being for their provisioning, regulating, cultural and supporting services. Human interference in ESS can intensify the benefits to human society. In this vein, the idea of _freedom of choice and action' is relevant and is briefly explained in the following sub-sections according to its various components and how these components are affected by changes in ecosystem services due to a range of drivers.

4.5.1 Freedom of Choice and Action

Freedom and choices are fundamentally founded on the numerous other features of well-being and are shaped by changes in ecosystems in terms of biodiversity and supporting, provisioning, regulating or cultural services. Human well-being can be enhanced by promoting the sustainable interaction between human and ecosystems which are further sustained by essential institutions including technology, organsations, and instruments. Formation of such institutons through participatory approaches and transparency can add to freedom of choice and may also enhance social, ecological, and economic security.

4.5.2 Physical and Economic Security

Changes in provisioning sevices can put a stress on economic and human security that may directly disturb provisions of food, fish and other goods and may lead to conflict over resource degradation. Similarly, variations in regulating services can affect the scale and frequency of landslides, droughts, floods, or other calamities. Additionally, alterations in cultural services can also affect human and economic security, for example, the degradation of spiritual and ceremonial attributes of ecosystems can lead to fragile social relations within a community. Ultimately, these kinds of variations can influence freedom of choice, material wellbeing, health, security and good social relations. Ecological security can be defined as the minimum level of ecological stock necessary to ensure sustainable flows of diferent ESS. Nonetheless, the tangible benefits enabled by scientific developments and various institutions are neither automatic nor equitably shared. In addition, some advances in technologies and institutions can cover or intensify responsible governance and different environmental challenges. In this regard participative decision-making is a vital component of governance accountability but it can be costly to sustain in terms of time and resources (Anggraeni et al. in press).

4.5.3   Good Life (Survival and Existence)

Provisioning services (production of fibre and food) and regulating services (natural purification of water) are crucial elements for a good life. Unequal access to these essential ESS can enhance the wellbeing of small segments of people at the cost of others who are denied access.

4.5.4   Health (Good Health and Enhanced Productivity)

Health has direct links with provisioning (production of fibre and food) and regulating ESS, including for example, those services that affect the dissemination of disease-transmitting pests as well as irritants and pathogens present in air and water. Recreational and spiritual benefits are another kind of cultural services.

4.5.5   Good Social Relations through Active Participation

Any changes in cultural services can affect social relations which are directly related to quality of life and human wellbeing. Table 4.3 links the impacts of the ESS of the different colours of water to human well-being. It is worth noting that although human well-being was stressed in the Millennium Ecosystem Assessment, the IPBES emphasises the concept of _more good life‘ (Díaz et al. 2015).

Table 4.3: Impacts of ESS of different colours of water on human well-being

Well-being	Provisioning (PS)							Supporting Services — Regulating (RS)							Cultural (CS)						
	R	Bs	Bg	Gn	Gr	BB	W	R	Bs	Bg	Gn	Gr	B	W	R	Bs	Bg	Gn	Gr	B	W
	+	+	+	+	+	±	+	+	+	+	+	?	?	+	+	+	+	+	-	-	+

Security

- **PS:** Adequate resources ensure safe access; where resources are low, the possibility & safety of access is influenced.
- **RS:** RS ensure safety of humans in general; however, climate change impacts like heatwaves, floods, and droughts will affect human security (Dallas and Rivers-Moore 2014) causing damage to life and property, migration and refugee flows.
- **CS:** CS are positive; however, where water bodies mark state boundaries and these are affected by climate change, this may affect inter-state relations; further where water is scarce, this may lead to conflict situations.

Good Life

- **PS:** PS are important for survival and livelihoods. However, pollution, salt water intrusion, salt concentration in soils, poor pollination and nutrient cycling affects food production; and oxygen depletion, acidic water bodies and river sediments suffocate aquatic life while sediments accumulate in the tissues of fish and shellfish. Affects related livelihood options.
- **RS:** RS are important for the good life. However, increased water temperature, decreased dissolved oxygen (Dallas and Rivers-Moore 2014), increased pollutants and pathogens, and changes in water flows affect aquatic ecosystems and ESS (Nellemann 2009: p.2009).
- **CS:** CS of water bodies provide aesthetic and recreation options. However, pollution, vector migration resulting from climate change and invasive species reduces water quality for recreation, biodiversity for tourism activities and the value for disadvantaged/indigenous communities (Dallas and Rivers-Moore 2014).

Health *(Freedom of choice and action)*

- **PS:** Clean water essential for good health; but contaminated drinking water leads to diarrhoea and intestinal parasites (Dallas and Rivers-Moore 2014). Harmful for both aquatic organisms and humans.
- **RS:** Increased contaminants affect human health.
- **CS:** The spiritual, recreational and cultural aspects of clean rainbow- and blue water can enhance the feeling of well-being of people.

Good Social Relations *(Freedom of choice and action)*

- **PS:** Good quality water supports social cohesion and mutual respect; while poor and reduced quality of water can create social stress and mutual disregard.
- **RS:** To the extent that climate change is seen as affecting water quality, this can affect good social relations and trust between countries (Hurlbert and Gupta 2015).
- **CS:** CS lead to social cohesion and mutual respect that have led to the institutionalisation of cooperative mechanisms through history (Dellapenna and Gupta 2009 (eds.)).

Source: Hayat and Gupta 2016

4.6   INFERENCES

This chapter (a) has shown that the direct drivers of poor water quality are agriculture, industry, households and demographic shifts. The indirect drivers include political drivers between and within states, culture and ethnic elements, non-water-related policies comprising poverty reduction policies and economic growth, the financial incentive of local industries, poverty of the local people, technological advances, international trade and climate variability and change and other natural factors. This implies that transboundary multilevel water governance needs to address the direct, if not, indirect drivers. This chapter (b) categorizes freshwater into atmospheric moisture, blue surface and groundwater, green water, white water, grey water, black water and. Further, (c) it demonstrates that each of these kinds of freshwater have different ecosystem services (see Table 4.2) with differential impacts on human well-being (see Table 4.3). These points imply that multilevel transboundary water governance requires to better understand the role of the different kinds of freshwater and their changing ecosystem services in order to achieve better management for human and natural purpose.

5

GLOBAL WATER INSTITUTIONS AND ITS RELATIONSHIP WITH INCLUSIVE AND SUSTAINABLE DEVELOPMENT

5.1   INTRODUCTION

This chapter presents a summary of global water institutions followed by a thorough analysis of five key global institutions that are relevant to the governance of transboundary water resources. It identifies some of the relevant principles (i.e. political, socio-relational and ecological) and instruments (regulatory, economic, suasive and management). It also recognises the obligation for building a River Basin Organisation (RBO) and using dispute resolution mechanisms for problem solving. This chapter intends to answer the research question: How have global institutions for transboundary water governance evolved and what are the implications of these institutions for governing transboundary river basins without a regulatory framework? To answer this question, this chapter also addresses some subsidiary questions: (1) How have the key global institutions for governing transboundary water resources evolved? (2) Which governance instruments (principles and instruments) are included and which are excluded? (3) How can the establishment of RBO's reduce conflict and promote cooperation and in a transboundary river basin? (4) How has power influenced the inclusion/exclusion of instruments (principles and instruments)? To answer these questions, this chapter first presents a summary of global water institutions (see 5.2) and then assesses the evolution of five key institutions at the global level (see 5.3). Second, it identifies different instruments (including principles) within these institutions (see 5.4). Third, it discusses the power and politics of inclusion/exclusion of certain instruments (see 5.4). Fourth, it briefly highlights the role of river basin organisation in dispute resolution (section 5.5). Fifth, the last section (section 5.6) draws inferences for transboundary water governance at multiple levels.

5.2   OVERVIEW OF GLOBAL WATER INSTITUTIONS

The codification of water rules started with the development of treaties over transboundary rivers, underground aquifers and lakes (Dellapenna and Gupta (eds.) 2009; Eckstein and Sindico 2014). These rules were included in court judgements over various water matters, as well as in regional and national agreements (Guzmán-Arias and Calvo-Alvarado 2016; Mehta 2016). Presently global water institutions are shaped to ensure efficient and equitable distribution of shared water resources between states (McKinney 2011). Furthermore, they also to ensure that no substantial transboundary harm should occur to those who utilise the resource (Drieschova and Eckstein 2014; McKinney 2011).

The codification of governance principles for transboundary rivers at the global level started at the beginning of the 20th century. The International Law Association (ILA) codified the rule on the utilisation of water in international river basins in the non-binding Helsinki Rules (International Law Association 1967) (Mechlem 2003; Dellapenna 2011). Following this, The Ramsar Convention on Wetlands of International Importance (UNESCO 1971) was adopted in 1971 to protect and conserve ecosystems of extremely sensitive wetlands of global importance. One year later in 1972, the Stockholm Declaration on the Human Environment (The Stockholm Declaration 1972) included water governance principles in the Preamble and Principle II. In 1977, the Mar del Plata Action Plan in the First World Water Conference set the framework for the management of water resources

(Worthington 1977). In this Action Plan, the water supply for human consumption and agricultural utilisation purposes was clearly specified (Worthington 1977).

To continue, in 1986, the ILA non-binding Seoul Rules were applied to transboundary surface and groundwater resources (The Seoul Rules 1986). In 1992, the UN Conference on Environment and Development (UNCED) produced three primary versions related to governing water resources including: (1) the Rio Declaration on Environment and Development (The Rio Declaration 1992) which recognized 27 governance principles to be addressed as vital principles on the environment and development; (2) the United Nations Framework Convention on Climate Change (UNFCCC 1992) which covers climate mitigation and adaptation; and (3) the Convention on Biological Diversity (The Convention on Biological Diversity 1992) which primarily aims to protect global biodiversity. The legally binding UNECE regional Convention for the Protection and Use of Transboundary Watercourses and International Lakes was adopted in 1992 (UNECE 1992). This Convention was originally a regional agreement, but because of the amendment to Articles 25 and 26 all UN member states were invited to become parties (UNECE 2003) making it a global agreement. The 1999 Protocol on Water and Health (UNECE 1999) of the UNECE Convention protects human health and enhances wellbeing. Similarly, the United Nations Convention to Combat Desertification (UNCCD 1994) was adopted to address the harmful effects of land cover changes and climate variability on socio-economic opportunities, food production, and water resources sustainability. In 1997 the United Nations Watercourses Convention, inspired by the Helsinki Rules, was adopted and has been in force since 2014 and is binding on all 36 ratifying parties.

The ILA's scholars wrote the non-binding Berlin Rules on Water Resources in 2004, which include principles of customary international law concerning the management of freshwater resources. The Berlin Rules replace the Helsinki Rules by adopting principles of human rights and international environmental law. The Berlin Rules consist of both codified and customary principles that relate to water resources (International Law Association 2004). The Human Rights Committee (UNHRC) as well as the General Assembly (UNGA) of the United Nations adopted two separate resolutions on the Human Right to Water and Sanitation in 2010 (UNGA 2010a; UNGA-HRC 2010). These resolutions recognise the human right to water and sanitation. The 2008 Draft Articles of International Law Commission on the Law of Transboundary Aquifers (International Law Commission 2008) have been reviewed three times by the member states of the United Nations. There continues to be disagreement among the participant states regarding the draft articles especially regarding whether it should be accepted as a policy statement, international framework agreement or non-binding text (Conti 2017). Following that, the member states of the UN adopted the Sustainable Development Goals (SDGs) in September 2015 (UNGA 2015). These are 17 goals and 169 targets which are built on the earlier construction of the eight Millennium Development Goals (UN Millennium Declaration 2000) where water is a cross-cutting theme in various goals.

5.3   KEY GLOBAL WATER GOVERNANCE INSTITUTIONS

In the overview of global water institutions above I have discussed 16 key institutions that were negotiated at the global level. This section emphasises five key institutions that are relevant to the analysis of transboundary water governance in this thesis. These include: (1) Customary International Water Law; (2) the 1992 UNECE Water Law; (3) the 1997 Watercourses Convention; (4) The Human Right to Water and Sanitation; and (5) the 2015 Sustainable Development Goals of Agenda 2030. Some of these institutions are binding global conventions (i.e., 2 & 3); non-binding declarations (i.e., 4) or goals (i.e., 5) agreed upon by countries. These five institutions are directly relevant for my analysis of the transboundary level (see Chap 6), national, and sub-national level of Afghanistan (Chap 7) and Pakistan (Chap 8) and multilevel (Chap 9) water governance in the Kabul River Basin.

5.3.1   Customary International Water Law

Customary international water law has been formally approved in the two UN water laws and codified in the 1966 informal Helsinki Rules and the 2002 Berlin Rules. Customary water law reflects existing state practice (Giordano and Wolf 2003; Rahaman 2009). It includes political principles (e.g., information exchange between riparians, obligation to cooperate, notifiying riprian states about emergency situationsand planned measures to avoid water flow interruption, limited territorial sovereignty, and dispute resolution); social-relational principles (e.g., equitable and reasonable utilisation and capacity building); and ecological principles (e.g. protecting and preserving groundwater recharge zones as well as ecosystems) (McIntyre 2011).

Customary law imposes substantive (using water in an equitable and reasonable manner) and procedural (obligations to cooperate in good faith) obligations on countries sharing watercourses (Shaarawy 2016). Principles of customary law can potentially play a vital role in dispute resolution concerning water resource sharing among those countries that are not yet party to the 1997 Watercourses Convention (McIntyre 2007).

The principles of customary international water law has evolved from various theoretical approaches such as:

a. **Absolute Territorial Sovereignty Doctrine:** It allows a riparian of the shared water resource to utilise the water freely which is flowing in its territory without any consultation and irrespective of the consequences (Correia and da Silva, 1999). This approach eventually encourages the upstream state to divert all water resources in a transboundary river basin without considering the dire needs of downstream state (McCaffrey, 1996). Most water experts and scholars are not in favour of this doctrine as it does not represent evolving international law (Birnie and Boyle, 1994; Salman and Uprety, 2002).

b. **Absolute Territorial Integrity Doctrine:** It is established on the claim that the lower riparian of a transboundary basin has the right to receive uninterrupted water flow of good quality. This indicates that the upstream riparian cannot divert or interrupt the flow without the permission of the downstream riparian. This approach reassures the lower riparian of a transboundary basin to demand the healthy and continuous flow of water from the upper riparian irrespective of the priority (Barandat and Kaplan 1998). States that are downstream riparians in a transboundary river basin always support this doctrine as it allows them to use the shared water resources in an unaltered way. Due to its limited role in state practice, scholars and water practitioners rejected this concept (Birnie and Boyle 1994).

c. **The Sovereign Equality Doctrine:** According to this concept, every riparian state is entitled to utlise water of a transboundary river inside its premises without affecting the legal rights and interests of other riparian states. Furthermore, all riparian states have mutual rights and obligations as well as entitlement to equitably share the benefits of a shared river basin. This concept is broadly known as the concept of sovereign equality or limited territorial integrity (Salman 2007; Schroeder-Wildberg 2002). Table 5.1 below presents the main principles of customary international law and their main objectives, while Table 5.2 contains information on incorporating the principles of customary law into international treaties and conventions.

5.3.2   The 1992 UNECE Water Law and its 1997 Protocol

Since the regional UNECE Water Convention (UNECE 1992) has now been opened for international participation, I discuss this here. The UNECE Water Law has been ratified by 43 Parties (including 42 states and those countries sharing transboundary waters in the region of the UN Economic Commission for Europe). It aims to improve the protection and management of transboundary surface waters and groundwaters. The main principles and instruments of the 1992 UNECE Water Law include political (information exchange, notifiying about planned measures as well as emergency situations, dispute resolution, and obligation to cooperate); social-relational (public access to data and information, equitable and reasonable utilisation); and ecological principles (Best Available Technology Not Entailing Excessive Costs, EIA, ecosystem conservation and protection, basin as the unit of management, prevention of pollution, monitoring, precautionary principle, polluter pays).

On March 17, 1992, the UNECE Water Law was made available for signature in Helsinki where it came into into force on October 06, 1996. The Convention was eventually converted into a global legal framework for enhancing transboundary water cooperation after the modification was adopted on February 06, 2013. If countries outside the UNECE region ratify the Convention, it would become universally applicable (UNECE 2016: p. 5). However, due to low participation of the non-UNECE countries in the negotiation process, the legitimacy of the Convention can be questioned. The Convention includes general provisions: to prevent, control and reduce transboundary impact (Art. 3); monitoring (Art. 4); research and development (Art. 5); information exchange (Art. 6); responsibility and liability (Art. 7); information protection (Art. 8) (UNECE 1992). Establishing a River Basin Organsation (RBO) to protect and manage transboundary water resources (Art. 10). The Convention is evaluated during the Convention's Meeting of the Parties while its implementation is overseen by the secretariat.

In the follow-up to the Convention, the parties adopted a Protocol on Water and Health (UNECE 1999) which is legally binding and entered into force in 2005. It aims to protect human health and enhance human well-being through improved water management. It includes the precautionary principle (to control water-related disease), the polluter-pays principle (to ensure that the polluter will bear the costs of pollution), the limited sovereignty principle (to ensure that the activities of one riparian state avoid harming another state or its environment), the intergenerational equity principle (to make sure that the ability of future generations are not compromised while meeting the needs of the current generation), protection of water resources protection, governing water resources at an appropriate geographic level, efficiently using water resources, awareness creation, ensuring public participation and access to information in decision-making regarding integrated water management, water and health, ecosystems conservation and protection, protection of vulnerable people against water-related disease and ensuring equitable access to water (Art. 5). The Protocol addresses issues of water resources and health of ecosystems, but primarily focuses on disease prevention (Art. 4, 5, 8, 12, and 13) in transboundary surface water and groundwater.

5.3.3   The 1997 Watercourses Convention

The ILC was requested in 1970 by the UNGA to formulate concrete international guidelines for transboundary water use similar to the 1966 Helsinki Rules (McCaffrey 1999; Dellapenna and Gupta 2008). This led to the adoption of the Convention on the Law of the Non-Navigational Uses of Transboundary Water Courses which was adopted in 1997 by states (McCaffrey 1999). As of 2019, this Convention has been ratified by 36 states. It takes the special situation and needs of developing countries into account. It describes an international watercourse as _a system of surface and groundwaters which crosses borders of states' (Art. 2). The Convention has both procedural rules and substantive norms for riparian states of shared watercourses. The key objective of the Convention is to _safeguard the conservation, protection, management, development, and utilisation of transboundary water resources as well as promote the sustainable and optimal utilisation of shared water resources.' Specifically, Article 5 and 6 of the Convention on equitable and reasonable utilisation and participation are very relevant in the context of transboundary water resources. Article 5 encourages the riparian states to develop and use transboundary water resources in an efficient and equitable manner with a view to achieve optimal and sustainable utilisation and benefits by taking into account the interests of the riparians states. It also educates riparian states about the rights of utlisation and the duty to cooperate with each other on the protection and development of the watercourse. Additionally, Article 6 describes factors which are related to the equitable distribution of water resources including geographic, hydraulic, climatic, economic, and social (including attention to dependent populations), existing and potential water usage, and its protection and usage, among others.

The Watercourses Convention includes some important principles including political (information exchange; notification about planned measure and emergency situations, dispute resolution, and obligation to cooperate); social-relational (prior informed consent as well as equitable and reasonable utilisation); and ecological principles (EIA, prevention of pollution, ecosystems conservation and protection, invasive species, and basin as the unit of management).

Each riparian state is obliged to provide data and information to co-riparians of a shared watercourse (Article 9) concerning the condition and proposed uses (Article 12) so that other states have sufficient time to study the intended water use and object if the use is considered harmful (Article 14). Thus it restricts states to avoid harming their co-riparian states by considering all appropriate measures while utilising water resources of an international watercourse in their territories (UN Watercourses Convention 1997: Art. 7) (Fitzmaurice 1997). The Watercourses Convention is limited in its scope to specify how priority should be given in terms of utilising an international watercourse (Art. 10). However, it sets out guidelines for dispute resolution among riparian states including: (a) a mediation or referral request from a third party; (b) negotiation; (c) setting up a commission of inquiry to establish a procedure; (d) providing information to the Commission; and (e) dispute submission in the International Court of Justice (Art. 33). The Convention calls upon states to take concrete steps to prevent pollution-related damage (Art. 21), the introduction of invasive species (Article 22), and to enforce a responsibility on states that damage to a shared watercourse should be remedied or compensated (Art 22). The Watercourses Convention has specific provisions for damages to international watercourses due to natural calamities such as drought or erosion. It encourages participating states to inform co-riparians about the watercourse related emergencies which might disturb them, for instance water borne diseases or floods (Art. 27). Since the time of its adoption, the Watercourses Convention has taken more than 17 years to enter into force and so far 36 states have ratified the Convention - while still many countries particularly the upstream ones - have not yet ratified it (Gupta 2016b). The Convention is recognized as a milestone which helps in the setting up of international water law (Krishna and Salman 1999).

5.3.4   The Human Right to Water and Sanitation

Most states have ratified either a human rights convention or signed on to a political declaration which identifies the human right to water and sanitation. This basic right proposes to guarantee access to water and sanitation services as a human right vital to the recognition of all human rights. Some countries enforce this right actively, while in others it has remained a normative principle.

The Resolution on the Human Right to Water and Sanitation of the UN General Assembly (UNGA 2010b) requests states and international organisations to encourage efforts to approve the right to water and sanitation and support its application through technology transfer, improved financing, and building capacities. The Resolution of the Human Rights Commission of the United Nations General Assembly (UNGA-HRC 2010) obliges states to ensure the application of the Resolution, even when dealing with third-party suppliers. Such application needs to be supported by other related principles e.g., accountability for human rights violations, mainstreaming gender equity and non-discrimination, EIA, transparency in development and application. Various states have supported these Resolutions. Although the Watercourses Convention does not explicitly prioritise the right to water and sanitation services as a right, the adoption of these two Resolutions makes clear that this should be seen as a priority (Salman 2012; Trigueros 2012).

5.3.5   The Sustainable Development Goals (SDGs) of Agenda 2030

The United Nations General Assembly approved the Sustainable Development Goals (SDGs) and Agenda 2030 in 2015 (UNGA 2015). The _plan of action' that SDGs offer include the integration of poverty alleviation, environmental sustainability, and economic development by 2030. The SDGs visualizes _a world where we endorse our obligations concerning the human right to clean and safe drinking water as well as improved sanitation and hygiene' (Paragraph 7). Goal 6 of the SDGs is the water-related goal that aims to cope with global water issues related to economic and human activities. Goal 6 of the SDGs is a cross-cutting goal that is linked to all other development goals. This particular goal ensures the accessibility and sustainable management of water and sanitation for all. In essence, it goes beyond drinking water, sanitation, and hygiene and emphasises the centrality of water resources to sustainable development, and the key role that clean and safe drinking water, improved sanitation, and hygiene play in achieving other goals related to health, education, and poverty reduction. Agenda 2030 was adopted by all UN Member States and specialised agencies to protect the planet, alleviate poverty, and make sure that people have freedom of choice, enjoy peace and prosperity.

It has eight targets to tackle the global water crisis and also addresses transboundary water issues. These include achieving: (1) support in building capacities of developing countries in different water- and sanitation-related programmes and activities; (2) efficient water use across all sectors as well as addressing water-scarcity related issues by promoting sustainable withdrawals and supply of freshwater; (3) the support and active participation of local people in enlightening governance of water and sanitation; (4) the application of Integrated Water Resources Management (IWRM) at all geographic levels; (5) enhanced water quality by reducing pollution; (6) reasonable and equitable access to safe and affordable drinking water worldwide; (7) access to sufficient and equitable sanitation and hygiene for all by taking into account the needs of girls, women, and other vulnerable groups; and (8) the conservation and protection of water-related ecosystems.

Agenda 2030 calls on all states to treat the different Goals in an inter-connected and inter-linked manner. Therefore, it is not unexpected to see that the other 16 SDGs and its various targets have strong direct and indirect connections with water. For instance, target 3.3 of Goal 3 aims to cope with water-borne diseases while 3.9 strives to decrease illness and deaths from water pollution. Similarly targets 11.5, 12.4, and 15.1 of Goals 11, 12 and 15 respectively ensure protection, conservation, restoration, and sustainable utilization of both terrestrial and inland freshwater ecosystems and their services. Target 15.8 strives to limit the impacts of invasive species on freshwater ecosystems. The review of SDGs specifies that non-water related goals are linked with water-related Goal 6 in one way or the other. For example, more investment in the water sector can lead to achieving Goal 1 on poverty alleviation (UNGA 2015); achieving food security to end hunger and promote sustainable agriculture (Goal 2); and enhancing access to reliable, affordable, modern, and sustainable energy for all (Goal 7). One of the necessary elements that can hamper sustainable development is the uncertain impacts of climate change. In this regard Goal 13 aims to tackle such challenges. Target 13.3 of Goal 13 is linked transboundary water governance where states are encouraged to include climate change into national level policies and planning.

Goal 16 of the SDGs is motivated by the concept of ‗good governance' which encourages sustainable social development by promoting inclusive and peaceful societies. It specifically emphasizes establishing global cooperation by creating, efficient, accountable, and transparent institutions at all geographic levels. Moreover, target 16.7 of the SDGs is closely linked to one of the important principles of inclusive development that aims at participatory, representative, and inclusive decision-making at all geographic levels. Such principles aim to enhance the legitimacy of decision-making and empower those without a voice, e.g., dam building by an upsteam riparian can negatively affect the downstream riparian because downstream states rely on the regular water flow for fisheries and agriculture. In such circumstances the participation of downstream riparians may enrich the quality of decision-making. Target 16.10 of the SDGs encourages the participation of the public in accordance with national law and international agreements (UNECE 1998). Public participation is one of the important elements of inclusive development which can only be effective if there is adequate and equitable access to information.

Table: 5.1: Global institutions governing transboundary water resources

Institution Ratification	Parties	Goals/Objectives	Principles & Instruments
Customary international water law	Only applicable to parties of the 1992 UNECE Water Law	To empower public, provide them appropriate resources, and to enable cooperation over the world's fresh water resources	political (information exchange, notification about planned measures and emergency situations, dispute resolution, obligation to cooperate, limited territorial sovereignty); social-relational (equitable and reasonable utilization, capacity building); and ecological (ecosystem conservation and protection)
The 1992 UNECE Water Law	43 Parties (including 42 states and all UNECE countries sharing transboundary waters	To enhance methods and improve national efforts for the safety and management of both surface and groundwaters in a transboundary river basin	political (notification about planned measures and emergency situations, information exchange, dispute resolution, and obligation to cooperate); social-relational (public access to information, equitable and reasonable utilisation); ecological (EIA, BATT, basin as the unit of management, pollution prevention, precautionary principle, monitoring, polluters pay, and ecosystem conservation and protection)
The 1997 Watercourses Convention	36 States	To create a framework for the utilisation, improvement, maintenance, administration, and safety of international watercourses, whilst encouraging optimum and sustainable usage thereof for current and future generations, and accounting for the distinct situation and requirements of	political (notification about planned measures and emergency situations, information exchange, dispute resolution, and obligation to cooperate); social-relational (prior informed consent, equitable and reasonable utilisation); ecological (pollution prevention, EIA, basin as the unit of managementecosystem conservation and protection)
The Human Right to Water & Sanitation (2010)	All states have ratified either a human rights convention or accepted a political declaration which recognises the HRWS	To identify the human right to clean water and improved sanitation and recognize that these are vital to the recognition of all human rights	The human right to water & sanitation
The Sustainable Development Goals of Agenda 2030	All UN member States	To protect the planet, erridicte poverty, and make sure that all people enjoy peace and prosperity	social-relational (human right to water and sanitation, poverty alleviation, food security, capacity building, participation of public, reduced inequality, intergenerational equity, sustainable urbanization, food security, rights of women, youth, & indigenous peoples); ecological (ecosystem conservation and protection, precautionary principle, pollution prevention, and invasive species)

Table 5.2: Major principles & instruments in the key global water institutions

	Customary International Law	1992 UNECE*	1997 UN WCC*	2010 UNGA	2015 SDGs
Political Principles					
Information Exchange	■	■	■		■
Notification of Emergency Situations	■	■	■		■
Notification of Planned Measures	■	■	■		
Obligation to Cooperate	■	■	■		■
Disputes Resolution	■	■	■		
Limited Territorial Sovereignty/ Do Not Harm	■	■			
Environmental Principles					
Basin as the Unit of Management		■			
BATT		■	■		
Conjunctive Use		■			
EIA		■			
Invasive Species		■	■		■
Monitoring		■	■		
Prevention of Pollution		■	■		■
Precautionary Principle		■	■		
Protected Areas for water			■		
Protected Recharge and Discharge Zones					
Ecosystem Preservation and Protection	■	■	■		■
Polluters Pay		■			
Social Principles					
Capacity Building	■				■
Equitable & Reasonable Utilisation	■	■	■		■
Human Right to Water & Sanitation				■	■
Intergenerational Equity				■	■
Poverty Alleviation					■
Prior Informed Consent			■		
Priority of Use					
Public Access to Information		■			■
Public Awareness & Education					■
Public Participation					■
Rights of Women, Youth, & Indigenous Peoples					■
Food Security					■
Human Well-being					■
Quality Education					■
Clean Energy					■
Economic Growth					■
Infrastructure					■
Reduced Inequality					■
Sustainable Urbanisation					■
Responsible Consumption & Production					■

*Legally Binding; Source:** Modified from Conti 2017

5.4   INVENTORY OF KEY PRINCIPLES IN GLOBAL WATER INSTITUTIONS

This section reviews governance principles for operationalising the fifth component of the theoretical framework (see 2.6). Global water institutions include many principles and are crucial for understanding the architecture of transboundary water governance. The term _principle' is applied by legal scholars to a theoretical value or law pertinent to specific cases (Alpa 1994). It is also applicable to an instrument (i.e. objectives of water quality) (ibid). As this thesis focuses on inclusive development as a goal of water governance, I emphasise the social/relational focusing on how power can be shifted to local people through, inter alia, the adoption of procedural and substantive principles (e.g. public participation and access to information, human right to water and sanitation, priority of use, public awareness and education, equitable and reasonable use, poverty eradication capacity building, priority of use, and intergenerational equity) and ecological (pollution prevention, monitoring, EIA, ecosystem conservation and protection, protection of water recharge and discharge) components. Although the standing of these principles in international law varies, they are considered as equally important. In this thesis, all the principles that are analysed were identified through both the literature review and a review of the important global water institutions. The following are the main categories of principles.

5.4.1   Political Principles

Political principles that were identified throughthe review of literature and key global water governance institutions include: the principle of sovereignty, obligation to cooperate, dispute resolution, exchange of information, notification about planned measures and of emergencies.

5.4.2   Socio-Relational Principles

Water law consists of various principles including both substantive and procedural, which I have grouped into socio-relational principles. These include: public awareness, education, participation, and access to information, equitable and reasonable use, the human right to clean drinking water and improved sanitation, rights of women, youth and indigenous people, priority of water use, and capacity building. The principle of equitable and reasonable utlisation obliges statesto distribute water (particularly transboundary waters) equitably and appropriately as per their purposes. The 1997 UN Watercourses Convention as well as the 2008 ILC Draft Articles include the principle of equitable and reasonable utlisation. The challenges related to the equitable and reasonable use of transboundary waters between the riparians of an international watercourse are specifically addressed by Articles 5 and 6 of the 1997 UN Watercourses Convention. The equitable use terminology explicitly refers to justice and fairness (including both distributive and procedural) whereas the reasonable utisation does not specifically translate into the most effective use). Moreover, Article 5 (2) exemplifies the notion of equitable participation of all the member states of a transboundary river as well as intergenerational equity and alleviation of poverty.

5.4.3   Ecological Principles

Ecological principles that are identified in this thesis review of literature and global water governance institutions include: ecosystem protection and conservation, monitoring, Environmental Impact Assessment (EIA), preventing transboundary rivers from pollution, and protection of water recharge and discharge zones.

5.5   ROLE OF RIVER BASIN ORGANISATIONS IN DISPUTE RESOLUTION

In recent decades, various River Basin Organisations (RBOs) were created by the riparian states of transboundary river basins to cope with the existing and potential challenges of shared water resources (Petersen-Perlman and Fischhendler 2018). These challenges include flood forecasting on the basis of sharing data and information among member states of transboundary river basins, such as the Mekong Basin (Wang et al. 2016); fixed and equitable distribution of shared water, for example, the Indus water sharing between India and Pakistan (Sattar et al. 2017); preventing waste from flowing downstream, e.g. in the catchment area of the Danube River Basin (Gasparotti 2012); improving water quality along the North Sea and entire Elbe River (Mangi 2016); or recovering the depleted fish stocks, for instance in Lake Victoria (Silsbe and Hecky 2008). In all the above examples, RBOs have contributed efficiently in settling disputes and bringing riparian states closer. However, the role of RBOs vary from basin to basin in strengthening water governance institutions, collaboration among riparian states and settling water-related disputes (Dinar 2009; Zawahri 2008).

There are two types of design characteristics which distinguish the design of River Basin Organisations: (a) organisational structure (b) and the governance mechanisms (Huitema and Meijerink 2017). The organisational structure further consists of seven categories: (1) the membership structure (e.g. to see whether all riparians states are members of the RBOs or just a portion) (Schmeier 2015); (2) scope of functions (e.g. whether RBOs focus exclusively on navigation or non-navigational issues as well including managing fisheries, monitoring of water utlisationand allocation) (Huitema and Meijerink 2017); (3) principles of international water law (e.g. whether the principles of international water law are customary and codified and therefore make a contribution to the sharing, regulation and protection of a watercourse) (Stoa 2014); (4) the degree of institutionalisation and legalisation of the RBO (e.g. the degree to which it is capable to create and device river basin management planes for the states and various other participants in the river basin) (Tir and M. Stinnett 2011); (5) the organisational structure of the RBOs (e.g. RBOs performance in river basin management) (Schmeier 2012); (6) the secretariat (e.g. the role of the secretariats in the institutionalised international environmental policy including water) (Saruchera and Lautze 2015); and (7) financing of the RBO for a more efficient governance of the watercourse (GIZ 2014).

Similarly, governance mechanisms also include the tools and instruments for river basin management to guarantee cooperation, cooperative management and transboundary water resources development. This includes: (i) policy-making tools (e.g. making alliances with co-riparians while governing transboundary water resources)   (Schmeier 2015; Schulze 2012); (ii) data and information-sharing mechanisms for exchange of data and information (which is crucial for transboundary water governance) (Gerlak and Schmeier 2016; Schmeier et al. 2016; Thu and Wehn 2016); (iii) monitoring mechanisms (e.g. compliance monitoring and environmental monitoring where the former denotes the monitoring of actors' performance and the latter discusses the activities that aim at seizing the condition of the river basin and its natural environment at a specific point or over a period of time) (Wingqvist and Nilsson 2015); (iv) dispute-resolution mechanisms (e.g., promoting cooperation among riparians of transboundary river basins) (De Bruyne and Fischhendler 2013; Huitema and Meijerink 2017); (v) mechanisms for stakeholder involvement (e.g. evaluating the role of various external actors involved in the distribution and allocation of water resources, for instance local support and Non-Governmental Organisations (NGOs), epistemic communities as well as other international or regional institutions that may have an influence (Carr 2015).

RBOs have a crucial role to play in the implementation of (particularly goal 6) e.g. implementing the IWRM approach and establishing transboundary water cooperation (Hooper 2003; Hooper 2017). Furthermore, RBOs contribute towards the equitable and reasonable utilisation of international watercourses and multilevel legal frameworks, including global conventions on water which define the general rules and principles for water cooperation (Mukhtarov and Gerlak 2014). There are three main areas that can particularly help in understanding the importance of RBOs. First, is RBO effectiveness i.e., to what extent RBOs actually influence transboundary water governance (Huitema and Meijerink 2017). While research on water governance has largely focused on whether and under which endogenous and exogenous situations RBOs help to properly manage shared water resources, the causes for why some RBOs improve the situations of a transboundary river basin while others have generally failed are still generally unknown (Schmeier 2012 2015). Second is the challenge of seeing extra and evolving actors in a transboundary river basin. Furthermore, the introduction of new actors makes water resources governance extremely complex (Seegert et al. 2014). Various water and non-water related actors from CSOs, NGOs/INGOs or large-scale institutionalised RBOs, as well as representatives of the academic community, the private sector, and international organisations need to be included in transboundary water governance. Understandings of the institutional design of RBOs can assist in developing a mechanisms for incorporating new actors into the activities of river basin governance (Schmeier 2015). Third, the environmental challenges and changes in the environment as well as the capability of an RBO to deal with them, place increasing demands on transboundary water resource management. RBOs are therefore required to recognise the task of measuring changes and developing appropriate reactions (Schmeier et al. 2016).

5.6   INFERENCES

These five institutions provide the global setting, definitions, principles, instruments and dispute resolution mechanisms within which transboundary water challenges between states need to be explored. The analysis of these five key institutions indicates that only two global institutions (i.e.

the 1997 Watercourses Convention and 1992 UNECE Convention) have binding instruments for their respective parties to make them effective in ensuring cooperation in transboundary river basins. In UNECE case, the EU parties already have close cooperation and a number of other binding agreements and mechanisms for dispute resolution on almost all bilateral issues among the members. Due to these strong cooperation mechanisms, the power of stronger countries has been neutralised and the UNECE Convention was signed despite some existing contentious issues. However, in global institutions such as the UNWC, differential power relations among states can influence its effectiveness, which is evident from the fact that only 36 states have ratified the Convention as of 2018.

Additionally, UNECE includes the polluters-pay principle (economic instrument) while the UNWC only uses a pollution prevention (the _no harm') principle (without any enforceable economic instrument). This step demonstrates the efforts to bring the majority of the countries to ratify the UNWC, which is an indicator of how powerful actors have been successful in influencing the Convention. Exclusion of economic instruments (i.e. polluters-pay) are usually beneficial for polluting and for powerful countries since they can continue to pollute Transboundary Rivers without any economic repercussions. Furthermore, although the UNECE mentions equity, it is the UNWC that unpacks the principle of equitable and reasonable sharing of water between states. This allows weaker and often downstream countries to gain more access to water – which may be one reason why upstream countries did not ratify the latter Convention. The lack of champions to steer the treaty such as the EU or G20 (Gupta 2016b), treaty congestion during the 1990s (Weiss 1992) and lack of awareness and capacity to take advantage of the Conventions are some of the prominent issues for the low rate of ratification.

Since international law provides mostly general guidelines without enforcement mechanisms, it creates space for stronger countries to use power for advancing their interests in case of shared water resources. Moreover, legal principles and clauses are ambiguous and contradictory in international water laws, which creates more diplomatic space for powerful states to disregard international water law without significant negative consequences (Dinar 2006; Sand 2016). From a neo-realist perspective, the UNWC is unlikely to reduce inter-state conflicts since it is not ratified by most upstream (and often powerful) countries.

However, despite these weaknesses and the influence of powerful states in not ratifying these agreements, there are a number of effective instruments within the UNWC that are useful in addressing transboundary water issues, especially in situations when transboundary river basins are governed without regulatory frameworks, such as the KRB. For example, since there is no existing transboundary treaty in the KRB, lessons from the UNWC can be learnt with respect to equitable and reasonable use (see 5.2.3), which can provide relevant guidance for a potential agreement as described in the following chapters on the KRB case study. As mentioned earlier (see 5.3.1 and 5.4.2), the principles of absolute territorial sovereignty and absolute territorial integrity are no longer useful in providing guidance for effective water cooperation in the KRB. However, the concept of limited territorial sovereignty upon which UNWC is based, stipulates that riparian states have equitable rights to shared water utilisation (Rieu-Clarke 2005). Articles 5 and 6 also reflect the concept of limited territorial sovereignty.

The identified principles and instruments from the five global institutions will be applied for analysing Afghanistan and Pakistan, as well as transboundary level water governance issues. However, the focus of analysis will be on Articles 5 and 6 in relation to the equitable and reasonable utilisation. The current analysis indicates that the notion of equitable and reasonable use of freshwater resources are very relevant in the KRB case study. Articles 5 and 6 aim to balance differing benefits across the political borders of the states, so that _all member states gain maximum benefit from the water uses without harming each other' (ILA 2001). Among the many advantages of Article 5 and 6, an important one is the recognition of the rights of both upstream and downstream states. These principles will be applied to analyse water governance mechanisms at transboundary level (chapter 6), in Afghanistan (chapter 7), Pakistan (chapter 8). The following chatpers will also explore how these principles and instruments can help in (re)designing a transboundary water cooperation mechanism in the KRB.

6

ANALYSIS OF INTERNATIONAL RELATIONS IN THE KABUL RIVER BASIN (KRB)

6.1   INTRODUCTION

This chapter describes, analyses and compares relations between Afghanistan and Pakistan in the KRB. It aims to answer the following questions: (1) how are various characteristics including ESS and drivers of freshwater problems taken into account at transboundary level in the KRB? (2) How have freshwater governance frameworks evolved at transboundary level in the KRB? (3) Which governance instruments address the drivers of freshwater problems at transboundary level in the KRB? (4) How does legal pluralism occur at transboundary level in the KRB? (5) How do power and institutions influence freshwater governance frameworks at transboundary level in the KRB?

This chapter draws on the methodology in Chapter 2 to answer these questions and proceeds as follows. First, it discusses the political organisation of governance in the KRB (see 6.2), ESS (see 6.3) and drivers of freshwater problems (see 6.4). Second, it investigates the evolution of transboundary level institutions in the KRB (see 6.5) and identifies the relevant goals (see 6.6.1), principles (see 6.6.2) and instruments (see 6.6.3) within these institutions. Third, it explains the instances of legal pluralism (see 6.7). Fourth, it discovers the correlation between principles/instruments and drivers to achieve inclusive and sustainable development (see 6.8). Finally, the last section draws inferences (see 6.9).

6.2   THE CONTEXT OF WATER GOVERNANCE AT TRANSBOUNDARY   LEVEL IN THE KRB

The Kabul River is an important tributary of the Indus River System (IRS) jointly used by two riparians - Afghanistan and Pakistan (Azizi 2007). There is no formal regulatory framework to equitably share the water resources and hence each country does what it wants in such an anarchic system. Overall, the key issues in the KRB are: (a) issues related to water sharing as (i) Afghanistan claims more water because 80% of the Kabul River is located in Afghanistan while Pakistan's claim is based on 50% water flow contribution in the Kabul River, (ii) Afghanistan is utilising only 10% while Pakistan 90% of the Kabul River water resources, and (iii) both countries claim more water explicitly and/or implicitly; (b) issues related to quality where urbanisation, industries including mining and manufacturing and increasing pesticide use in both commercial and subsistence agriculture is increasing pollution in the river.

Although, Pakistan is utilising a major portion of the water resources in the KRB, it is worried about the future access to water as Afghanistan is planning new dams and irrigation infrastructure (Hessami 2017; Majidyar 2018). The unilateral development of the shared river basin by an upper riparian (e.g. Afghanistan) to misuse its geographic advantage or the deployment of other forms of power (i.e., material, bargaining and ideational) by a lower riparian to restrict the upper riparian from doing so can have serious repercussions for hydro-relations and regional peace (Zeitoun and Warner 2006). The already tense environment and hostile situation in the KRB can exacerbate the existing security situation in the region (Hanasz 2011b; Kakakhel 2018). Favre and Kamal (2004: p. 107) argue that riparian issues in the KRB are very complicated given the ongoing conflict in the basin and the border dispute between Afghanistan and Pakistan. The long standing border dispute has become especially sensitive since the emergence of the war-on-terror in 2001 (Hussain 2011). The irrigation development and hydro-power generation of Afghanistan can possibly prompt tensions between the two neighbours (Ahmad, 2010). The new conflict can trigger the old and unsettled *Durand Line* border dispute between the two countries (Ahmadzai and McKinna 2018; Renner 2009).

It has been assumed that lack of water in many parts of Afghanistan has caused civil unrest, including the rise of the Taliban and other militant groups, because water scarcity severely affected subsistence farmers and herders leading to militancy and violence as a coping strategy (Ahmad 2016; Habib 2014). Combined with shrinking water availability in both countries due to intensive agriculture and mining practices, population influx, urbanisation and climate change, the hydro-relations between Afghanistan and Pakistan may be further deteriorated. Hence, water security has been linked to national security narratives in the region. (Azizi 2007; Habib 2014). The water experts and diplomats in Pakistan are of the opinion that new hydro-power generation and irrigation infrastructure projects would significantly affect the water-related infrastructure and economy of the Khyber Pakhtunkhwa (KP) province of Pakistan (Kiani 2013; Pervaz and Khan 2014). Reduction in the freshwater flow to Pakistan in the KRB area can potentially have adverse impacts on the livelihoods of many poor people combined with the uncertaninities linked to climate change (Azizi 2007; Hanasz 2011b).

6.3   CHARACTERISTICS AND ECOSYSTEM SERVICES OF THE KRB

6.3.1   The Kabul River

The Kabul River rises 72 km west of Kabul city in the Sanglakh Range of the Hindukush Mountains and is situated in north-western Pakistan and eastern Afghanistan (Akhtar 2017; SIWI 2015). The total length of the Kabul River is 700 km where only 20% of the river length is located in Pakistan while the remaining 80% in Afghanistan. Before entering the Khyber Pakhtunkhwa province of Pakistan - some 25 km north of the Durand Line near Torkham, it passes through various important cities of Afghanistan such as Sarobi, Jalalabad, and the capital city of Kabul (Ramachandran 2018). Kunar, Swat, and Bara in Pakistan and Alingar, Panjsher, and Logar in Afghanistan are some of the major and important tributaries of the Kabul River (SIWI 2015: see Figure 6.1). The largest tributary is the Kunar River (contributing more than 50% of the water flow) which starts out as the Mastuj River in District Chitral, Pakistan (Kiani 2013) and joins the Kabul River near the city of Jalalabad (Kakakhel 2018). The Kabul River contributes approximately 26% to the total annual water flow in Afghanistan whch covers about 12% of the total land area (IUCN 2013).

The river remains as the Kabul River predominantly for historical and political reasons although the Kunar River contributes more water than the Kabul (Kakakhel 2018). The Kabul River connects 11 provinces of Afghanistan with one province of Pakistan (Khalid et al. 2013). The town of Asadabad is the first main inhabited area on the Kunar River on the Afghanistan side of the border, while

Kabul City is situated on the Kabul tributary (IUCN 1994). The city of Jalalabad is the last major town of Afghanistan before the Kabul River enters into Pakistan. Jalalabad is basically situated at the convergence of the Kunar and Kabul Rivers (Thomas 2014). After entering Pakistan, the Kabul River passes through various settled and populated areas of the Khyber-Pakhtunkhwah Province (Majidyar 2018), including the city of Peshawar which is located close to the Shah Alam tributary of the Kabul River. The cities of Nowshera and Charsadda are two other densely populated cities situated close to the Kabul River (Iqbal 2017; IUCN 1994).

The KRB has the potential of reaching 21 Billion Cubic Meters (CBM) of water per year and can potentially generate nearly 23,000 Megawatts of hydroelectric power from the rivers such as the Kabul River, which can produce up to 3,100 megawatts of electricity (Kiani 2013; Yousaf 2017). The Kabul River monthly discharge data (average montly discharge 38,120 cusecs) shows high seasonal variability (Ahmad et al. 2009) where the low flow period is September to April and the high flow period is from May to July (Khan and Khan 1997). This seasonal snow and glacial melt is contributing largerly in this variation (Yousafzai, Khan, and Shakoori 2008b). The entire area of the KRB is highly arid and any rainfall impact is basically covered by glacial inputs (Rasouli et al. 2015; Iqbal et al. 2018). In addition, the tributaries of the KRB in Afghanistan are also situated in low rainfall areas. The River Swat is the main tributary of the Kabul River below Warsak Dam before it joins the Indus (Yousafzai, Khan, and Shakoori 2008b).

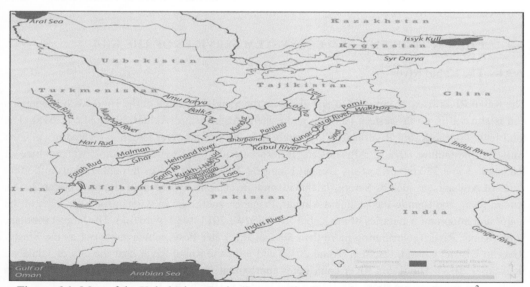

Figure 6.1: Map of the Kabul River Basin (Source: The Center for Afghanistan Studies)[3]

---

[3]https://www.unomaha.edu/international-studies-and-programs/center-for-afghanistan-studies/academics/transboundary-water-research/DLM12/DLM12.php.

6.3.2   Freshwater Biodiversity in the KRB

The entire stretch of the Kabul River hosts about 35 commonly known species of fish (Nafees et al., 2011). Among these, the most famous species is called Masheer (king of river fish) whose population is on the decline since 1990 (Nafees et al. 2011; Saeed 2018). This decline is linked to large scale commercial fishing activities and municipal pollution in the river near Nowshera and Peshawar which, disturb fish migration (Nafees et al. 2011). The raw municipal and industrial sewage is frequently discharged through irrigation canals and urban drainage channels due to lack of wastewater treatment facilities. This results in high absorption of pathogenic microorganisms and severe microbial pollution in the KRB (Murtaza and Zia 2012; Zaryab et al. 2017). An enormous number of seasonal bird species (i.e. cranes, waterfowl and waders) migrate to the Kabul River wetlands in the winter. These birds arrive through the Chitral District and then travel through diverse routes. Some fly to Kashmir via Shandour pass while some follow the Chitral River, entering Afghanistan. Observations in winter have identified a number of ducks including pintail, shoveller, widgeon, mallard, and ruddy shelduck. Other birds include gulls, egrets, lapwings, herons, and terns that are commonly spotted. Cranes were once frequent visitors in the past. They are rarely sighted these days as their numbers seem to have considerably dwindled over time (IUCN 1994; Nafees et al. 2016). Turtles are commonly found along various sites of the Kabul River but are predominantly abundant near the junction of two Peshawar sewage drains (IUCN 1994).

6.3.3   Supporting Services

The Kabul River supports soil formation by transporting minerals (sand/gravel) (Tunnermeier and Himmelsbach 2005). The organic nutrients enhance fertility of the floodplains and support its reconstruction (Nafees et al. 2016). The flow in the river continues to provide other ecosystem services. The KRB provides supporting services to numerous plant and animal species, and supports vegetation which further helps in erosion control (Frischmann 2012). Supporting services provided by the KRB are presented in Table 6.1.

6.3.4   Provisioning Services

The Kabul River and all its tributaries collect water from their catchment areas through rainfall, snow melt and glacial melt (Ahmad et al. 2009; Rasouli et al. 2015). On their way to the ocean after joining the Indus River System, this water becomes available to humans and the environment (Ahmad et al. 2009). The common uses of the freshwater in the Kabul River include sewage disposal, irrigation, watering livestock, fishing, transportation, washing, bathing and recreation. The livelihoods of many local people in the KRB are dependent on the regular and healthy flow of the Kabul River (Nafees 2004). Navigation by boats and vessels, extraction of sand and gravel and plant resources from the floodplain support numerous people (Frischmann 2012). Water itself in the KRB is considered to be the most important provisioning service. The Kabul River contributes a fourth of Afghanistan's freshwater needs (Kakakhel 2018). In Afghanistan, 28% of the Afghan households are connected to the power supply system most of which is produced from the Kabul River and its tributaries (Kakakhel 2017, 2018). Most of the tributaries of Kabul River originate in the mountains and form a potential source of energy (Nafees et al. 2016). Provisioning of groundwater through infiltration in muddy floodplains is another important service of the Kabul River (Tunnermeier and Himmelsbach 2005). The 35 different fish species throughout the KRB is a source of livelihoods and diet for many poor people living in the KRB. Additionally, the Indus waters fish can also be considered as part of the provisioning services because both the rivers merge near Attock District of Punjab Province in Pakistan (Nafees, Ahmad, and Arshad 2011). The Kabul River and its tributaries also provide, through annual flooding of the floodplains, a variety of plant resources which are utilised as food, fuel, timber, fibre and forage (Shroder and Ahmadzai 2016). Local communities and tourists utilise the Kabul River for navigation (Abbas et al. 2018). Before the construction of Warsak Dam, the Kabul River was the main source for transporting timber from Afghanistan to Pakistan (ibid). There are hundreds of villages on the right and left bank of the Kabul River in Pakistan. However, very few bridges are serving these people to cross the river. Therefore, people use taxi-boats, chairlifts, and cable cars to cross the river which provides earnings for some poor people. This enhances tourism in the area while lowering energy demand and saving on infrastructural costs. Provisioning services provided by the KRB are presented in Table 6.1.

6.3.5   Regulating Services

One of the most important regulating services of the Kabul River is hydrological regulation (Tunnermeier and Himmelsbach 2005; Akhtar 2017). From June to August the flow in the Kabul River is mostly high due to fast melting snow and glaciers which causes flooding (Iqbal et al. 2018). However, the flood water is sometimes diverted and used for irrigation in other arid areas downstream (Ahmad 2010). A big part of the KRB consists of floodplains that help in recharging groundwater (Tunnermeier and Himmelsbach 2005). This is done when some of the flood waters are held in the floodplain which infiltrates into the ground. The extra water is gradually released back into the river. The connectedness of the Kabul River with the groundwater recharge is another regulating service which helps in improving the quality of groundwater by removing nutrients and pollutants while passing through soil layers (ibid). However, increasing pollution in the KRB contributes to groundwater pollution if the sediments are absorptive. The capacity to assimilate organic waste is another regulating service of the Kabul River (Nafees 2004; Zaryab et al. 2017). Some of the organic waste is removed by direct consumption by certain aquatic fauna, such as fish and turtle. Aquatic and wetland plants in and around the KRB play a critical role in sequestering nutrients as well as many other pollutants (IUCN 1994). Forested floodplain barriers prevent nonpoint source pollutants (i.e., pollution from agricultural lands, urban runoff and energy production) from inflowing into small streams and enhance in-stream handling of both nonpoint and point source (e.g., pollution from air, water, thermal, noise or light) pollutants (Khuram et al. 2017; Zaryab et al. 2017). Another regulating service of considerable interest of the Kabul River is climate regulation through carbon sequestration in the floodplains and surrounding forests (Khuram et al. 2017). Regulating services provided by the KRB are presented in Table 6.1.

6.3.6   Cultural Services

KRB has been providing a variety of social, cultural and religious activities for centuries (USAID 2017a). A number of activities including sport fishing, swimming, rafting, tourism and recreational boating are common throughout the world though their importance varies greatly between cultures (Kakoyannis and Stankey 2002; Khan 2005). There are mosques at various point on the Kabul River shore where worshipers can access freshwater to clean themselves (doing _wadu' or bathing) before praying.[4] Mass bathing in the Kabul River which occurs mostly in summers provides an opportunity for social gathering and communication. Similarly, the KRB also provides a cultural opportunity for women to gather and do laundry on the banks of River.[5] These activities do not harm perceptible degradation of the river as long as flows are not regulated, and the natural biodiversity helps assimilate these _wastes' as described above. The waterfowl shooting is a famous sport activity for tourists as well as local people in the KRB. The waterfowl shooting activity takes place between December and April when they migrate along the Indus flyway (IUCN 1994). Hunting and fishing for recreation and food are currently the main recreational uses of the Kabul River (Yousafzai, Khan, and Shakoori 2010; Mohammad Nafees et al. 2018). The increasing number of riverside restaurants near Nowshera and Charsadda serve Kabul River fish. Additionally, the local people enjoy walking and appreciate the quiet and peaceful environment on the river banks (IUCN 1994; Yousafzai, Khan, and Shakoori 2008a). In some of the areas in Swat and Chitral people conduct various other activities such as kayaking, canoeing, and white water rafting (IUCN 1994). The local people sing various songs about the Kabul River as this River has stimulated the thoughts of local communities (ibid). There are various famous poems and songs about the Kabul River which are sung at different local and national festivals and ceremonies.[6] Cultural services provided by the KRB are presented in Table 6.1.

---

[4] Interviewee 47, 55
[5] Interviewee 1, 2, 13
[6] Interviewee 1, 2, 10, 24, 33

Table 6.1: Major ecosystem services provided by the Kabul River

Kinds of freshwater in the KRB	**Supporting Services** Formation of soil (gathering of organic matter and sediment retention); habitat provision (provision of habitat for wildlife feeding, shelter, and reproduction; nutrient cycling (processing, storage, recycling, and gaining of nutrients)		
	Provisioning	**Regulating**	**Cultural**
Rainbow water	Huge storage of water on Earth; habitat for birds & insects	Climate regulation, hydrological regulation	Aesthetic (inspiration for art), spiritual (rain Gods/ Gods of thunder), inspiring knowledge)
Blue surface & groundwater	Water collection from catchment areas through rainfall, snow & glacial melt; irrigation, finishingwaste disposal, recreation, navigation, shelter, medicine, bathing, washing, water for power suply system; provisioning of groundwater through infiltration in muddy floodplains; services such as indicators of proper land management, and land use	Hydrological regulation through flood water usage for irrigation in arid zones; groundwater recharge through infiltration of flood water into the ground; water quality improvement by removing nutrients and pollutants through soil layers; assimilation of organic waste by certain aquatic fauna such as fish and turtle; nutrients sequestration; climate regulation through carbon sequestration	Recreational uses such as fishing, hunting waterfowl sports, canoeing, kayaking, white water rafting, sport fishing, swimming; opportunity to worshipers to access water and clean themselves; mass bathing and opportunity for social gathering mostly in summers; riverside restaurants attract tourists and provide peace and quiet environment; inspiring the imagination of the local people through songs
Green water	Fodder, food, pastureland, herbs and shrubs	Evaporation (flowing downwind to fall as precipitation later), and aquifer recharge	Forests &landscapes for tourism, spiritual needs and education
Grey water	Rice and vegetable production, fodder crops, energy production, mining	Climate and water regulation, evaporation flowing downwind to later fall as precipitation	Education services regarding the negative impacts of chemicals in water
Black water	Animal fodder, insects and worms as birds' food	Spreads disease unless managed	Educational services regarding its negative effects
White frozen water/glaciers	Habitat for markhor and snow leopard, storage of water	Albedo effect	Preserving data for humans, information about CO_2 in the past, preserving life forms frozen in the past

Source: IUCN 1994; Kakoyannis and Stankey 2002; Khan 2005; Tunnermeier and Himmelsbach 2005; Ahmad 2010; Yousafzai et al. 2010; Frischmann 2012; Rasouli et al. 2015; Shroder and Ahmadzai 2016; Khuram et al. 2017; Akhtar 2017; Zaryab et al. 2017; USAID 2017a; Iqbal et al. 2018; Abbas et al. 2018; Kakakhel 2017, 2018; Nafees et al. 2018.

6.4   DRIVERS OF FRESHWATER PROBLEMS AT TRANSBOUNDARY LEVEL IN THE KRB

6.4.1   Direct drivers

The key direct drivers of KRB related to water conflict at transboundary level are (see Table 6.2): (a) agriculture development (e.g., commercial agricultural practices including animal husbandry, the extractive sector and water use in energy)[7]; (b) industry (including services and infrastructure);[8] and (c) demographic shifts (i.e., migration, population growth, increase in population density, urbanisation, population growth).[9] The demand for water has dramatically increased in both Afghanistan and Pakistan due to unplanned population growth (Majidyar 2018). The returning and settling of more than three million Afghan refugees may put additional stress on the freshwater resources (BBC 2016). The KRB covers nine provinces in Afghanistan and two in Pakistan. Additionally, around 25 million inhabitants reside in the KRB. (Paula-Hanasz 2011; Ramachandran 2018). Agriculture development in both countries is vital to the livelihoods of millions of people where it contributes 50% to GDP in Afghanistan and 22% in Pakistan greatly (Ahmad 2010; Aziz 2013). Agriculture provides direct and indirect employment to 85% of the population in Afghanistan and 70% in Pakistan (Paula-Hanasz 2011). The conflict between Afghanistan and Pakistan which is currently negligible can potentially be exacerbated in the near future due to the growing demands of growing populations of both countries (Kakakhel 2017; Pervaz and Khan 2014). Subsistence and commercial agriculture as well as mining practices provide livelihood opportunities to many poor people in the region (The World Bank 2014; USAID 2017b).

6.4.2   Indirect Drivers

The key indirect drivers of the freshwater problems in the KRB at transboundary level are (see Table 6.2): (a) Political dynamics between states;[10] (b) culture and ethnic elements (such as wasteful use of resources and behaviours concerning access and allocation etc.);[11] (c) non-water-related policies (e.g. land tenure and land use, agriculture & food security, as well as economic development);[12] (d) economy (economic growth);[13] (e) poverty;[14] (f) technological advances (agriculture intensification);[15] (g) global trade (e.g. virtual water trade or _globalisation' ;[16] and (h) natural change and variability in weather – droughts, floods, landslides and tectonic movement.[17] Political dynamics between states is an important transboundary level driver where _Durand Line' as an

---

[7] Interviewee 2, 13
[8] Interviewee 6, 10, 11
[9] Interviewee 9, 12, 22, 33, 34, 35, 46
[10] Interviewee 5, 25, 26, 27, 41, 52
[11] Interviewee 7, 10, 32, 50
[12] Interviewee 1, 13, 21, 67, 71
[13] Interviewee 1, 2, 10, 29, 31
[14] Interviewee 7, 8, 28
[15] Interviewee 54
[16] Interviewee 2, 10, 17, 49
[17] Interviewee 15, 18, 37, 64

internationally recognised border is not acceptable to Afghanistan. The Durand line – apart from separating Afghanistan from British India in 1893 – also divides water resources in the KRB. It allows Pakistan to claim the water flows from the Hindukush Mountains of Pakistan. Afghanistan, on the other hand, argues that sources of the Kunar River which originates in Pakistan actually belongs to them. The political context of extremism, Taliban proxies and Pakhtunistan inhibits collaboration affects and is affected by water related issues between the two neighbours. Similarly, freshwater resources in the KRB at transboundary level is threatened by the rise of commercial agriculture practices and the mining industry which may be further aggravated with the completion of US\$62 billion worth of China-Pakistan-Economic Corridor Projects (see 8.2.1).

Freshwater resources in the KRB are also under severe threat from climate variability and change which has resulted in droughts and floods in the recent past. The KRB is predominantly fed by the Hindukush glaciers which are vulnerable to earthquakes and the negative impacts of climate change and weather variability. Changes in the melting of glaciers or climate variability can also directly influence freshwater resources.

Table 6.2: Driver of freshwater challenges at transboundary level in the KRB

Direct Drivers	Key References
Agriculture development (e.g., commercial agriculture practices including animal husbandry, the extractive sector and water use in energy)	Lashkaripour and Hussaini 2008; Mack 2010; F. Akhtar 2017
Industry (including services and infrastructure)	Nafees 2004; Rasouli et al. 2015; Akhtar 2017
Demographic shifts (i.e., migration, population growth, increase in population density, urbanisation)	Paula Hanasz 2011; Akhtar 2017; Najmuddin, Deng, and Bhattacharya 2018
Indirect Drivers	
Political dynamics between states (e.g. on Durand line)	Mack 2010; Paula Hanasz 2011; Pervaz and Khan 2014; Najmuddin, Deng, and Bhattacharya 2018
Culture and ethnic elements (attitudes regarding inefficient use of resources as well as behaviourial approaches towards water access and allocation, etc.)	Lashkaripour and Hussaini 2008; Frischmann 2012; Pervaz and Khan 2014; Shroder and Ahmadzai 2016; UNAMA 2016
Non-water-related policies (land tenure and land use planning, economic development, food security and agriculture, China-Pakistan Economic Corridor related projects)	D'souza and Jolliffe 2013; Gohar, Ward, and Amer 2013; The World Bank 2014; Najmuddin, Deng, and Bhattacharya 2018
Economy (economic growth)	Qureshi 2002; Kawasaki et al. 2012; Habib 2014; Ahmadzai and McKinna 2018
Poverty	Lashkaripour and Hussaini 2008; Akbari et al. 2008; King and Sturtewagen 2010; Mack 2010; Frischmann 2012; Kakakhel 2017
Technological advances (agriculture intensification)	King and Sturtewagen 2010; Ghulami 2017; F. Akhtar 2017; Najmuddin, Deng, and Bhattacharya 2018
Global trade (e.g. trade in virtual water or ‚globalisation‘)	Lashkaripour and Hussaini 2008; Renner 2010; King and Sturtewagen 2010; Vick 2014a
Natural change and variability in weather, Droughts; Floods; Earthquakes; Landslides, tectonic movement	Ahmad 2010; Vick 2014a; Shroder and Ahmadzai 2016; Akhtar 2017; Masood and Mushtaq 2018; Iqbal et al. 2018; Akhtar et al. 2018

6.5   EVOLUTION OF TRANSBOUNDARY LEVEL INSTITUTIONS AND PRACTICES IN THE KRB

6.5.1   Overview of Transboundary Level Institutions and Practices in the KRB

Both Afghanistan and Pakistan are facing severe security and governance issues which can hinder their future development and growth prospects.[18] In the last two decades, relations between the two neighbours has been marred by acts, rumours and cross-border incursions (Pervaz and Khan 2014; Thomas et al. 2016). The current geopolitics in the region along with the existing issues of extremism, intolerance, violence, cross-border terrorism, and the deteriorating security situation make bilateral relations worse (Pervaz and Khan 2014; SIWI 2015). Both states face common issues of population increase, urbanisation, food and energy security, economic growth and agricultural productivity, but securing and utilising freshwater resources in efficient ways can be the key determinant to achieving social, ecological and economic well-being as well as peace and stability (Majidyar 2018; SIWI 2015). The significance and importance of freshwater resources of sufficient quality and quantity makes the Kabul River an important transboundary resource (Habib 2014; Kakakhel 2017). Due to the non-existence of formal mechanisms, freshwater resources in the KRB are currently shared based on historic patterns. However, the future development of such sharing could take into account international customary water law (see 5.2.2), religious water practices (see 5.2.3) and/or be inspired by global rules including the 1992 UNECE Water Convention 1992, UN Watercourses Convention 1997, ILC Draft Articles 2002 and the Sustainable Development Goals 2015 (Lead Pakistan 2017; Pervaz and Khan 2014). While neither country is party to the two international water law treaties, both Afghanistan and Pakistan have ratified the binding global agreements on biodiversity (UNCBD 1992), climate change (UNFCCC 1992), and desertification (UNCCD 1994). The normative obligations arising from water and related agreements could provide inspiration for developing transboundary water governance on the Kabul.[19]

There have been various efforts by concerned ministries and authorities in both countries to formalise freshwater governance in the KRB. These efforts include: the 2003 meeting between Pakistan's federal flood commission and the Ministry of Energy and Water (MEW) in Afghanistan to share flood related data where talks collapsed due to lack of information and data sharing; the 2005 discussions between the Water and Power Development Authority (WAPDA) Pakistan and the Provincial Government of Khost in Afghanistan regarding joint hydroelectric power which is yet to be implemented; the 2006 World Bank initiative to draft a Pak-Afghan transboundary agreement over the KRB; the 2009 Islamabad Declaration to further regional collaboration; the 2013 discussion between Afghanistan and Pakistan's finance ministers related to finance for a joint-power project on the Kabul River; the 2014 Afghanistan-Pakistan Joint Chamber of Commerce (APJCC) pledged to develop a joint hydropower-sharing agreement on the Kabul River; the 2014 discussions for a

---

[18] Interviewee 1, 40
[19] Interviewee 51, 59

proposed formal KRB governance structure including the World Bank as well as foreign ministries of Afghanistan and Pakistan in Dubai; the 2015 meeting in Dubai, organised by the Global Water Partnership (GWP), for academia, practitioners, experts, and engineers from Afghanistan, India and Pakistan to improve cooperation in combating climate change in the Himalaya-Karakoram-Hindukush (HKH) region; the 2015 trilateral meeting between Afghanistan, China, and Pakistan for a proposed 1500 megawatt joint hydropower project; and the 2015 meeting between MEW-Afghanistan and the climate change ministry in Pakistan to cope with Glacial Lake Outburst Floods (GLOF). For the evolution of formal/informal transboundary-level water governance frameworks in the KRB, along with their included principles, see Figure 6.2 which shows the accumulated inclusion of different categories of principles over different eras. Similarly Figure 6.3(a) presents the number of adopted principles for each category over time; Figure 6.3(b) explains the trends of different categories of principles over time; and Figure 6.3(c) presents the actual progress and regress of different categories of principles over time.

6.5.2 Water Governance in the Colonial Era

Before the creation of Pakistan in 1947, the whole KRB was an integral part of the Kingdom of Afghanistan (Kaura 2017; Kayathwal and Kayathwal 1994; Omrani 2009). Freshwater governance in the KRB gradually emerged from traditional irrigation practices as well as principles of Islam which were merged with modern conceptions of water management during the British colonialisation in India (Abderrahman 2000). The 2,430 km (1,510 miles) long Durand Line was created in 1896 tbetween Afghanistan and the British India by the colonial adminsitration to reduce the spheres of influence of both countries and improve bilateral relations including trade in the region (Kaura 2017; Omrani 2009). The Durand Line cuts through the Pakhtun tribal areas, politically dividing not only humans but also the natural resources including land and water.[20] According to some scholars (Kayathwal and Kayathwal 1994; Omrani 2009; Walker 2011) having geopolitical and geostrategic perception, the Durand Line is still termed as one of the most unsafe borders in the world in terms of ideological-based ongoing conflicts. Pakistan inherited the 1893 Durand Line agreement and the subsequent 1919 Treaty of Rawalpindi. Both Kabul and Islamabad never signed a formal agreement or ratification regarding border issues (Janjua 2009; Biswas 2013; Yousafzai and Yaqubi 2017) nor on water resources in the KRB (Ahmad 2010; Azam 2015; Kakakhel 2017). Pakistan claims that agreements passed on to successor states remain valid and do not need to be renegotiated as per the principle of _uti possidetis juris_ (UPJ)‛[21] (Biswas 2013; Brasseur 2011; Omrani 2018; Warraich 2016).

Throughout the history of the KRB, only two formal transboundary water governance frameworks were created by British Colonialists which have no legal enforcement since the partition of India. One of these two frameworks was the 1873 Frontier Agreement between British-administered

---

[20] Interviewee 10, 12

[21] Uti possidetis juris or uti possidetis iuris (i.e., _as you possess under law') is a principle of customary international law that preserve the boundaries of colonies emerging as States. The policy behind the principle has been explained by the International Court of Justice (ICJ) in the Frontier Dispute between Burkina Faso and Mali.

Afghanistan and Russia on the Amu Darya Basin, although this agreement was not directly linked to the KRB but lay the foundation for transboundary water governance with the inclusion of the sovereignty principle to treat transboundary rivers as international boundaries. However, at that time no agreement was made regarding a water resource sharing mechanism. The 1873 agreement led to another treaty between British-India and the Afghan Government in 1921. Although its reference to water was very limited, there was a clear water utilisation mechanism (SIWI 2015; Saeed, Hassani, and Malyar 2016). This treaty comprised of 14 Articles and 2 Schedules and was valid for three years from the date of signing. According to Article 2 of the treaty, the British Colonialists agreed to permit the residents of Torkham (a border village and crossing point between Afghanistan and Pakistan) in Afghanistan to draw water through a pipeline (Treaty 1921: Art. 2). In return, Afghanistan would allow the British officers and tribesman to utilise the water resources of Kabul River for navigation and sustain the current irrigation rights (Favre and Kamal 2004).

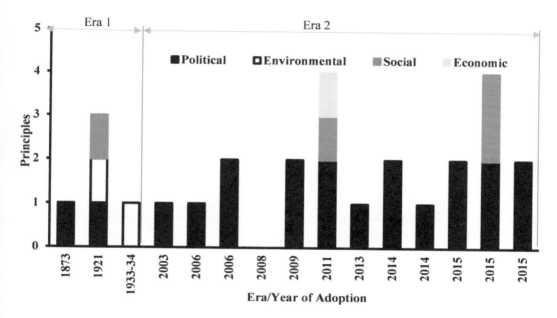

Figure 6.2: Evolution of water governance in the KRB (accumulated)

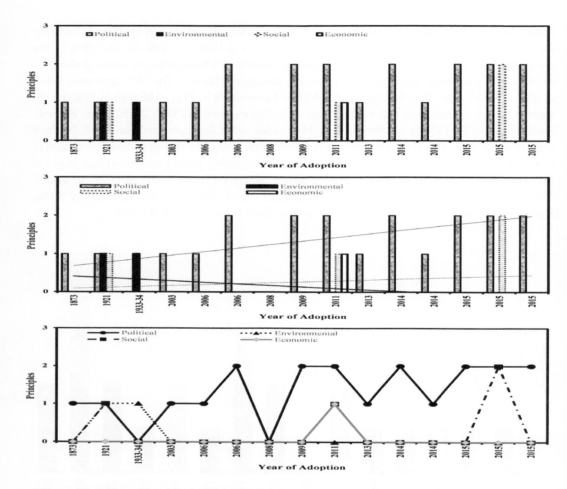

Figure 6.3: (a) Number of included principles (b) overall progress (c) trend analysis

6.5.3   Water Governance Practices in the Modern Era

Since 2000, Afghanistan, Pakistan, and the international community have been struggling to formalize water governance in the KRB.[22] The creation of Water Sector Strategies (WSS) in 2000 by Afghanistan was such an effort to initiate transboundary water cooperation in the KRB with Pakistan. This effort eventually gave birth to the Kabul River Basin Council (KRBC) which provided an opportunity to build capacities of water professionals and improve various issues within Afghanistan (such as allocation and distribution, monitoring water quality and quantity and

---

[22] Interviewee 1

enhancing storage) before going for transboundary collaboration.[23] In 2003, the federal minister of Pakistan for Water and Power also proposed an idea to Afghanistan's Ministry of Foreign Affairs for a possible Kabul River Treaty to efficiently and peacefully utilise shared water resources. However, Afghanistan declined this offer because it lacked technical and financial capacities to sign a bilateral treaty with Pakistan (Kakakhel 2017; Vick 2014b). In 2004, Pakistan restated an offer for a transboundary agreement over the KRB, enabling them to make long-term water availability projections (Kakakhel 2017). In 2006, USAID actively supported the idea of formally sharing water resources in the KRB and initiated dialogues between the two neighbours. In 2006 the World Bank also offered to arbitrate a dialogue over water issues between Afghanistan and Pakistan, but it was not successful as the Afghan officials refused to collaborate or share water-related data.[24]

The Islamabad Declarations in 2009 were also an attempt to strengthen bilateral relations, but collaboration has not materialised in the water sector. Technical water experts from Afghanistan who were present at the time confirmed that as per the State's policy it was not possible to share any hydrological data, joint flood-protection strategies or joint dam feasibility studies with any of its neighbours.[25] After several meetings, the World Bank offered to make the Kabul River Basin Management Commission (KRBMC) functional in 2011 for conflict resolution and Kabul River joint management. The core purpose of making KRBMC functional was to enhance collaboration, build institutional capacity, inititate water-related data sharing, and use the hydraulic, hydrological, and economic modeling to improve the Kabul River management plan (Ahmadzai and McKinna 2018; Burki 2013). This idea was inserted within World Bank funded capacity-building projects in Afghanistan, such as the Afghanistan Water Resources Development (AWARD) project (The World Bank 2016a). The Government of Pakistan has urged the US and World Bank time again to mediate dialogues over water sharing in the KRB to draft an agreement or at least develop an information sharing mechanism to avoid potential disputes over shared water resources (Ahmad 2010; Kakakhel 2017; SIWI 2015).

In August 2013, the federal finance ministers of both countries signed a Memorandum of Understanding (MoU) to mutually develop a hydropower project on the Kunar River (SIWI 2015; Malyar 2016; Saeed, Hassani, and Malyar 2016), however it is yet to be implemented. Similarly, the water and foreign affairs ministries of both Afghanistan and Pakistan were invited to Dubai by the World Bank in late 2014 to mutually develop techniques for exchanging water-related data as well as water allocation mechanisms in the KRB.[26] However, a follow-up meeting has not yet been called. A trilateral meeting in 2015 was conducted between Afghanistan, China and Pakistan in Kabul where China agreed to finance a joint hydropower project of 1500 megawatt on the Kabul River. However, progress has been slow, and despite multiple attempts, officials from both countries

---

[23] Interviewee 40, 43
[24] Interviewee 6, 16
[25] Interviewee 2, 3, 15, 23
[26] Interviewee 42, 50.

have failed to operationalise this initiative.[27] During a UNDP-hosted event in 2015 in Islamabad, MEW Afghanistan and the Ministry of Climate Change in Pakistan agreed to collaborate in GLOF[28] related flood events. Similarly, in June 2016, the Chief Executive of Afghanistan, Abdullah-Abdullah, expressed his interest to formalise water sharing in transboundary river basins with Afghanistan's neighbours so that all riparian countries can benefit. During the event, the Chief Executive expressed a vision for Afghanistan to enhance its water storage capacity.

Currently, in business-as-usual scenarios, Afghanistan only consumes 25% of the surface water and less than 30% of groundwater resources in the KRB under the informal governance structure (Malyar and Hearns 2014). Furthermore, Afghanistan has a very limited water storage capacity (i.e., approximately 100-110 m3/capita/year - one of the lowest in the world), limited hydropower generation, and aims to expand irrigated land and irrigation networks (IUCN 2013; Malyar 2016; Hessami 2017; Jain 2018). On the other side of the border, Pakistan is over-exploiting water resources. Pakistan's groundwater extraction rate far exceeds the average recharge. Similarly its water withdrawal to water availability ratio is 77% (IUCN 2013; Saeed, Hassani, and Malyar 2016). There are visible imbalances between the two countries in many other fields, such as Pakistan's greater technical expertise (human capital); institutional capacity; water-related data; security challenges; public participation; and awareness about customary international water law. Afghanistan lacks capacities and resources in all these fields. Both face problems due to funding limitations; water flow and quality challenges; and competing stakeholders in transboundary waters (Kakakhel 2017; Thomas et al. 2016; Vick 2014b). In addition, various existing and potential challenges persist that may complicate transboundary water issues in the KRB, such as population growth, increasing urbanisation and industrialisation on both sides, climate change and the zero-sum mind-set in the region with absolute winners and losers (Ahmad 2010; Vick 2014b; Kakakhel 2017). The evolution of both informal and formal governance frameworks which have references for the transboundary Kabul River or its tributaries are provided in Annex G.

Currently there is no formal mechanism of transboundary water governance in the KRB between Afghanistan and Pakistan (Ahmad 2010; Vick 2014b). The 1921 Treaty between Afghanistan and Great Britain was the only applicable bilateral treaty in the KRB but it was not specifically a water treaty. According to this Treaty, the British Government allowed the residents of the border village of Tor Kham to take water through a pipe for domestic use. Responding to the British offer, Afghanistan also allowed British officers and tribesmen to navigate and preserve the existing irrigation rights in the Kabul River (Favre and Kamal 2004; SIWI 2015). Both Afghanistan and Pakistan have been trying since the Taliban decline to discuss a mechanism concerning water sharing in the KRB (Aziz 2013; Thomas et al. 2016; Vick 2014b). However, this has not been successful despite the support of the international community. In 2003, friends of both Afghanistan and Pakistan drafted a treaty between Afghanistan and Pakistan in the KRB via a joint technical

---

[27] Interviewee 33, 46.
[28] Interviewee 7, 11.

committee but all attempts were unsuccessful due to mistrust and lack of sharing data.[29] The World Bank once again offered support in 2006 to start negotiation over the water sharing in the KRB but their efforts were not successful.[30] Advances were made at a high level meeting in 2013 when the then President of Afghanistan, Mr. Hamid Karzai and Prime Minister of Pakistan, Nawaz Sharif signed an engagement to discover options for mutually developing the Kunar tributary of the Kabul River for hydropower generation. However, details of this agreement were not clear and therefore very limited progress was made since the announcement.[31]

6.6   GOALS, PRINCIPLES AND INSTRUMENTS

6.6.1   Goals of Transboundary-level Freshwater Governance Framework in the KRB

There is no formal regulatory framework at transboundary level in the KRB, which means there are no goals on social and ecological inclusion at this stage which can be achieved. To set up such goals there is a need to design an institution at the transboundary level by establishing a river basin commission, and developing a water allocation/distribution mechanism and the inclusion of relevant principles and instruments.

6.6.2   Governance Principles at Transboundary Level in the KRB

This section identifies governance principles in the existing and former practices to operationalise the fifth component of the conceptual framework (section 2.6). Furthermore, based on the content analysis, literature review, and interview data, it also analyses governance principles. In line with my conceptual framework I have explained and discussed the key principles (see 5.3) under three main categories of inclusive development: (a) political principles (see 5.3.1), (b) social-relational principles (see 5.3.2), and (c) ecological principles (5.3.3). Table 6.3 shows which of the relevant principles for transboundary water governance are currently applicable to the KRB.

[29] Interviewee 2, 10.
[30] Interviewee 13.
[31] Interviewee 3, 17, 18, 45.

Table 6.3: Major principles and instruments the transboundary water

	The 1873 Frontier Agreement	The 1921 Agreement between Afg. & Russia	The 1933 Kunar Agreement	The 2003 dialogues for Proposed KRB Treaty	The 2006 Discussions for Joint Hydro-power	The 2006 World Bank Sponsored Dialogues	The 2008 Kunar Cascade Dams Project	The 2009 Islamabad Declaration for KRB	The 2011 Proposal for KRBMC	The 2013 Discussions for Joint Hydropower	The 2014 Joint Hydropower Over Kabul River	The 2015 Regional Stakeholders Meeting	The 2015 China Proposal for Joint Hydropower	The 2015 Discussions on GLOFs Events
Political Principles														
Information Exchange				X			X	X	X	X	X	X	X	X
Warning about Emergency Situations														
Warning about Proposed Planned Measures														
Duty to Cooperate				X	X			X	X		X		X	X
Dispute Resolution													X	
Limited Territorial Sovereignty/ No Harm	X	X												
Environmental Principles														
Basin as the Unit of Management														
BATT														
Conjunctive Use of Water														
EIA														
Invasive Species														
Monitoring														
Prevention of Pollution														
Precautionary Principle														
Protected Areas for water														
Protected Recharge & Discharge Zones														
Ecosystem Protection & Preservation		X	X											
Polluter Pays														
Social Principles														
Capacity Building										X	X		X	
Equitable & Reasonable Use														
Human Right to Water & Sanitation		X												
Intergenerational Equity														
Poverty Eradication													X	
Prior Informed Consent														
Usage Priority														
Public Access to Information														
Public Awareness &														

Education							
Public Participation							
Rights of Women, Youth, & Indigenous Peoples							
Food Security							
Human Well-being							
Quality Education							
Clean Energy							
Economic Growth							
Infrastructure							
Reduced Inequality							
Sustainable Urbanisation							
Responsible Consumption & Production							

Source: Modified from Conti 2017

6.6.3   Governance Instruments at Transboundary Level in the KRB

The analysis of goals, principles and instruments in this section indicates that there are no existing regulatory frameworks with effective instruments to address transboundary level water challenges in the KRB. Although, Afghanistan and Pakistan have ratified the Climate Convention and its Protocols (which binds them to reduce their GHGs emissions and adapt to potential climate change); the Biodiversity Convention (UNCBD 1992) and its Protocols (which obliges them to protect their biodiversity); and the Convention on Combating Desertification (UNCCD 1994) (which binds them to manage their lands), they have yet ratified the two global water law treaties (the 1992 UNECE and the 1997 UNWCC). These laws can provide starting points for addressing transboundary water issues in the KRB by highlighting different kinds of ESS of the transboundary waters (as identified in section 6.3) on which the population of both the countries depends. Despite not being parties to the UNWCC, Articles 5 & 6 on equitable and reasonable use can provide guidance and a way forward to initiate dialogues and cooperatively manage water resources in the region. Table 6.3 reveals that both countries have adopted principles concerning cooperation and information exchange, but are not moving forward to establish genuine cooperation over water issues. Although they have accepted the principle of limited territorial sovereignty in two treaties, by not coming to agreement on water issues, they do not respect this limitation. However, there is a menu of principles that the two countries could adopt to enhance their capabilities and jointly manage the shared water resources. As signatories to the SDGs, both countries have committed to a common agenda.

Given that there are limited transboundary agreements and policies affecting these two countries, it does not make sense to undertake a legal pluralism analysis. Furthermore, since very few principles and instruments have been adopted, the key drivers affecting transboundary cooperation as explored earlier in this chapter are not addressed. The most challenging driver of conflict is the issue of the border which affects water related collaboration. However, the formation of the Kabul River Basin Management Commission could explore how the SDGs, the water conventions, the human right to water and sanitation, and other environment related Conventions can be combined to develop a framework to address water issues. My analysis of direct (agriculture, industry, demographic shifts) and indirect drivers (political dynamics, culture, non-water policies, economic growth, poverty, technological advances, trade, and natural changes) shows that these are difficult to be addressed through the current practices in the KRB and governance principles in the existing frameworks. However, the analysis of global institutions (Chapter 5) indicates that there are a variety of applicable instruments (e.g. in UNWC and UNECE) to address a majority of the identified drivers, and can provide building blocks and guidelines for working towards a cooperation mechanism within the KRB. Moreover, the unaddressed principles are covered by the SDGs, which are universal, if both the countries are committed towards achieving these goals based on their national development priorities. The SDGs and other global institutions can provide an effective basis for working towards a transboundary water sharing mechanism since the SDGs have a specific water related goal (Goal 6) with a target on transboundary water sharing. Although the SDGs are voluntary and not binding, less developed countries have incentives to achieve these targets with the support from international cooperation. The SDGs and global water law instruments (e.g. in UNWC and UNECE) can inspire the design of a treaty for transboundary cooperation to accomplish the objectives of inclusive and sustainable development.

6.7   POWER ANALYSIS OF THE KRB RIPARIAN STATES IN TERMS OF GEOGRAPHIC AND THREE ELEMENTS OF POLITICAL POWER

6.7.1   Introduction The section compares and analyses the power dynamics of Afghanistan and Pakistan in the context of the KRB. Power not only drives riparian states of a shared river basin towards hostility but it also influences the transboundary institutional architecture (Zeitoun and Mirumachi 2008). The role of power in influencing Pak-Afghan water relations can be analysed by considering the four key elements of hegemonic struggle (see 3.4), while the influence of power over the institutional context can be analysed in terms of donor driven policies in the KRB. According to the literature on hydro-hegemony, four basic elements of power (i.e., material, geographic, bargaining, and ideational) influence power asymmetries and the making of a hydro-hegemon. The strength of each element matters and adds to the collective power of a country. The power play between Afghanistan and Pakistan is discussed below followed by highlighting the role of donors in power play.

6.7.2   Afghanistan and Pakistan

a. **Geographic Power:** Afghanistan is the upstream riparian and located at the top of the Kabul River watershed, with two main tributaries, the Panjshir and Ghorband rivers, including a number of glacier and snowbank water sources in the Hindukush located to the north of Kabul City; and the Logar River, having its sources in the Paghman Mountains and Kohi Baba Ranges to the west of Kabul (Shroder and Ahmadzai 2016; The DAWN 2018; Akhtar 2017; The World

Bank 2010).[32] However, Afghanistan does not possess the headwaters of the Chitral and Kunar tributaries of the KRB which is located in Pakistan (Kakakhel 2017). The important Swat River tributary to the Kabul River originates and ends within Pakistan's territory (Kakakhel 2017; Khuram et al. 2017). Although Afghanistan has much of the discharge of the Kabul River within its borders, Pakistan is considered to be an upstream hegemon of the Chitral River, which is one of the main sources of the Kunar River tributary to the Kabul River located within Afghanistan (Kakakhel 2017; Sedeqinazhad et al. 2018; Shroder and Ahmadzai 2016). The Chitral River then, is also a main water contributor to the lower Kabul River through the Kunar tributary, which gives Pakistan some important geographic power over the Kabul River system as well (Sedeqinazhad et al. 2018; Shroder and Ahmadzai 2016). In addition to the Chitral-Kunar tributary, Pakistan also has the entire drainage system of the Swat tributary to the Kabul River drainage basin (Kakakhel 2017; Iqbal et al. 2018), which adds additional water into the system and more positional geographic power to Pakistan. This creates some kind of balance in geographic power.

b. **Material Power:** The material power of Afghanistan is very low in the region because it has been engaged in war and near-constant insurgency for three decades leading to low capacities and capabilities particularly in the water sector (Shroder and Ahmadzai 2016). Various indicators of material power reveals Afghanistan's low material power including a low literacy rate, low technical capacity, low population, low GDP per capita, as well as a limited number of combat troops (The World Bank 2016b). However, it is financially and politically supported by India, USA, most of the NATO countries within the UN system, as well as the World Bank and other important donor agencies[33]. In comparison to Afghanistan, the material power of Pakistan is quite high where it has technological and military competencies, nuclear power, a stronger educational base about water, a comparatively strong and stable economy, as well as more political and financial support in Southwest Asia than Afghanistan (Majidyar 2018; Shroder and Ahmadzai 2016). Therefore, in terms of hydro hegemony, Pakistan's influence must be viewed as objectively high in comparison to Afghanistan.[34]

c. **Bargaining Power:** The bargaining power of Afghanistan is comparatively low except for its position as the upper riparian in the KRB. Afghanistan has minimal water infrastructure compared to Pakistan (Hessami 2018; Salahuddin 2010; Shroder and Ahmadzai 2016). Pakistan has been utilising the water resources of the Kabul River for a long time (Ahmadullah and Dongshik 2015; Kakakhel 2017; Thomas 2014). Therefore, a potential reduction of water supplies in Pakistan because of dam construction in Afghanistan would certainly upset Pakistan (Shroder and Ahmadzai 2016). Besides manipulation by Pakistan there are various other factors that contribute to the lack of water sector development in Afghanistan including the four decades of ongoing ideological-based insurgency, lack of investment in the water sector, as well

---

[32] Interviewee 5
[33] Interviewee 1
[34] Interviewee 25

as low capacity.[35] Comparatively, Pakistan has advanced knowledge about the the rules of the
water game because of its bargaining with India over water sharing in the Indus River Basin
since its partition in 1947 (Shroder and Ahmadzai 2016). It is better able to set the agenda for
negotiations, work with complex negotiations, provide incentives to encourage the weaker party
to comply, and to apply influence by associating non-water issues to regional security or other
areas.[36] In terms of hydro hegemony, this provides enhanced power.

d. **Ideational Power:** Afghanistan would probably be proficient to enforce particular ideas and
    narratives about water in the region (Shroder and Ahmadzai 2016). The issue of the Durand
    Line and the self-constructed ideology of _Greater Pakhtunistan' is used by Afghanistan to
    balance its power relations with Pakistan (Gall 2014). Afghanistan could argue that it has been
    cheated by its neighbours, Iran and Pakistan, who used the waters to their own benefit while
    Afghanistan was facing security challenges (Mashal 2012; Salahuddin 2010). No substantive
    ideas about water seem to be emerging in Afghanistan, other than the worrisome factor that
    Pakistan might have been able in some fashion to ban Afghanistan from being able to develop
    its water resources (Mashal 2012; Hanasz 2017; Hessami 2017). At the present time in Pakistan,
    many water statistics are low-level state secrets, which is a negative form of ideational power.

This shows that Afghanistan has more geographic power, but Pakistan has more of the other forms
of power (see Table 6.4). Pakistan can promote hegemonic stability in the region in a positive way
by assisting Afghanistan in the construction of dams and other power generation projects. This can
also be done with channelling assistance from China (King and Sturtewagen 2010; Vick 2014b). It is
also important that Pakistan uses its influence to control the Taliban insurgency. After the Taliban
murdered children of the Army Public School in Peshawar, the Government of Pakistan increased
pressure through its strong military spy agency-ISI[37] to control the Taliban. However, due to
historical reasons of cross-border militancy this is quiet a challenge.[38] Pakistan needs to look toward
Afghanistan in a less colonialist or materialistic manner and more in a developmentally helpful and
kind way to support Afghanistan in irrigation development, hydrpower generation, and flood
control. This will most probably be helpful in achieving a more legitimate collaboration.[39] However,
this is yet to happen.

---

[35] Interviewee 32, 33, 38
[36] Interviewee 10, 11, 16
[37] ISI is the military spy agency of Pakistan, responsible for collecting, processing, and analysing information concerning national security
[38] Interviewee 10, 11, 19, 33
[39] Interviewee 8, 9, 25, 30

Table 6.4: Relationship between geographic power & elements of political power in the KRB

<table>
<tr><td rowspan="2"></td><td rowspan="2"></td><td colspan="4" align="center">Elements of Geographic Power</td></tr>
<tr><td>Elements</td><td>Afghanistan (Upstream)</td><td>Pakistan (Downstream)</td><td>River length & drainage area</td></tr>
<tr><td rowspan="10">Elements of Political Power</td><td rowspan="5">Material</td><td>Economy</td><td>Low GDP per capita</td><td>Higher GDP</td><td>Pakistan's full control over the Chitral & Swat tributaries have enhanced irrigated areas which increase its GDP</td></tr>
<tr><td>Military</td><td>Negligible, no causal relationship</td><td>Negligible, no causal relationship</td><td>Negligible, no causal relationship</td></tr>
<tr><td>Population</td><td>Negligible, no causal relationship</td><td>Negligible, no causal relationship</td><td>Negligible, no causal relationship</td></tr>
<tr><td>Technology</td><td>Negligible, no causal relationship</td><td>Negligible, no causal relationship</td><td>Negligible, no causal relationship</td></tr>
<tr><td>Pol. Stability</td><td>Negligible, no causal relationship</td><td>Negligible, no causal relationship</td><td>Negligible, no causal relationship</td></tr>
<tr><td>Bargaining</td><td>Strategic relations with powerful states</td><td>Financial and political backing from the US, India, and most of the NATO countries; support from the UN agencies and the World Bank, Minimal water infrastructures, proposed dams' construction The lack of water sector development</td><td>International political and financial support and skilled in knowing the rules of the water game due to Indus Waters Treaty</td><td>Negligible, no causal relationship</td></tr>
<tr><td>Ideational</td><td>Power of ideas</td><td>Due to lack of technical capacity no water/climate information can be gathered to shape ideas and norms of downstream riparian</td><td>In Pakistan, numerous statistics about water are low-level state secrets; Influence over non-actors in Afghanistan</td><td>Negligible, no causal relationship</td></tr>
</table>

6.7.3   Donors' Influence and Institutional Context

The current political will and struggle of both Afghanistan and Pakistan to initiate collaboration in the KRB can be a window of opportunity for both countries (Kakakhel 2017; Kerry et al. 2011; Pervaz and Khan 2014; Price 2014; Razzaq 2018). According to the UN-Water (2012) report, the political will of States, at all geographic levels, is the primary condition for successful transboundary water cooperation. The international community and donors cannot foster cooperation or influence dialogues unless there is political will and interest of the governments themselves (Hanasz 2017). Indeed, it is vital for improved governance of water (Vogtmann and Dobretsov 2006). In the last two decades, the international community and donors' involvement in Afghanistan is exceptional (Bjelica and Ruttig 2018; European Commission 2012). Donors such as the World Bank, Asian Development Bank, USAID, European Union, GIZ and USAID are already prioritising transboundary water related work in Afghanistan. The World Bank has already shown considerable interest in the KRB to bring Afghanistan and Pakistan closer to each other through bilateral dialogues and negotiations. For instance, studies like the

_Investment Plan for the Kabul River Basin (2013)', and _Scoping Strategies Options for Development of the KRB (2010)' are some of the efforts by the World Bank.[40]

Donor organisations can potentially remove various important barriers which obstruct interstate cooperation.[41] In addition, they can encourage riparian states to view cooperation as a win-win situation (Mostert 2005). Donors have access to a diverse range of techniques which can be utilised for improved transboundary water governance e.g., building capacities of riparian states, exchange of expertise, conditional funding, loans and debt relief, financial support for non-water related activities, direct intervention, convening, facilitation and mediation (ibid). Donor organisations can develop consent, outline the negotiating agenda and create a problem-solving atmosphere (Mostert 2005; Yasuda et al. 2017). Keeping all these tools and mechanisms in mind, donors have much to offer in the KRB (as UNDP did in the Mekong River negotiations, and the mediating role of World Bank in the Indus River), be it the provision of technical expertise, financial support for infrastructure development or assisting dialogues between Afghanistan and Pakistan to foster collaboration. Currently, there are more transboundary focused efforts by donors in Afghanistan than before, which can ultimately bring Afghanistan closer to its co-riparian States. Donors have initiated various capacity building programmes to train Afghan officials in transboundary water management and in strengthening transboundary water institutions.[42]

The initiation of bilateral dialogues between Afghanistan and Pakistan by donor agencies can play an influential role in making the transboundary governance framework work (IUCN 2013; Malyar 2016; Thomas et al. 2016; Vick 2014b). Due to the institutional capacity and experience of arbitrating the 1960 Treaty in the Indus Basin (of which the Kabul River is a tributary), the World Bank might have the ability to facilitate another bilateral agreement between Afghanistan and Pakistan in this Basin or include Afghanistan as a party to the existing IWT. In addition to the formal support by donor agencies, there are activities that could further help the understanding of transboundary water issues between Afghanistan and Pakistan. For instance, some joint studies on transboundary waters by Afghan and Pakistani experts are conducted by Heinrich-Bol-Stiftung (HBS) (Saeed, Hassani, and Malyar 2016), and the International Water Management Institute's media dialogues (IWMI 2016) has been held. These activities allow experts from both countries to work together and understand each other's needs, challenges and limitations. Some of these activities have already happened while some are under consideration.[43] Furthermore, the World Bank funded an initial scientific study which revealed that development in the upper KRB would have limited impacts on Pakistan, as there is limited potential for irrigation development, and Afghanistan's needs are more related to hydropower projects (Frischmann 2012) which generally involves the non-consumptive use of water. One of the opportunities therefore is to show that development and benefit sharing can occur in the basin if done properly and collaboratively. This is a basin where the benfits of cooperation can outnumber the consequences of conflict. Taking into

---

[40] Interviewee 28.
[41] Interviewee 22, 30.
[42] Interviewee 27, 28, 43.
[43] Interviewee 28, 43, 48.

consideration, the exisiting political turmoil between Afghanistan and Pakistan, water cooperation could be an opportunity for greater regional stability (Ahmadzai and McKinna 2018; Price 2014). The spill over benefits into other areas could be significant.[44] The influence of donors supporting both Afghanistan and Pakistan can be a way forward as they have the power and resources to bring both the riparian countries to the table for dialogues. Joint initiatives of research, capacity building and institution building on water with the support of donors can build trust and enable application of Articles 5 & 6 of the Watercourses Convention on equitable and optimal utilisation in the KRB.

6.8   INFERENCES

This chapter has described and analysed the water-based relations between the two riparians of the KRB in order to answer the question of how power and institutions influence international relations between the two countries and how they obstruct or contribute to achieve inclusive and sustainable development. It has done so by i) discussing the overall political context, ii) identifying key ESS including biodiversity, iii) recognising direct and indirect drivers of freshwater challenges, and iv) providing a detailed overview of the transboundary level institutions by analysing recent and historic practices including the pre-colonial agreements.

This chapter draws four conclusions. First, there is no formal regulatory framework to equitably share the KRB water resources between the two riparians, and hence each country does what it wants in such an anarchic system. There are various issues in the KRB including: (a) issues related to water sharing as (i) Afghanistan claims more water because 80% of the Kabul River is located in Afghanistan while Pakistan's claim is based on 50% water flow contribution in the Kabul River, (ii) Afghanistan is utilising only 10% while Pakistan 90% of the Kabul River water resources, and (iii) both countries claim more water explicitly and/or implicitly; (b) issues related to quality where urbanisation, industries including mining and manufacturing, and increasing pesticide use in both commercial and subsistence agriculture is increasing pollution in the river. Although, Pakistan is utilising a major portion of the water resources in the KRB, it is worried about the future access to water as Afghanistan is planning new dams and irrigation infrastructure. The government of Pakistan thinks developing such projects would significantly affect the water-related infrastructure and economy of the already socially and politically deprived province of Khyber Pakhtunkhwa (KP). Additionally, riparian issues in the KRB are very complicated given the ongoing conflict in the basin and the Durand Line dispute between Afghanistan and Pakistan.

Second, acknowledging the variety of BESS at transboundary level within the Kabul River Basin can provide the basis for similar problem framing as well as highlight the benefits for local livelihoods. For example, the basin provides a winter habitat for migratory birds, enables soil formation (enhancing fertility) and is fed by similar sources of water in KRB (snow melt, rainfall and glacial melt), and enables hydro-power generation and holds more than 35 fish varieties. All

---

[44] Interviewee 10, 11, 19, 33.

these not only provide water but also nutritious food and livelihood opportunities for people across the border in both countries. Similarly, regulating services of the Kabul River include connected and improved groundwater recharge by removing nutrients and pollutants, climate regulation through carbon sequestration in the floodplains and surrounding forests. The River also provides a cultural opportunity for women to gather and do laundry on the banks of the River. These activities cause no perceptible degradation of the river as long as they do not use chemical detergents and the natural biodiversity helps assimilate these _wastes'. There are various famous poems and songs about the Kabul River which are sung at different local and national festivals and ceremonies.

Third, acknowledging and identifying similarities in the key direct and indirect drivers can be a first step towards addressing transboundary water issues in the KRB. Key direct driver's include: (a) agriculture development; (b) industry; and (c) demographic shifts. In particular, agriculture in both Afghanistan and Pakistan rely on surface freshwater which is supporting the well-being of millions of people in both countries. Agrictulture contributes 50% of the GDP in Afghanistan and 22% in Pakistan. The main indirect drivers of the freshwater problems in the KRB at the transboundary level are: (a) political dynamics between states; (b) culture and ethnic elements; (c) non-water-related policies; (d) economy; (e) poverty; (f) technology; (g) international trade; and (h) natural change and variability in weather. In particular, political dynamics between states is an important transboundary level driver where the Durand Line is not recognized by Afghanistan as an international border. The Durand line – apart from separating Afghanistan from British India in 1893 – also divides water resources in the KRB. It allows Pakistan to claim the water flows from the Hindukush Mountains of Pakistan. Afghanistan, on the other hand, argues that sources of the Kunar River which originate in Pakistan actually belong to them. The issue of the Durand Line also needs to be urgently resolved and this requires the establishment of a fact-finding mission by including concerned authorities and international legal experts. The political context of extremism, Taliban proxies and Pakhtunistan inhibits collaboration and is affected by water related issues between the two neighbours. The Kabul River is predominantly fed by the Hindukush glaciers which has significant threats from the climate change impacts and weather variability. The fast melting of glaciers or variability in the climate can directly influence freshwater resources.

Fourth and finally, the absence of formal institutions has enabled power politics to prevail between the two countries. While pre-colonial agreements did not deal with the allocation issue, the three colonial agreements allocated water for drinking and navigation purposes among the villages of riparian countries, but these agreements became void in 1947 after the partition of British India. In the post-colonial era there has been a return to anarchy – each country does what it wants. The analysis of power (see 6.9) and institutions (see 6.5) shows that the lack of formal institutions enables the more powerful country to behave as a hydro-hegemon. Although Afghanistan has a geographic advantage, its low capacity and resources make it less influential to exploit freshwater resources whereas Pakistan possesses higher material, bargaining and ideational power consuming about 90% of water from the KRB and is less motivated to change the status quo, which has influenced the formation of the transboundary institutional architecture significantly (Zeitoun and Mirumachi 2008). Due to these power asymmetries, Pakistan has adopted a security oriented foreign policy agenda towards Afghanistan and India. Pakistan has been able to exclude relevant domestic actors (i.e. ministries of water, agriculture, environment, IRSA) except for defence and foreign ministries and has kept the role of international actors in Afghanistan (NATO, other donor countries) minimal in negotiations about transboundary water sharing and other socio-relational issues (Taliban proxies, Durand Line, Pakhtunistan). Sovereignty politics is practiced in the KRB by excluding the voices of the less powerful, key non-water related actors, and international and domestic actors at different levels. Inadequate attention to these actors and identified characteristics can prevent institution building and hinder the effectiveness of informal or formal water cooperation in the KRB. As a consequence, no goals on social and ecological inclusion can be achieved.

7

ANALYSIS OF WATER GOVERNANCE IN AFGHANISTAN

7.1   INTRODUCTION

This chapter describes and analyses multilevel freshwater governance in Afghanistan and aims to answers the following questions: (1) How are the various characteristics including ESS and drivers of freshwater problems taken into account at multiple geographic levels in Afghanistan? (2) How have freshwater governance frameworks evolved at multiple geographic levels in Afghanistan? (3) Which governance instruments address the drivers of freshwater problems at multiple geographic levels in Afghanistan? (4) How does legal pluralism occur at multiple geographic levels in Afghanistan? (5) How do power and institutions influence freshwater governance frameworks at multiple geographic levels in Afghanistan?

This chapter uses the methodology in Chapter 2 to answer these questions, and continues as follows. First, this chapter describes the political organisation of water sharing within Afghanistan (7.2), various ESS (7.3), and drivers of freshwater problems (7.4) at multiple geographic levels in Afghanistan. Second, it discusses the evolution of freshwater governance (7.5) in Afghanistan. Third, it discusses goals, principles and instruments (7.6). Fourth, it conducts an analysis based on legal pluralism (7.7), and explores the relationship between governance instruments and drivers and their contribution in achieving inclusive and sustainable development (7.8). At the end, this chapter draws inferences (see 7.9) about the interplay of power and institutions in influencing freshwater sharing within Afghanistan.

7.2   THE CONTEXT OF WATER GOVERNANCE IN AFGHANISTAN

The current population of Afghanistan is more than 31 million which is expected to reach approximately 56 million by 2050 (Yıldız 2015). This 80% increase in population will put severe stress on already stressed water resources. Some studies (see Aich et al. 2017; Yıldız 2015) reveal that the precipitation patterns will change with global climatic changes and variability. The impacts of changing climate and weather variability will also affect snowfall in higher elevations affecting water flows in rivers. Being one of the world's poorest countries and with an economy largely based on subsistence agriculture (Yıldız 2015), Afghanistan will not be able to cope with such challenges. Farmers in the country are greatly relying on fresh surface and groundwater resources for irrigation and watering their livestock due to the highly arid climate (Campbell 2015). Glacial and snow melt are the primary sources for feeding seasonal streams, rivers, and aquifers. These sources of freshwater provide drinking water mainly to cities(Mack 2010; Zaryab et al. 2017). Due to the regular water supply to cities people in rural villages were once encouraged during the drought period from 1999 to 2005 to move to the larger cities and abandon their land (Campbell 2015; Yıldız 2015). These abandoned areas which were hit by severe droughts can still be found throughout the country (Campbell 2015).

Afghanistan has historical conflicts with its neighbours over the flow of water due to its land-locked status (Ahmadzai and McKinna 2018; Gadgil 2012; Salahuddin 2010). Due to its geographic location, the mountain snow runoff passes through Afghanistan into its neighbouring countries, Iran,

Pakistan, and other Central Asian States. In the past, Afghanistan had built various water reservoirs to store its surplus water but most of those reservoires were damaged during the four-decades long conflict (Zaryab et al. 2017). Today hardly 30-35% of the freshwater stays in Afghanistan due to lack of water infrastructure (Gadgil 2012). These damaged reservoirs and other water infrastructure are challenging to repair and reconstruct in short time due to continuous militancy and unrest (Pervaz and Khan 2014; Ramachandran 2018). The estimated cost of repairing or re-constructing such nationwide projects is approximately USD 11 billion. Investors and donors hesitate to invest such a huge amount of money in projects where workers and project-related infrastructure are vulnerable to the attacks of militants and terrorists (Salahuddin 2010).

It is quite challenging to have reliable and regular supplies of clean and safe freshwater as well as freshwater-related data due to the longstanding conflict since the 1980s (Campbell 2015; Yıldız 2015). Besides damage to infrastructure including water monitoring devices, the unrest in Afghanistan has severely affected the abilities of water scientists and the institutional knowledge of Afghanis (Yıldız 2015). In comparison to other neighbouring countries of South Asia, the water-related expertise of Afghanistan is inadequate (King and Sturtewagen 2010; Qureshi 2002; The World Bank 2018). Various Afghan scholars particularly trained water scientists are not much aware of global technological developments and are disconnected from the international scientific community (Campbell 2015; Mack 2010). The United States of America attacked Afghanistan in October 2001 (Campbell 2015) and by as early as 2002 the US and NATO forces in partnership with concerned Afghan water experts started reconstructing water-related infrastructure and important civic institutions (Perry 2015; Qureshi 2002). This has resulted in steadily building scientific knowledge, the institutional capacitiy of the NGOs and Afghan water and other related ministries, as well as understanding of the water-related needs of the country (Yıldız 2015).

7.3   ECOSYSTEM SERVICES OF FRESHWATER IN AFGHANISTAN

The natural resources of Afghanistan such as water, forests and minerals are a key source for the country's peaceful and prosperous future (Ahmadzai and McKinna 2018). A large percentage of the country's population (around 70-80%) directly relies on artisanal excavation, animal husbandry, and agriculture for their livelihoods (Sharifi et al. 2016). Afghanistan must harness these assets for job and revenue generation in order to improve its place on the Human Development Index and to supply basic services to its citizens. The rural Afghan population uses freshwater ecosystems for their livelihoods. The deterioration of these natural ecosystems due to variability in freshwater flow or degradation in quality may negatively affect people (Saba 2001). Following are the four types of ESS provided by freshwater in Afghanistan (see Table 7.1).

7.3.1   Freshwater Resources in Afghanistan

Afghanistan has a series of snow covered mountains including Baba, Wakhan, and Hindu Kush which makes it rich in terms of freshwater resources despite the fact that it is located in an arid zone (Shroder and Ahmadzai 2016; Yıldız 2015). Among the three mountain ranges Hindu Kush is located at an elevation of 2000 meters and hosts approximately 80% of Afghanistan's freshwater resources. The Hindu Kush mountains function as natural storage for Afghanistan which deliver permanent water flow to major rivers by snow melt round the year (Qureshi 2002).

Afghanistan is surrounded by large rivers, for instance the Amu Darya Basin is situated in the North while in the East it is bounded by the Indus River (Ahmad and Wasiq 2004; King and Sturtewagen 2010). Afghanistan can be divided into four main basins based on the watershed units as well as on the morphological and hydrological systems (Qureshi 2002). Approximately 75 Billion Cubic Meters (BCM) of potential water resources exist in Afghanistan in various forms where the amount of groundwater is about 20 BCM and surfacewater is 55 BCM (Ahmad and Wasiq 2004; Qureshi 2002). Approximately 20 BCM of freshwater annually is consumed by the agricultural sector which is roughly equal to 99% of all freshwater used. In addition, approximately 3 BCM of the groundwater is extracted each year (Ahmad and Wasiq 2004). It is estimated that out of the total annual water consumption in Afghanistan, approximately 9% comes from the groundwater aquifers, 7% from the natural springs, and roughly 85% from rivers and streams (Angelakis et al. 2016; Qureshi 2002).

Approximately 2500 cubic meters of freshwater per capita is available in Afghanistan which is comparatively the highest in the region. For example, per capita water availability in Iran is 1400 cubic meters while Pakistan's per capita water availability is less than 1200 cubic meters (ibid) (Qureshi 2002). The freshwater resources in Afghanistan is projected to be mainly under-utilised (Qureshi 2002; The World Bank 2014; USAID 2002). However, it is not clearly known how much of the underused water resources can be retrieved to avoid damage to ecosystems and people.

7.3.2   Freshwater Biodiversity in Afghanistan

Seven of eight exclusive biogeographical areas of Afghanistan, belong to the Palaearctic Empire (Palka 2001). The Indi-Malayan region consists of a very small area in the lower Kabul River area (UNEP 2008). According to a recent classification, the country can be divided into 15 smaller ecoregions, four of which are considered critical / vulnerable, eight are vulnerable and only two are relatively stable and intact (Khan 2006). The species composition of all regions was significantly influenced by fuelwood collection, overgrazing, and exploitation by large herbivorous animals (Government of Afghanistan 2014). Deciduous and evergreen forests can be found in the eastern monsoon part of the country and once accounted for approximately 5% of the land area (Akhtar 2017). Some studies suggest that only 5% of these original forests are left (Carberry and Faizy 2013; Government of Afghanistan 2014). Pistachios and juniper once accounted for about 38% of the territory of Afghanistan (Government of Afghanistan 2014; NEPA-Afghanistan 2008). It is also known that the country was one of the most significant centres for the origin and development of food crops (UNEP 2008) such as wheat and other crop varieties (Government of Afghanistan 2014). In Afghanistan there are about nine indigenous breeds of sheep, eight breeds of cats and seven of goats (Thomson et al. 2005).

7.3.3 Supporting Services

There is limited literature on the supporting services of nutrient flows and soil fertility. However, there is some information on biodiversity including habitat availability and genetic diversity, ensure the functioning of all other ecosystem services. The Government of Afghanistan (2014) highlighted the importance of biodiversity in freshwater, in particular Koh-e Baba for the provision of ESS in the region. The country has a large number of breeding and migratory birds and is home to several species of plants and animals, such as wolves, foxes, wild cats, rabbits, deer, bats and many other birds. Afghanistan has a considerable genetic diversity of wild relatives of wheat and other plants that can provide genes for disease resistance (ibid). These different ESS are of particular importance in predominantly agricultural areas.

7.3.4 Provisioning Services

Agricultural activities in Afghanistan highly depend on freshwater from precipitation, ice, and melting snow (NEPA-Afghanistan 2008). Similarly, production of different crops greatly vary from year to year depending on rainfall and other weather conditions. The Hindukush mountain is situated at an altitude of more than 2000 meters and stores about 80% of Afghanistan's freshwater resources (Ali and Shaoliang 2013; The World Bank 2018). The mountains act as natural reservoirs and sources of water, where snow accumulates in the winter, snow melts and rain falls in the spring, and frozen water emerges from the glaciers in the summer to maintain the water flow in rivers (Hanasz 2011b). The large river basins of Afghanistan such as Amu Darya, Helmand and Kabul offer the greatest potential for irrigated agriculture and hydropower (King and Sturtewagen 2010). It is important to preserve the natural functions of these river basins of future generations and for prosperity of the land (Ahmadzai and McKinna 2018). The water cycle can be affected by climate change and land degradation, which can have serious consequences for communities living in the downstream areas (Hanasz 2011b; Kakakhel 2018; Vick 2014b). Soil erosion and forest degradation are largely caused by overgrazing which hinder forest regeneration (NEPA-Afghanistan 2008). Farmers note that changes in vegetation and productivity, including changes in weather patterns such as rain and snow, force them to shift grazing from traditional areas to higher areas (Savage et al. 2009). As a result, these practices increase the pressure on alpine ecosystems where large areas of vegetation have been converted into grazing-resistant cushion shrub lands (NEPA-Afghanistan 2008). In addition, increased cross-border cooperation with neighbouring countries is needed to ensure that water system infrastructure development does not harm local communities in Afghanistan or neighbouring countries (Hanasz 2011b; Mashal 2012; Nafees et al. 2016; Ramachandran 2018).

7.3.5 Regulating Services

It is estimated that approximately 16% of Afghanistan's land mass is affected by human-made activities. Afghanistan is one of the most desertified areas in the world (75% of the country is vulnerable to desertification) (NEPA-Afghanistan 2008; Savage et al. 2009). The topographic, geological and climatic conditions of the country escalate the susceptibility to soil erosion. However, growing on steep slopes and unsustainable use of pastures can lead to a significant deterioration of rangelands (Government of Afghanistan, 2014; NEPA-Afghanistan 2008). Agriculture in rainfed zones is particularly detrimental to soil preservation, but is not practiced due to the low availability of irrigated land (Rao et al. 2016). Freshwater can potentially preserve the natural vegetation on the slopes, decrease soil erosion and enhance the productivity of the land that provides the key livelihood source (Government of Afghanistan, 2014; Rao et al., 2016). Freshwater can potentially preserve natural vegetation on the slopes, reduce soil erosion and increase land productivity, which enables livelihoods (The Government of Afghanistan, 2014).

Floods and landslides can be detrimental to mountains and valleys, especially in spring and summer, when ice-covered glacial lakes and snowy lakes start melting leading to devastating floods (Savage et al. 2009; The World Bank 2017). Dust storms and frequent droughts can also cause substantial harm. Furthermore, earthquakes are very common as the country is in a high seismic zone (The World Bank 2017). The depletion of natural resources and rapid urbanization are key factors that increase Afghanistan's vulnerability to hazards (Savage et al. 2009; Tami 2013). Various socio-economic development elements such as umemployment, land tenure practices, and migration have enhanced the vulnerability of some segments of society where they are forced to live in disaster-prone locations (Cordesman 2010; Loschmann et al. 2015). The exclusive dependence on natural resources as part of daily life, and catastrophic events, puts great pressure on freshwater ecosystems (Savage et al. 2009). Competetive demands and unsustainable consumption have undermined the safety net that freshwater provides and reduces the resilience of communities (Government of Afghanistan 2014; Rasul and Sharma 2016; The World Bank 2016b). Pollination is another important and indirect regulatory service of freshwater and supports local incomes (Government of Afghanistan 2014).

7.3.6   Cultural Services

Afghanistan's various regions have substantial cultural value for diverse local communities. For example, it is believed that the snow line on Koh-e-Allah Mountain expresses the word _Allah' and, therefore, has great cultural significance (The Government of Afghanistan 2014). Similarly, one of the hot springs provides recreational opportunities for the local communities apart from its direct use for consumption (Erfurt 2011). The local population reveals that there were many springs before the start of the existing long ongoing conflict (ibid). These areas still offer many opportunities for ecotourism, hiking, and skiing which can be of great benefit to local communities (The Government of Afghanistan 2014; UNAMA 2016). In mountain areas, there are a number of shrines that commemorate important events and represent important parts of the landscape that should be considered in future land use plans (The Government of Afghanistan 2014). Both formal and traditional institutions govern an unequal and insecure land ownership system in the country (Pfeiffer 2011; Khan 2015). Recently, the Government of Afghanistan has tried to initiate land reforms and modernize the land management system. Since 2011, there has been a new regime for the management of land rights in the form of the Afghan Land Authority, but these regulations are still in the early stages (Saltmarshe 2011; The Government of Afghanistan 2014). Due to the continuous deterioration of the soil, the demand for fertile land is increasing (Shrestha 2007; Gaston and Dang 2015).

Table 7.1: Major ecosystem services provided by freshwater in Afghanistan

Kinds of Freshwater in Afghanistan	Supporting Services		
	Availability of habitat and inherited variety confirm the functioning of all other services; freshwater biodiversity of Koh-e Baba supporting the provision of all ESS in the area; home to a large number of breeding and migratory birds; supporting diverse plant species, and wildlife such as wolves, fox, wild cats, rabbit, deer, bats, and numerous birds; genetic diversity of wild relatives of wheat and other flora provide genes of resilience and resistance to disease		
	Provisioning	Regulating	Cultural
Rainbow water	Huge storage of water on Earth; habitat for birds and insects	Climate regulation, hydrological regulation	Aesthetic (inspiration for art), spiritual (rain Gods/ Gods of thunder), inspiring knowledge)
Blue surface and ground water	Natural storage facility; supporting the vibrant flow in rivers through snow accumulation in the winter, melting of snow and precipitation during spring, and discharge of frozen water from glaciers in summer; prevalent potential for agriculture and hydropower gemeration; maintaining watersheds for the future prosperity; freshwater in the KRB can enhance cross-border cooperation with neighbour countries	Freshwater in Afghanistan naturally increases susceptibility to the practices of soil erosion; rainfed farming is especially harmful to soil preservation; freshwater preserve the natural vegetation and prevent soil erosion which enhance the productivity; pollination is indirect regulating service which provides an alternative source of income for local communities	The snow line on the top of the Koh-e Allah spell out the word _Allah' is of traditional importance; holy natural hot springs appealed many tourists before the war; provides occasions for trekking and backcountry skiing as well as ecotourism to local communities and tourists; many caves in the KRB on Afghanistan's side have earliest animal statues that locals visit; monuments in the KRB memorialise important events; a mix of formal and informal institutions in the KRB governs an variable and indeterminate land tenure system
Green water	Fodder, food, pastureland, herbs and shrubs	Vaporization (flowing downwind to later fall as precipitation); undergraound aquifer recharge	Forests and landscapes for tourism, spiritual needs and education
Grey water	Rice and vegetable production, fodder crops, energy production, mining,	Climate and water regulation, vaporization flowing downwind to later fall as precipitation	Educational services regarding its potential uses in agriculture
Black water	Animal fodder, insects and worms as birds' food	Spreads disease unless managed	Educational services regarding its negative effects
White frozen water/glaciers/frozen polar regions	Habitat for markhor and snow leopard, storage of water	Albedo effect	Preserving data for humans, information about CO_2 in the past, preserving life forms frozen in the past

Source: Modified from the Government of Afghanistan 2014

7.4   DRIVERS OF FRESHWATER PROBLEMS AT NATIONAL & SUB-NATIONAL LEVEL IN AFGHANISTAN

There are various drivers of freshwater problems at national and sub-national levels in Afghanistan. In this section I discuss national and sub-national level direct (see 7.3.1) and indirect (7.3.2) drivers which also include drivers recognized through literature review, case studies, as well as through literature that is not case specific.

7.4.1   Direct Drivers

The four direct drivers of freshwater problems in Afghanistan are (see Table 7.2): (a) agricultural development including commercial agricultural practices and animal husbandry, the extractive sector and water use in energy;[45] (b) industry including services and infrastructure;[46] (c) municipal supply of clean water and improved sanitation services including water usage at household level i.e., drinking water, water for sanitation and hygiene as well as subsistence agriculture for survival;[47] (d) demographic shifts including migration, population growth, increase in population density and urbanisation[48]. The four decades long war has destroyed water infrastructure and the agriculture sector which had employed more than 80% of the population. Moreover, arrival of more than three million registered and almost the same number of unregistered Afghan refugees from Iran and Pakistan, coupled with unsustainable population growth, can put enormous pressure on available freshwater resources in the KRB where the population density is comparatively high than other river basins in the region. Increasing demand for improved sanitation practices anddrinking water particularly in urban areas as well as water uses in subsistence agriculture and localised mining industry is expected to put significant stress on available freshwater resources.

7.4.2   Indirect Drivers

The key indirect drivers of freshwater problems are (see Table 7.2): (1) political dynamics among administrative units within Afghanistan;[49] (2) culture and ethnic elements including approaches concerning access and allocation, behaviour towards wasteful use of resources, etc.;[50] (3) the non-water-related policies on food security, agriculture, land trnue and use, as well as economic development;[51](4) economy i.e., economic growth;[52] (5) poverty;[53] (6) technological advances including agriculture intensification;[54] (7) international trade including ‗globalisation‘ or trade in

---

[45] Interviewee 53, 57, 65.
[46] Interviewee 42.
[47] Interviewee 42, 66, 70.
[48] Interviewee 41, 42, 43, 46.
[49] Interviewee 43.
[50] Interviewee 43, 47, 57, 58, 67.
[51] Interviewee 47, 52.
[52] Interviewee 41.
[53] Interviewee 47, 70.
[54] Interviewee 51, 52, 64, 70.

virtual water;[55] and (8) natural change and variability in weather, including droughts, floods, landslides, tectonic movement.[56] In Afghanistan privately owned water pumping stations control the flow and provide clean drinking water to approximately 80% of the population. This may be also due to lack of water-carrying infrastructure due to the four-decade long ongoing conflict. In addition, security challenges are also a disincentive to building infrastructure. Massive flooding caused by river overflow and glacial lake outbursts are in the western region and central belt, while drought in the southwest and northern regions have put farmers out of work and degraded water quality. In addition, earthquakes and landslides are also of concern in the northern regions.

Table 7.2: Drivers of freshwater problem in Afghanistan

Direct Drivers	Key References
Agriculture development (e.g., commercial agriculture practices including animal husbandry, the extractive sector and water use in energy)	Interviewee 53, 57, 65
Industry (including services and infrastructure)	Interviewee 42
Municipal water and sanitation services e.g., household uses (drinking water, sanitation and hygiene) and subsistence agriculture	Interviewee 42, 66, 70
Demographic shifts (i.e., migration, population growth, increase in population density, urbanisation, population growth)	Interviewee 41, 42, 43, 46
Indirect Drivers	
Political dynamics within states	Interviewee 43
Culture and ethnic elements (approaches concerning access and allocation, wasteful use of resources, etc.)	Interviewee 43, 47, 57, 58, 67
Non-water-related policies (food security, agriculture, land tenure and land use, as well as economic development)	Interviewee 47, 52
Economy (economic growth)	Interviewee 41
Poverty	Interviewee 47, 70
Technological advances (agriculture intensification)	Interviewee 51, 52, 64, 70
Global trade (e.g. globalisation' or trade in virtual water)	Interviewee 45, 59
Natural change and variability in weather, Droughts; Floods; Earthquakes; Landslides, tectonic movement.	Interviewee 56, 60, 63

7.5   EVOLUTION OF THE FRESHWATER & RELATED INSTITUTIONS IN AFGHANISTAN

7.5.1   Overview of Water Governance Institutions & Practices in Afghanistan

The institutional structure in Afghanistan has three diverse parts: customary tribal law, Sharia law, and state legal codes. Additionally, the legal system in Afghanistan is a combination of both informal and formal structures (Bassiouni and Rothenberg 2007; Christensen 2011; Hashimi 2017; Singh 2015; Strand et al. 2017; Wardak 2004). Customs, religion and state sectors define their own exclusive share of authority but also ally with other sectors (Barfield 2003). In various regions of the

---

[55] Interviewee 45, 59.
[56] Interviewee 56, 60, 63.

world, where state power has replaced all other authorties, in Afghanistan the power of each competitive part is diminished (Barfield 2008). At some point of time, customs and religious norms were dominant while on other occasions the state has effectively enforced its power (Barfield 2003). Each sector claims to be autonomous and to have exclusive legal authority in theory, but in practice none has ever been able to completely replace others (ibid). It has been observed that these systems continuously work side by side, and it is the nature of particular conflicts which will determine which system will dominant in comparison to the others (Lau 2003). In this sense, the formal water governance structures will, from time to time, integrate principles from the informal sector, which is then applicable to both systems. Informal water rules are mostly verbal and non-written as these rules vary from region to region and even from community to community (Barfield 2003, 2008).

In Afghanistan, water is formally governed according to the Constitution, Civil Code, as well as Water Laws (Constitution 2004; UNAMA 2016) while informally water is governed as per the local customs and principles of Islamic law (ALEP 2011; Qureshi 2002). The past generation of customs, laws and policies in Afghanistan declares that freshwater in the rivers is _public property‘ (Constitution 2004) and everybody has a legitimate right to consume it as long as it is not _contradicting to public wellbeing or special laws‘ (ALEP 2011; UNAMA 2016). These prinicples of _water as a basic right‘ and _public property‘ were practiced informally for more than four decades and were eventually formally included into Afghanistan‘s current water and environmental laws. These rights are further protected by various versions of the Constitution of the Islamic Republic of Afghanistan and are also being promotedby an Independent Human Rights Commission (Ginsburg and Huq 2014; UNAMA 2016). Historically, informal _water rights‘ in Afghanistan go back to ancient settlements near Kandahar some 4500 years ago (Qureshi 2002), and to the customary practices of Mehergarh and Indus Valley Civilisations (see 7.5.2) in present day Pakistan. The current system of codified water rights in the form of land rights in Afghanistan started during the reign of Abdur Rahman (1880-1901) which can be best described as hybrid rights. Over the past four decades, parallel to informal practices, various formal water governace laws and policies were developed. The management and governance of water resources has also been integrated centrally into the Constitution and various other laws related to environment, mining, forests, pastures and agriculture (see Annex H). For the evolution of formal/informal multilevel water governance frameworks in Afghanistan along with included principles see Figure 7.1 which shows the accumulated inclusion of different categories of principles over different eras. Similarly Figure 7.2(a) presents the number of adopted principles for each category over time; Figure 7.2(b) explains the trends of different categories of principles over time; and Figure 7.2(c) presents the actual progress and regress of different categories of principles over time.

7.5.2   Water Governance before Common Era (BCE)

Early settlers (around 3000 BCE ago) in Afghanistan were linked to the Indus Valley Civilisation through trade and culture where Mundigak (an early city near Kandahar) was a colony of the Indus Valley Civilisation (Dupree 1977). The earlier inhabitants in Northern Afghanistan were Indo-Iranians (Shroder 2006). The strong linkages of Afghanistan with the Indus Valley civilisation were further revealed when - Shortugai in Northern Afghanistan - an Indus Valley site was found on the Oxus River (Kenoyer 1998) (see 8.5.2 for water governance in ancient civilisations). In Afghanistan, water governance has traditionally been undertaken by community-based management structures with elected or, more often, selected water masters[57] (called Mirabs), who supervised construction and maintenance of the water infrastructure, implementation of rules and peaceful resolution of conflicts[58] (McCarthy and Mustafa 2014). Water allocation or rights to access water in Afghanistan is primarily linked with land ownership, the size of the land and the level of the contributions to the construction and maintenance of water infrastructures[59] (ALEP 2011; Qureshi 2002). Although, customary practices can offer an important foundation for shaping indigenous water governance, it is to be noted that customary structures are not always inclusive and are often subject to elite capture (Weinthal et al. 2014).

7.5.3   The Islamic Way of Water Governance

Water management in Islam represents cultural views on the relationship between water and people in various Islamic states which helped in the creation of the Sharia law (Hamid 2013). Understanding the Islamic principles of water governance is useful for obtaining information on existing freshwater governance practices and defining how the Islamic way of water management can be reformed (Faruqui et al. 2001). Water has played a prominent role in key religious teachings including those of the Hadith and Quran. This is due to the fact that the Islamic faith has emerged in the desert (Gilli 2004; Jamil and Haddad 1999).[60] The Quran stresses the significance of water in a frequently quoted verse that _We made every living thing from water' (Holy Quran: 652). In the Quran, the word _water' is repeated more than 60 times, with several references to other related words such as _rivers', _the sea', _fountains', _springs' and _rain' (Haleem 1989: p. 34). The Quran pronounces that water is a gift from God and further declares that: _We may give life to a dead land, and give it for drink to cattle and many people that we have created' (Holy Quran: 724). In addition to the obvious benefits for the preservation of life, water is of particular importance in Islam, as faith attaches importance to the cleanliness and central role of water for the cleansing services prior to praying (Faruqui et al. 2001). Because of their importance, Islamic sources require an equitable and efficient distribution of freshwater resources, and to accomplish this goal, water must be shared justifiably (ibid). The need for an equitable distribution of water is clarified in the verse of the Quran: _and inform them that the water is shared between them; every share of water shall be attended' (Holy Quran: 1039).

A series of Hadith offers an additional perspective when affirming that _God denies his favour to people who deny others the use of water when they have excess water' (Haleem, 1989). These declarations concerning irrigation indicate the need for unbiased circulation of wealth in society.

---

[57] Interviewee 41, 43, 44.
[58] Interviewee 68.
[59] Interviewee 56, 63, 70.
[60] The documented sayings and actions of the Prophet Muhammad (PBUH).

Faruqui explains that, ‗basically all hadith stresses on the provision of equity, and no exemption for those dealing with water' (Faruqui et al., 2001: p. 2). One of the techniques to make sure that water is not monopolized is its position as a publicly owned resource. Public ownership of water is justified by the testimony of Prophet Muhammad in that ―Muslims have a common share in three (things): grass, water and fire" (Caponera, 2001: p.95). Hamed notes that Muhammad treated monopoly or competition by deciding that ‗essential means, such as ‗grasslands, wildlife, forests, certain minerals and especially water should not be privately owned in their natural state' (Hamed, 1993, p.154). In addition to these common principles of Sharia regarding water resources, the Quran and the Hadith provide clear instructions that polluting or exploiting water and other natural resources are strictly forbidden. Thus, Islam offers humankind the opportunity to sustainably consume natural resources by not harming others or the natural environment - in line with the Quranic verse, ‗make not mischief in the land' (Holy Quran: 11).

The mechanism to protect the environment is the responsibility of Khalifah or the resource manager of the land (Abdelzaher et al. 2017). This concept is derived from Islamic philosophy and is founded on the notion that humans - as the greatest creation of God - are responsible to protect and preserve the earth's resources (Hassan et al. 2010). This interpretation is reinforced by the Quran's claim that ―we made you rulers in the land after them, so that we might see how you act" (Holy Quran: 435). The notion of Khalifah basically presents environmental ethics to protect the excessive use of water resources (Gada 2014). For example, in one of the hadith, the Prophet Muhammad restricted the water amount that could be utlised to water the plants to the depth of the ankle (Wilkinson, 1990: p. 61). This facility, which limits the irrigation water use to the amount necessary for sufficient soil moisture sets a clear example to avoid overexploitation of water resources (Faruqui et al. 2001). This hadith creates a general rule for Muslims to protect the environment through concrete actions. This general context of the water-human relationship is principally resilient as it is rooted in religion itself. Islam as a religion is more than a set of ethics to which followers must adhere.

7.5.4   Water Governance in the Colonia Era

During the 18th century, Afghanistan was under attack by the two super powers from two different directions.[61] Britain was occupying the Indian subcontinent between 1757 and 1857, while Russia was expanding its control near Afghanistan's border by 1828.[62] Britian's efforts during the 19th century to protect its Indian empire from Russia and colonise Afghanistan led to a number of Anglo-Afghan Wars (1838-1842, 1878-1980 and 1919-1921).[63] Britiain's eyes were on capturing the Afghan territory between the Hindu Kush and the Indus Basin and on pushing back Russia (Fitzgerald and Gould 2009). Hence, in 1878 after the invasion, the British overthrew the king and formed a new British colony (Visalli 2013). The British created the Durand Line in 1893 to consolidate its gains, separate British India from Afghanistan, and divide the Pakhtun people (Kaura

---

[61] Interviewee 47, 50, 55, 62.
[62] Interviewee 41, 45.
[63] Interviewee 41, 57, 58, 59.

2017; Omrani 2009). Since then the province of Khyber Pakhtunkhwa (where the majority of Pakhtun people are settled) has been considered an integral part of Afghanistan (Hanauer and Chalk 2012). This created a deep hostility among the Pakhtun who survived, which still exists today (Hanauer and Chalk 2012; Omrani 2009). It is to be noted that neither Britain during the British Raj nor Pakistan after the independence ever attained full control of the northwest province which later became the source of Islamic radicalism that produced both Al-Qaida and the Taliban (Visalli 2013). This hostility has its origin in the drawing of the Durand Line[64] for water governance practices during the colonial era; see 8.5.4).

7.5.5   Water Governance in the Post-Colonial Era

Afghanistan's Constitution aims to achieve a ‐prosperous life and sound living environment" for all Afghans. The Constitution requires that ‐no law shall breach the beliefs and provisions of the holy religion of Islam" (Constitution 2004: Art. 3). It encourages governance of all natural resources including water, declaring that the ‐management, adequate utilization, and protection of public properties as well as natural resources shall be regulated by law" (Constitution 2009: Art. 9). It confirms that the government will take essential actions to ‐enhance forests as well as the living environment" (Constitution 2004: Art. 15) and the government will act, ‐within its financial means" to ‐design and implement effective programmes to develop agriculture and animal husbandry" (Constitution 2004: Art. 4). Agriculture is identified by the Constitution as an essential element to enhance the economic as well as socio-ecological means of improving the living conditions‘ of herders, farmers, and other citizens (ibid).

Furthermore, Afghanistan's Civil Code claims to be the special source of appropriate legitimate authority for the entire country (ALEP 2011) . It discusses that water from rivers and their tributaries are _public property‘ (Civil Code: Art. 2347). Each person has the legitimate right to consume water for irrigation of private lands including for irrigation of crops and trees, as long as the usage is not ‐contrary to public interests or special laws" (Civil Code: Arts. 2346-2347). However, the Civil Code is unable to explain which types of uses are opposed to public interests except perceiving that the ‐usage of water from public streams and its distribution shall be exercised with due observation of prevention of harm to public interests and proportionate to the lands that it is intended to be irrigated" (Civil Code: Art. 2349). Without legitimate rights, no one is allowed to build an irrigation canal or a watercourse (Civil Code: Art. 2353). In addition, if a person builds an irrigation canal on his own property, he has the legal right to use it in any way he wishes and so that no one can utlise it without the builder‘s prior permission (Civil Code: Art. 2348). The Afghan Civil Code is primarily based on one of the Sharia principles where the right to breathe fresh air is comparable with the right to freshwater.[65] Each individual has the right to drink water from any source, both from private or public sources.[66] For example, in case of a private water source the right to use water comes with

---

[64] Interviewee 69, 70.
[65] Interviewee 45, 46.
[66] Interviewee 61.

obligation not to harm the water source.[67] The Civil Code has a number of other indirect references to water rights regarding the obligation not to harm water sources and the need for reciprocity in case of shared water sources (Nijssen 2011).

Current water laws in Afghanistan maintain that the rights of water users, including the right of way to water resources, are understood in accordance with the _principles of Islamic jurisprudence' (Water Law 2009: Arts. 1 and 8/8) as well as the _traditions and local customs' of the Afghan people (Water Law 2009: Art. 1). However, as with water laws mentioned above, local customs are not adequately defined and differ from region to region and even from community to community.[68] Water distribution in rural Afghanistan among the local communities is in the hands of local _Mirab Bashis' (local official) or _Mirabs' whose verdicts are mostly valued.[69] Relying on the local customs for conflict resolution however is quite challenging as there is no unique set of customary principles which can be codified and applied in an identical manner (Barfield 2003; 2008). The system rather works more as a voting process than as a decision-making process (Barfield 2003: p.42). This flexibility permits conflicts to be fixed in a context specific way that is sensitive to local distresses (ibid).

The 2009 Afghanistan Water Law was implemented to enforce Article 9 of the Constitution (Constitution of 2004) through provisions that promote _the equitable distribution, conservation, and the efficient and sustainable use of water resources' (Water Law 2009: Art. 1). The purpose of this law is to strengthen the economy and protect the rights of water users in accordance with the principles of Sharia and local customs (Water Law 2009: Art. 1). Similar to the Civil Code and Constitution, the 2009 water law stipulates that _water belongs to the public' and the _people of Afghanistan' (Water Law 2009: Art. 2 and 8/1). Water may be used in accordance with customary practices to meet the needs of consumption, income, agriculture, industry, public services, energy generation, navigation, shipping, fisheries and the environment (Water Law 2009: Art 6)

The 2009 Afghan Water Law - among many other uses - gives priority to use water for _drinking and livelihoods' purposes (Water Law 2009: Art. 6).[70] As per the 2009 water law, the use of water shall be free of cost (Water Law 2009: Art. 7). However, service suppliers may charge fees for the storage, supply, diversion, transmission, treatment, operation, water supply maintenance, irrigation systems, and other related activities (Water Law 2009: Art. 7).

A private water supply company can disconnect users from the water tap if they do not pay the required fees or abuse the water supply (Water Law 2009: Art. 28). The government is committed to conserve and manage freshwater resource (Water Law 2009: Arts. 2 and 8).

---

[67] Interviewee 44, 47, 50, 51.
[68] Interviewee 50.
[69] Interviewee 53.
[70] Interviewee 46, 52, 59

118

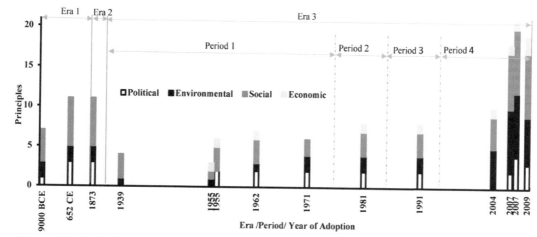

Figure 7.1: Evolution of water governance in Afghanistan over three eras

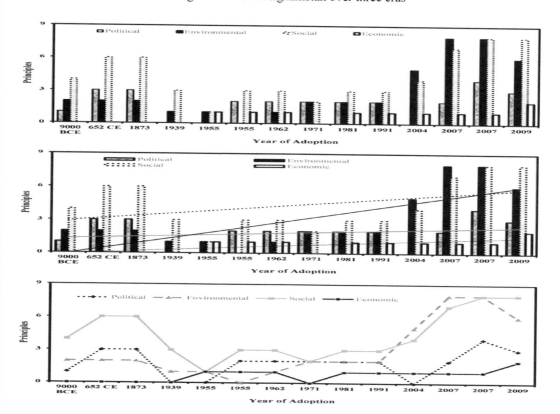

Figure 7.2: (a) Number of included principles (b) overall progress & (c) trend analysis

7.6   GOALS, PRINCIPLES AND INSTRUMENTS

This section assesses goals, principles and instruments in multilevel freshwater governance frameworks in Afghanistan to operationalize the fifth element of the conceptual framework (see 2.5). The analysis of principles is based on the content analysis, literature review and interview data of the national level water governance framework. In line with my conceptual framework, the key principles are discussed under three main categories of inclusive development: (a) political principles (see 5.3.1), (b) social-relational principles (see 5.3.2) and (c) ecological principles.

7.6.1   Goals of Freshwater Governance Framework at Multiple Level in Afghanistan

The Afghan water law aims to enforce Article 9 of The Constitution of Afghanistan. The Afghan water law aims for i) preservation, fair distribution, as well as the effective and justifiable utilization of water resources, ii) protecting the water user rights and making the national economy stronger as per the fundamental principles of Sharia as well as the local customs and traditions.

7.6.2   Governance Principles in Afghanistan's Multilevel Freshwater Governance Frameworks

As Table 7.3 shows, the legal framework consists of a number of laws and policies listed in row one. These include political principles which contain the warning about planned measures and emergency situations, information exchange, dispute resolution, and duty to cooperate; social-relational principles which comprise public involvement, the human right to clean water and improved sanitation, equitable and reasonable use, priority of water use, rights of women, youth and indigenous peoples, public awareness and education, intergenerational equity, poverty alleviation, capacity building; and ecological principles which consist of pollution prevention, monitoring, precautionary principle, ecosystem protection and preservation, water as a finite resource, EIA, and protected recharge and discharge zones. Table 7.3 also reveals the kinds of principles necessary for water management that are not included in Afghanistan's water law framework. The 2009 Water Law has replaced all previous water laws, although it is built on previous laws and provisions including the Water Law 1955, the 1962 Law Fixing the Price and Sale of Water, Afghan Water Law 1981, Afghan Water Law 1991, Constitution of Afghanistan (1931; 1964; 1977; 1980; 1987; 1992; 1994; 2004), Afghanistan Water Sector Strategies 2007, and other relevant policy processes and provisions. My analysis of these legal frameworks shows that the identified drivers through this research are not addressed by the principles in Afghan water laws. The latest Water Law 2009 is greatly influenced by the donors' agenda of considering water as an economic good and promotion of private companies for water provision that benefits the private sector including large multinational companies. Despite increased investment in agriculture by the EU in the agriculture and irrigation sectors for livelihood improvement (see 7.9), the agricultural yield has declined by two percent (European Union, 2019; Leao et al., 2018).

Table 7.3: Principles & instruments in Afghanistan's multilevel water governance frameworks

	Local Customs	Sharia Law	The 1873 Canal & Drainage Act	1939 Law regulating the sale of land under dams & rivers	The 1955 Commercial Code of Afghanistan	The 1955 Law of Ownership & Price of Water	The 1962 Law Fixing the Price & Sale of Water	The 1971 Law on Pasture & Grazing Land	The 1981 Afghan Water Law	The 1991 Afghan Water Law	The 2004 Constitution of Afghanistan	The 2007 Afghanistan Environmental Law	The 2007 Afghanistan Water Sector Strategies	The 2009 Afghanistan Water Law
Political Principles														
Information Exchange			■									■	■	
Warning of Emergency Situations	■		■											
Warning of Planned Measures		■				■								
Duty to Cooperate		■	■						■	■		■		■
Conflict Resolution		■	■			■	■		■	■		■		■
Limited Territorial Sovereignty/ Do Not Harm														
Environmental Principles														
Basin as the Unit of Management			■											
BATT														
Conjunctive Use													■	
EIA												■	■	■
Invasive Species												■		
Monitoring									■	■		■	■	■
Prevention of Pollution	■								■			■		■
Preventive Principle			■				■							
Protected Areas for Water			■											
Protected Recharge and Discharge Zones								■			■			
Ecosystems Protection and Preservation					■									■
Polluters Pay			■											
Water as a Finite Resource	■	■							■					
Social Principles														
Capacity Building				■		■			■	■		■	■	■
Equitable & Reasonable Use				■		■	■		■	■		■	■	■
Human Right to Water & Sanitation		■										■		■
Intergenerational Equity												■		
Poverty Alleviation														
Prior Informed Consent			■											
Priority of Use		■			■	■	■							■
Public Access to Information							■						■	
Public Awareness & Education		■					■					■		■
Public Involvement	■									■		■		■
Rights of Women, Youth, &		■		■		■			■					■

Indigenous Peoples		▬		▬		▬ ▬		▬	▬
Food Security								
Human Well-being								
Quality Education								
Clean Energy								
Economic Growth								
Infrastructure								
Reduced Inequality								
Sustainable Urbanisation								
Responsible Consumption & Production								

Source: Modified from Conti, 2017;   ***Bold laws & policies** in Row 1 of the table are legally binding

7.6.3   Governance Instruments in Afghanistan's Multilevel Freshwater Governance Frameworks

These instruments in the Afghanistan governance framework can be divided into four categories including: regulatory instruments (6.6.1); economic instruments (6.6.2); suasive instruments (6.6.3); and management instruments (6.6.4).

At the national level there are five regulatory instruments in the freshwater governance frameworks in Afghanistan which can regulate freshwater uses in industries including commercial agriculture, manufacturing and mining through licenses penalties and fines. Similarly mapping and zoning can restrict urban settlements near the water recharge and discharge zones, and river beds to prevent pollution. Regulatory instruments in Afghanistan's national frameworks allows line ministries and departments to prepare policies and strategies for sustainable water use. Water Usage Licenses regulate water use for sustainable use and to supply water to the growing population. The economic instruments include rights, permits and fees. Currently water is a right, which will be gradually converted into an economic good through permits. However, water rights can be suspended in case of non-payment of fees and misuse of water. River Basin Councils (RBCs) have the authority to issue, modify or cancel permits for under- and over-utilisation, lack of water availability, and in the national interest. RBCs can also impose and collect penalties. For drinking, livelihood, navigation and fire extinguishing purposes water can be used without permits. Similarly, there are fines for triggering financial damage to others, abusing a water right, threat to a downstream community in terms of life, health, and the livelihoods. National freshwater governance frameworks do not have suasive instruments to empower communities about the importance of freshwater. Capacity building of water-related professionals would have been meaningful in the long run. Additionally, there are two management instruments in Afghanistan freshwater governance frameworks at national level. These are: (1) the devolution of management functions and self-regulation by forming River Basin Agencies (RBAs) containing advisory boards involving substantial institutions for equitable and participatory water distribution; (2) Water User Associations for participatory management, operation and maintenance and equitable water distribution at local level.

Despite the creation of RBCs, their effectiveness is limited since they are unable to influence the political nature of inter-provincial coordination by addressing the interests of different powerful groups (e.g. farmers, Taliban and other militant groups, private sector, international organisations and donors). Similarly, the existing regulatory and economic instruments in the form of licences, permits, fees and fines are not strong enough to change the behaviour of users especially the powerful actors and large businesses. The RBCs are unable to play any role on the transboundary level due to the absence of treaties; but could play a part if treaty negotiations should begin.

7.7   LEGAL PLURALISM ANALYSIS

It is important to note that before the formation of the Durand Line that divides the Pakhtun population on both sides—the entire KRB region (including tribal areas and the KP province of Pakistan) were part of Afghanistan, which means that the practices of Sharia and local customs for water management applies to both countries (see 8.7).

The evolution of freshwater governance frameworks in Afghanistan reflects that Afghanistan has three parallel governance systems that govern natural resources including water: the local customs, sharia law and the state legal codes (vertical between local customs and Sharia). However, the relationship between these laws is complicated (Barfield 2003). Local customs dominate in Afghanistan among the three governance systems.[71] With regard to water and land rights, the civil code identifies the implication of local customs. It is prohibited as per the Constitution of Afghanistan to implement laws which contradict the basic principles of Islam.[72] At the same time, the Constitution is quiet on local customs. Customary practices are allowed in Sharia if they do not contradict the basic values of Sharia (Senier 2006). Similarly, the non-water laws at the national level are also analysed (horizontal analysis), which also address some of the direct and indirect drivers.

An analysis of freshwater governance frameworks in Afghanistan reveals first, that there is the integration of some instruments from informal systems (e.g., local customs and Sharia Law) into formal governance structures (water and non-water laws). These informal rules (particularly customs) are mostly verbal, non-written and vary from region to region as well as from community to community. Second, at the national level, both water governance frameworks as well as non-water laws address the issues linked to sustainable water use.

Legal pluralism occurs when freshwater resources are governed through multiple laws, policies and strategies. Generally, Afghans believe that their customary practices are not opposing the basic values and standards of Sharia, however this is not always the case.[73] Indeed, various customary practices contradict both Islamic law as well as Afghan statutory law.[74] This inconsistency is not visible to many people because a religious figure, _the mullahs' is most often included in the informal judicial system (e.g., Jirga and Panchayat) because of his religious affiliation and trust of

---

[71] Interviewee 49
[72] Interviewee 45, 52, 55
[73] Interviewee 43, 47, 52
[74] Interviewee 53, 61

the people.[75] With the inclusion of the mullah, people come to believe that the ruling complies with the basic principles of Sharia although the ruling of the Jirga opposes the basic values of Sharia (Brohi 2016).

Water governance frameworks contain instruments that contradict each other (type 2). For example, private ownership is forbidden in Sharia but in local customs it is a regular and accepted practice. Similarly, the Sharia instructs Islamic states to avoid pricing water or applying wastewater treatment or reuse methods. However, in local customs it is the other way round. Furthermore, priority of use is encouraged in Sharia, whereas it is prohibited in local customs. Subsidies for deep tube wells are promoted in Afghan non-water laws which can undermine the sustainable use of groundwater resources. The formal laws of the State of Afghanistan contemplate water as an economic good, whereas it is a basic human right in Sharia Law. In terms of indifference (type 1), water types are defined in Sharia but are not specified in local customs and the Afghan water laws. Moreover, conservation and efficient water use are well defined in Sharia but are not part of local customs and water use practices. There are some elements in the water governance frameworks that mutually support each other (type 4). For example, water quality monitoring, mandatory reporting and the polluters pay principle for water use in industry can achieve water goals separately and recognise the importance of other instruments. Similarly, no permits are required for domestic purposes in the Afghan water law but promotion of metering in non-water laws can help in sustainable water use and water conservation. Both water and non-water laws suggest fines and imprisonment for illegally diverting water and tampering with water infrastructure and permit systems are incorporated for drilling of deep tube wells for drinking purposes to increase access to drinking water. Licences are used in both water and non-water laws at the national level for addressing direct drivers of demographic shifts. For controlling pollution, both water laws and Sharia enforce penalties. Moreover, the Afghan water law recommends suspension of water rights in the case of non-payment which is supported by water pricing mechanisms in non-water laws of the country. This is depicted in Table 7.4. What stands out is that there are few participation principles and instruments that would enable accommodation of local policies except through the informal judicial system.

My analysis of the legal frameworks shows that different laws for water, energy, land, food and agriculture treat water differently, which in turn reduces their effectiveness and implementation. Despite the presence of formal legal frameworks for water, the local customs and Sharia laws are grounded in the local contexts historically which undermines some elements within the formal laws that contradict the local customs. Furthermore, weak organisational capacity and governance mechanisms provides ample manoeuvring space for powerful actors (at local level as well as large private sector organisations at national level) to circumvent these laws in their favour at the expense of low-income subsistence farmers and other water users.

---

[75] Interviewee 43, 49

Table 7.4: Legal pluralism analysis of Afghan water governance framework

Quality / Intensity	Weak relations	Strong relations
Contrary	*Type 1: **Indifference** - lack operational relationship among principles/instruments* - Types of water defined in Sharia, but not specified in local customs and water law - Conservation & efficient water use well defined in Sharia but not in local customs & practices	*Type 2: **Competition** - contradiction among principles/instruments* - Subsidies for deep tube wells can undermine sustainable groundwater use - Private ownership forbidden in Sharia but allowed in customs - Water as an economic good (state law) & human right (Sharia) - Priority of use in Sharia but not in local customs - Water reuse in customs but prohibited in Sharia
Affirmative	*Type 3: **Accommodation** - recognition of principles/instruments* - Informal judicial system (Jirga) may try to link local customs to Sharia and national law - River Basin Councils when established could perhaps enable accommodation of different policy approaches - Participatory water management through formation of FOs and WUAs at local level including informal conflict resolution mechanism (Jirga)	*Type 4: **Mutual support** - principles/ instruments support each other* - Water quality monitoring, mandatory reporting & polluters pay principle for water use in industry - No permits for domestic purposes (water law) but metering required (non-water law) - Fines & imprisonment for illegal diverting and tempering - Permits for deep tube wells for drinking purposes to increase access to drinking - Water use licences for addressing demographic shifts - Penalties for pollution in both water & Sharia law - Water pricing (non-water law) & suspension of water rights (water law)

Source: Modified from Bavinck and Gupta 2014

7.8   PRINCIPLES AND INSTRUMENTS ADDRESSING DRIVERS AND ACHIEVING INCLUSIVE AND SUSTAINABLE DEVELOPMENT

7.8.1   Principles and Instruments Addressing Drivers through National Water Law in Afghanistan

Afghanistan's legal and governance structure is based on a democratic national government. The three higher level branches (executive, legislative and judicial) are the foundation of the national level government. However, other formal entities including various ministries, the Afghan National Security Forces, and other commissions formed for this pupose also serve different government functions. Under the national government, the public sector is composed of provincial-level government departments, municipalities at the city level, and district-level governments at the lowest level. In terms of water governance, the same conditions apply i.e., all the water laws and policies are designed by the central government. Provinces are not allowed to make their own policies or laws. Afghanistan's Constitution encourages the government to make policies which do not contradict the local customs of Afghan's and Sharia.

The 2009 Afghanistan National Water Law lists a number of water related functions including new constructions, repair and rehabilitation, development, and monitoring of irrigation infrastructure, including agricultural development in order to mitigate risks of floods (an indirect driver), and protect and maintain the environment. It also promotes research activities and the economic effectiveness of the irrigation system and appropriate irrigation technologies to improve water quality and reduce water losses (addressing a direct driver). However, these measures are not supported by any regulatory or economic instrument such as subsidies. Although the policy recommends formation of River Basin Councils to include representatives of water users, concerned federal and local departments, and other groups of stakeholders in the river basin, the provinces are not empowered, and all matters fall directly under the administrative authority of the Federal Ministry of Energy and Water (MEW). These arrangements can further strengthen the centralised form of government where the provinces are not given decision making powers on water management and governance.

Although water is described as a human right, it is clearly considered as an economic good through the development of a number of policy instruments. The 2009 Afghanistan Water Law relies on regulatory, economic and management instruments but lacks suasive policy measures to achieve its stated goals. Through regulatory instruments, ministries and departments can prepare strategies and policies for sustainable water use to address direct drivers. Permits have been introduced for regulation of water use to cater to the demands of a growing population. Industries are required to report on regular water quality monitoring where pollution prevention is ensured through the imposition of penalties and fines. In addition to financial penalties, there are strict punishments in the form of imprisonment for blocking, diverting and re-routing water, tampering with equipment and signs, interference in water distribution, and encroaching upon the right of way of a number of water resources for the wider public (such as river banks, streambeds, canals, ditches, springs, Karezes groundwater sources, swamps and wetlands). Permits are a mandatory requirement to be approved by the ministry of water and electricity in agreement with the ministry of Agriculture and Land Reforms for construction of deep wells for agricultural use, as well for commercial, industrial and urban water supply.

Since water is an economic good, rights can be suspended in case of non-payment of fees and misuse of water by utilising economic instruments within the water law which can address most of the direct drivers (agriculture, industry and municipal water use). The River Basin Councils are authorised for issuance, modification and cancellation of permits for water use in the cases of under- and over-utilisation, lack of water availability, and in the national interest. These Councils have authority for imposition and collection of penalties for agriculture and industrial uses. The only exception for water use without permits is limited to drinking purposes, livelihood, navigation and fire extinguishing which address the indirect driver of poverty reduction. The law has proposed a number of fines for misuse of a water right, causing financial harm to another, threat to life, health and the livelihood of the downstream community.

There are a number of management instruments within the water law for ensuring devolution of management functions and self-regulation by establishing River Basin Agencies including advisory boards consisting of relevant institutions for participatory and equitable water distribution. Formation of Water User Associations (WUA) for participatory management, operation and maintenance (O&M) for equitable water distribution at the local level are central features. It also includes a dispute resolution mechanism through River Basin Councils (RBCs) over water distribution and use in the respective basins to address the indirect driver of political dynamics between/within states.

In terms of addressing the indirect driver of natural change and variability (e.g. flood and drought forecasting through data collection and analysis), normative management principles are applied without any effective policy instruments as mentioned above. Although it can improve equitable distribution at the canal level, the permit system is likely to provide benefits to the existing water bureaucracy by providing avenues for financial benefits and control over water recourses decision making (Lee 2006). Table 7.5 below briefly explain the instruments in Afghan Water Laws that address various multilevel drivers.

Table: 7.5: Instruments in Afghan water laws addressing drivers at multiple level in Afghanistan

Direct Drivers	Instruments			
	Regulatory	Economic	Suasive	Management
Agriculture development (e.g., commercial agriculture including animal husbandry, the extractive sector and water use in energy)	Imprisonment and fine for diverting and tampering; Permit for deep tube wells; Fines for misuse	Suspension of water rights for non-payment	-	Water User Associations (WUAs) for O & M, participatory management
Industry (including services and infrastructure)	Penalties and fines for pollution prevention; Water quality monitoring, mandatory reporting; Permits for water use in mining; Fines for misuse	Suspension of water rights for non-payment	-	-
Municipal water supply and sanitation services e.g., household uses (drinking water, sanitation, and hygiene) and subsistence agriculture	Penalties and fines for domestic pollution; Permits for urban water supply; Fines for misuse	Suspension of water rights for non-payment	-	-
Demographic shifts (i.e., migration, population growth, increase in population density, urbanisation, population growth)	Water use licences	-	-	-
Indirect Drivers				
Political dynamics between/within states	-	-	-	Dispute resolution mechanism (RBCs)
Culture and ethnic (attitudes for access and allocation, wasteful use of resources, etc.)	-	-	-	-
Non-water-related policies (agriculture & food security, land use, land tenure, economic development)	-	-	-	-
Economy (economic growth)	-	-	-	-
Poverty	-	No permits required for subsistence agriculture	-	-
Technological advances (agriculture intensification)	-	-	-	-
International trade (e.g. globalisation' or trade in virtual water)	-	-	-	-
Natural change and variability in weather, Droughts; Floods; Earthquakes; Landslides, tectonic movement.	-	-	-	Early warning system

7.8.2   Principles and Instruments Addressing Drivers through Non-Water Laws/Policies in Afghanistan

This section analyses the non-water related laws and policy frameworks in terms of the four policy instruments about water governance and management.

Under the Constitution of Afghanistan, the subterranean resources including water are the property of the state which can formulate laws and policies for their management and protection. The 2007 Environment Law has a number of regulatory instruments for environmental protection that can affect water quality and quantity. The National Environment Protection Agency (NEPA) of Afghanistan is established under the law which defines its powers and functions such as an apex body for policy formulation, implementation, environmental regulation and proper monitoring of environmental laws as well as coordination of international environmental agreements. For integration and coordination of environmental matters with other government agencies, the Environment Law has established Afghanistan's Committee for Environmental Coordination and the National Environmental Advisory Council. The law includes regulatory provisions such as fines and the polluter pays principle to address direct drivers (agriculture and industry) and licences for addressing demographic shifts. There are also some economic instruments such as water pricing to address direct drivers (agriculture and industry); subsidies for solar tube wells for agriculture and industry; and metering for domestic/municipal use. In terms of the indirect drivers, zoning and mapping are prescribed measures for flood control as well as conservation of ecological hotspots. Subsidies for tube wells are also suggested measures in terms of agricultural intensification (indirect driver), which can also lead to unsustainable groundwater abstraction. The non-water related laws and policies do not rely on suasive and management instruments for addressing water related issues. Table 7.6 below briefly explain the instruments in Afghanistan's non-water laws and policies that address various multilevel direct and indirect drivers.

Table 7.6: Instrument in non-water laws & policies addressing drivers at multiple level of governance in Afghanistan

Direct Drivers	Instruments			
	Regulatory	Economic	Suasive	Management
Agriculture development (e.g., commercial agriculture including animal husbandry, the extractive sector and water use in energy)	Fines Polluters pay principle	Water pricing Subsidies for solar tube well	-	-
Industry (including services and infrastructure)	Fines Licences for industrial use Polluters pay principle	Water pricing	-	-
Municipal water supply and sanitation services e.g., household uses (drinking water, sanitation, and hygiene) and subsistence agriculture-	-	Subsidies for domestic use Metering	-	-
Demographic shifts (i.e., migration, population growth, increase in population density, urbanisation, population growth)	Water use licences	-	-	-
Indirect Drivers				
Political dynamics between/within states	-	-	-	-
Culture and ethnic (attitudes for access and allocation, wasteful use of resources, etc.)	-	-	-	-
Non-water-related policies (agriculture & food security, land use, land tenure, economic development)	-	-	-	-
Economy (economic growth)	-	-	-	-
Poverty	-	Subsidies for solar tube well	-	-
Technological advances (agriculture intensification)	-	-	-	-
International trade (e.g. globalisation' or trade in virtual water)	-	-	-	-
Natural change and variability in weather, Droughts; Floods; Earthquakes; Landslides, tectonic movement.	Mapping and zoning for flood protection	-	-	-

7.8.3   Principles and Instruments Contribution in Achieving Inclusive and Sustainable Development

The analyses and contribution of principles and instruments included in the freshwater governance frameworks within Afghanistan as per the dimensions of inclusive and sustainable development reveals that inclusion of political and economic principles are very few while social-relational (seven) and ecological principles (eight) are in considerable numbers. Economic instruments dominate regulatory, suasive and management instruments. The inclusion of more economic instruments indicates the influence of international donors on Afghanistan's freshwater governance. Overall, the water governance structures within Afghanistan indicate that local priorities and political dynamics as well as donor-driven global development agendas and the neo-liberal economic system are influencing the governance of freshwater resources in Afghanistan from period to period. These findings suggest that there are weak linkages among drivers of freshwater problems and instruments to address these drivers.

The fact that most of the governance framework in all the four major periods have remarkably similar principles and instruments in most of the dimensions of inclusive and sustainable development shows that there are hardly any context specific designs for water governance frameworks within Afghanistan. For instance, arid or semi-arid areas in the country can be better managed by protection of recharge and discharge zones. The lack of context specific designs in water governance frameworks is likely to limit contribution of water governance frameworks to inclusive and sustainable development within Afghanistan.

In terms of inclusive development, relational inclusiveness is challenging to achieve even though the principle of advance notification of planned development measures was part of almost all governance frameworks within Afghanistan prior to the 2009 water law, which is not implemented. Throughout all periods of water governance frameworks, there is much stress on the inclusion of management instruments in terms of wide-ranging stakeholder participation and access to information which can contribute towards achieving relational inclusiveness with proper implementation. Moreover, social inclusiveness has relatively moderate potential due to emphasis on instruments of cost recovery; capacity building, poverty eradication and public education and awareness and can help Afghanistan to re-focus on human as well as natural capital development. However, this is also subject to implementation and developing a transparent and accountable governance structure at all levels. Additionally, environmental inclusiveness has very limited potential due to the absence of instruments to protect ecosystem services by regulating flow, preventing pollution, monitoring and giving significant attention to EIA.

Table 7.5 shows that the Afghan water law has a list of regulatory (e.g. imprisonment, fines, permits, fines, penalties, water use licences) and economic instruments (water charges, subsidies) to address the identified direct drivers, such as growing population, unplanned urbanisation and migration. However, there is no clear adoption of goals such as from the SDGs, and no clear prioritisation in terms of human rights to water and sanitation or ecological instruments as in international law. Natural changes in freshwater quality and quantity in the context of Afghanistan (see Table 7.2), such as climate change and variability, floods, earthquakes and landslides are not addressed by Afghan Water Law.

To address the direct drivers of agricultural and industrial development as well as indirect drivers including economic development the country's water law relies on regulatory instruments by suggesting punishments for locking, diverting, re-routing water and tampering with equipment and signs that can interferewith water distribution. It also suggests a number of management instruments (formation of River Basin Councils and WUAs) for participatory management, operation and maintenance (O&M) for equitable water distribution, a dispute resolution mechanism through River Basin Councils (RBCs) and use in the respective basins. However, agricultural policy is not developed based on a consideration of water conditions or how climate change can affect it. The provinces are not empowered through RBCs and a dispute resolution mechanism over administrative boundaries is not addressed. The country could benefit by examining the implications of the SDGs, the equity and ecological principles and instruments in the international Water Laws, the human right to water and sanitation and integrating the implications of the Climate, Desertification and Biodiversity Conventions.

The analysis of Afghanistan's legal frameworks reveals that there are various laws and policies for governing natural resources. Each of these laws treat water differently, which in turn reduces their effectiveness and implementation. The existing institutional structure is a mix of informal and formal rules which are often contradictory at both sub-national and national levels. These developments have caused the overlapping of local customs, Sharia and modern rules of freshwater use in Afghanistan and created a legal pluralistic form of governance. Relying mainly on traditional practices to resolve water disputes can be problematic because there is hardly any set of rules and institutions in traditional practices to be systematically codified for application in a uniform way. While customary water management practices can provide an important foundation for fashioning locally appropriate water governance, it is also important to acknowledge that traditional structures are not necessarily inclusive and are often subject to elite capture. Similarly, water is of particular importance in Islam where the Quran and the Hadith provide clear instructions that mankind must protect water and other important natural resources from overutilisation and degradation. In a hadith, the prophet Mohammed issued instructions to limit the amount for watering plants to the depth of the ankle. This provision establishes a clear precedent that prohibits unsustainable consumption of water resources.

7.9   POWER ANALYSIS OF FRESHWATER GOVERNANCE IN AFGHANISTAN

Afghanistan's aid dependence has influenced its decision-making powers with negative consequences economically as well as politically.[76] With the fragile government institutions and

---

[76] Interviewee 42, 44 45

lack of technical capacity, the country is not yet ready to negotiate the terms of aid (Samim 2016 2017).[77] The low technical capacity, limited and fragile institutions could be some of the key reasons for a two percent decline in the agricultural productivity despite the European Union's support for agriculture and irrigation development (European Union 2019).

With a mixed and inter-linked legal framework for water governance and allocations (i.e. Sharia and local customs at local/sub-national level and formal laws at the national level), a fragile institutional setup due to four decades of conflict, and donor influence, the role of power becomes central in water governance in Afghanistan. Since some principles of Sharia contradict local customs as well as the formal water and non-water-related policies, the existing water governance frameworks are unlikely to address the identified drivers and other issues, which can lead to an asymmetrical power structure at sub-national and national levels. Due to Afghanistan's total dependence on foreign aid, weak organisational capacity, limited and fragile institutions and the long-term presence of a large number of donor organisations as well as states (also in the context of Pakistan, Iran and China), it is unable to address the pressing and real socio-economic and environmental issues including water challenges in contrast to issues of security and counter-terrorism. Likewise, the influence of donor organisations is manifested in the Water Law 2009 where water is treated as an economic good by empowering large private companies, commercial farmers and private groundwater providers at the cost of the low-income population, which are mostly dependent on subsistence agriculture. These resulting power imbalances at the local and national levels is used to include some people and exclude others, and can hinder effectiveness of water cooperation at the transboundary level.

The institutional analysis of Afghanistan shows that local customs and principles of Sharia include equity principles which promote community-based water sharing principles and benefit large land-owners as opposed to weak actors by disregarding regulations of the central government. The focus of the Colonial era shifted towards a market economy, giving more individual rights regarding the utilisation of natural resources including water, which resulted in disproportional income gaps between the rural and urban population as well as between small and large landowners. Post-colonial legal systems comprise the Constitution of Afghanistan, civil codes and water laws where the constitution clearly directs to include sharia principles and local customs in all legal codes and policies. However, the existing institutional context and water related policies are often contradictory at both sub-national and national levels. These developments have caused the overlapping of local customs, Sharia, and modern rules of freshwater use in Afghanistan and created a legal pluralistic form of governance where: (a) at the local level some principles of Sharia (included in civil codes) contradict local customs, and (b) freshwater governance is disaggregated across multiple texts and often contradict non-water-related policies. Therefore, the existing water governance frameworks do not address direct and indirect drivers and other issues which can lead to an asymmetrical power structure, difficulties in achieving equitable access and allocation, and hinder the protection and conservation of ecosystem services at sub-national and national levels. The

---

[77] Interviewee 43, 48

weak institutional framework and governance mechanisms have provided space to powerful actors to advance their interests for water usage and control at the cost of small and subsistence farmers.

7.10   INFERENCES

This chapter has described and analysed multilevel freshwater governance in Afghanistan. It has done so by looking at (1) how different characteristics and drivers of freshwater problems are taken into account at multiple geographic levels in Afghanistan; (2) evolution of freshwater governance frameworks at different geographic levels in Afghanistan; (3) which governance instruments address the drivers of freshwater problems at multiple geographic levels in Afghanistan; (4) how legal pluralism can be observed at various geographic levels in Afghanistan; and (5) how power and institutions influence freshwater governance frameworks at multiple geographic levels in Afghanistan. Through these sub-questions, this chapter draws four conclusions.

First, widespread poverty, weak institutional and human capacity due to long-lasting conflicts and instability, and lack of knowledge and capacity to manage water resources have contributed to water challenges in Afghanistan. Afghanistan has enough water resources, approximately 2500 cubic meters, as compared to its neighbours, which are still largely underused. However, increasing population coupled with the changes in global climate will put immense pressure on the existing water resources. Due to Afghanistan's land-locked status and important geographic location, it has had historical disputes with its neighbours over water shares from its highlands which are very important for its survival. Currently, only 30-35% of the water is stored due to the lack of water storage and irrigation infrastructure in Afghanistan. It is challenging to get safe and reliable water supplies because water-resources data collection was stopped in the 1980s during the unrest and the Soviet invasion. Consequently, institutional knowledge, water related technology and abilities of water scientists were severely affected.

Second, since a large segment of the population in Afghanistan directly and indirectly depends on a variety of BESS provided by Kabul River, degradation of these vital services due to variability of water flows can have negative consequences. Afghanistan is directly dependent on its vast natural resources such as water, forests and minerals for its development and prosperity. Based on a recent classification, the country has 15 smaller eco-regions (four are considered as endangered, eight are categorised as vulnerable and only two can be considered as stable). The composition of species in these regions has been negatively affected due to unchecked overgrazing, wood collection for fuel and deforestation and overusage of land due to increasing large herbivorous animals. About 70-80% of the population is directly dependent on animal raising, subsistence agriculture, and small-scale mining for their income generation and livelihoods. Afghanistan must develop these assets for job and revenue generation in order to fund primary government services and improve the human wellbeing in terms of Human Development Index (HDI).

Third, it is vital to identify and understand key direct and indirect drivers of water issues for evidence based policies in a changing geopolitical and climatic scenarios for sustainable water use in the country. Key direct drivers of water issues in Afghanistan are: (a) agriculture and industrial development; (b) demographic shifts; (c) increasing demand for clean drinking water and improved sanitation; and (d) natural changes due to climate variability and changes such as flooding and glacial lake outbursts and droughts. The four decades long war has destroyed water infrastructure and the agriculture sector which was providing employment opportunities to more than 80% of the population. Indirect drivers include: (a) political drivers where (i) the State's weak regulation encourages privately owned water abstraction (ii) the damaged water-carrying infrastructure (iii) security challenges creating disincentives to building infrastructure (in 2017, the World Bank sponsored Kamal Khan Dam and India-sponsored Salma Dam were attacked by the Taliban), (iv) donor countries continue to prioritise security and strategic priorities; (b) social drivers including poverty where about half of the population has access to clean drinking water and just about 35% use improved sanitation facilities; (c) economic drivers; and (d) cultural drivers including, wasteful behaviour towards water consumption and pollution, low technical knowledge and low education, which places Afghanistan at 169th on the HDI. Fourth, due to a pluralistic legal framework and weak organisational capacity, Afghanistan is unable to address the freshwater drivers.

Existing institutions are unable to achieve equitable access and allocation, and hinder the protection and conservation of ecosystem services (ESS) at sub-national and national level. Moreover, donor organisations are able to define and push their agenda in Afghanistan by prioritising their strategic and security interests (in the context of neighbouring China, Iran and Pakistan with a focus on cross-border terrorism, security and other geo-strategic issues) as opposed to addressing the socio-economic and environmental issues of the larger population. In Afghanistan, water resources are formally governed by the Constitution, Civil Code, and Water Laws while informally through the local customs and principles of Sharia. Based on the principles of Sharia, water governance practices in the pre-colonial era did not include equity principles which could have strengthened cooperation among local communities. Additionally, these principles would have discouraged favouring large land-owners against small-landholders. The focus of the Colonial era shifted towards a market economy, giving more individual rights regarding the utilisation of natural resources including water, which resulted in income disparities between the rural and urban population as well as between small and large landowners. The existing Constitution clearly directs to include Sharia principles and local customs in all legal codes and policies. However, the existing institutional context and water related policies are often contradictory at both sub-national and national levels. These developments have caused the overlapping of local customs, Sharia and modern rules of freshwater use in Afghanistan creating a legal pluralistic form of governance where: (a) at local level some principles of Sharia (included in civil codes) contradict local customs, and (b) freshwater governance is disaggregated across multiple texts and often contradict non-water-related policies.

8

ANALYSIS OF MULTILEVEL
FRESHWATER GOVERNANCE
IN PAKISTAN

8.1   INTRODUCTION

This chapter describes and analyses multilevel freshwater governance in Pakistan and intends to answer the following questions: (1) How are the various characteristics including BESS and drivers of freshwater problems taken into account at multiple geographic levels in Pakistan? (2) How have freshwater governance frameworks evolved at multiple geographic levels in Pakistan? (3) Which governance instruments address the drivers of freshwater problems at multiple geographic levels in Pakistan? (4) How does legal pluralism occur at multiple geographic levels in Pakistan? (5) How do power and institutions influence freshwater governance frameworks at multiple geographic levels in Pakistan?

To answer these questions and using the methodology in Chapter 2, this chapter describes the political organisation of water sharing within Pakistan (8.2), various ESS (8.3) and drivers of freshwater problems (s8.4) at multiple geographic levels; discusses the evolution of freshwater governance (8.5); discusses goals, principles and instruments (8.6); conducts a legal pluralism analysis (8.7); explores the relationship between governance instruments and drivers and their contribution in achieving inclusive and sustainable development (8.8) and draws inferences (8.9) about the hybrid role of power and institutions in influencing freshwater sharing within Pakistan.

8.2   THE CONTEXT OF WATER GOVERNANCE WITHIN PAKISTAN

Pakistan's water sector has numerous challenges. Colonial legacies, outdated laws and insufficient management capacity of water related disasters have failed to address the challenges of droughts and floods (Young et al. 2019). Poor sanitation and hygiene practices have led to the high child mortality rate (Sarfaraz 2014). Predictions regarding glaciers melting remains worrying in the Himalayas. Furthermore, water storage and generation of hydro-power are essential for irrigation development and to cope with energy crisis in Pakistan (Hanasz 2011). Despite these threatening issues, still various opportunities exist in the water governance field in Pakistan. For instance, the impacts of climate change empowers decision makers to observe water scarcity issue within the context of changing climate (Yousaf 2017). The currently changing local, regional, and global political dynamics as well as the peace processes in Afghanistan encourages Pakistan to seek for greater cooperation within the country and among Pakistan and its nehbouring countries.

The National Water Policy of Pakistan (2018) addresses that mananging and developing water resources in Pakistan have severe challenges to resolve many associated issues. The sustainability and expansion of irrigated agriculture is being exposed to: (a) the growing demand of water to meet the requirements of a increasing population in addition to socioeconomic demands; (b) historical and spatial disparities in available water resources (e.g. 65% of rainfall and approximately 81% of flow in the river occurs during the monsoon season for about three months; similarly, groundwater quality varies with respect to depth and location); (c) reduction in freshwater availability due to silt deposits in dams; (d) poor maintenance of the irrigation canals; (e) salinsation and waterlogging in various canals; (f) lack of interests from relevant organisations to establish a drainage network for the existing irrigation network; (g) depletion and pollution of aquifers; (h) poorly disposing off saline effluent, industrial & household waste as well as contaminants from fertilizer and pesticides; (k) inadequate   participatory practices; (l) increase frequency of droughts and floods; (m) lack of consensus-building as well as mistrust among the four provinces over equitable water sharing, poor pricing, valuation of water; and degrading water quality.

8.2.1   The China-Pakistan Economic Corridor (CPEC) and Freshwater Challenges

The China-Pakistan Economic Corridor (CPEC) project aims to enhance the strategic relationship as well as trade and economic cooperation between China and Pakistan. CPEC was formally launched during the Chinese Premier's visit to Pakistan in May 2013 where he emphasised the construction of the 2033 km long trade corridor (The Diplomat 2014). The then Government of Pakistan and all other subsequent governments have also shown considerable enthusiasm for this project. This corridor will link the deep-sea port of Gwadar in Balochistan Province of Pakistan to Kashgar city of North-western China. Since February 2013, the port of Gwadar is undertaking a major extension to make it one of the biggest commercial port (South China Morning Post 2014). The economic corridor – through Pakistan – will link China with the Middle East, Europe and Africa. Besides fulfilling China's energy needs and alleviating poverty in its western region, CPEC also intends to improve Pakistan's economy and enhance economic cooperation among South Asian countries (China Daily 2013).

However in 2017, when the Master Plan of CPEC was publicly disclosed, the Chinese plans and primacies in Pakistan for the next two decades became acknowledged to the Pakistani citizens (Husain 2017). For instance, some of the important activities include: (1) leasing out of thousands of acres of agricultural land to the Chinese enterprises; (2) round-the-clock video recordings and surveillance from Peshawar to Karachi on roads and in commercial areas to maintain law and order; (3) and building of a nation-wide fibre optic for internet traffic and digital TV. The plan predicts a deep saturation of Pakistan's economy by Chinese enterprises and culture. The large scale commercial agricultural projects by Chinese enterprises may cause displacement of small local farmers leading to social and political unrest. In addition, the landowners in southern Punjab and Sindh would risk losing political influence if they sell their land (ICG 2018). Furthermore, water resources may be further depleted by the extensive CPEC activities as warned by the then Governor of Balochistan which could be very serious as the water table in many parts of Baluchistan has been falling for decades (The DAWN 2016).

8.3   ECOSYSTEM SERVICES OF FRESHWATER IN PAKISTAN

This section highlights freshwater biodiversity and the four categories of ecosystem services (ESS) provided by freshwater in Pakistan.

8.3.1   Freshwater Biodiversity in Pakistan

Pakistan has a variety of the world's animals, plants and birds due to its location between three zoogeographical zones: The Palearctic, the Oriental and the Ethiopian. However, Pakistan also has experienced rapid changes in attitude that has a significant impact on its flora and fauna (IUCN 1997; Akhtar, Saeed, and Khan 2014). In the North of Pakistan, three mighty mountain ranges meet – the Himalaya, Karakoram and Hindukush (HKH). These snow and ice covered mountains host the endangered snow leopard (biological name: Panthera uncia) which has survived due to its spotted coat and survival skills (Jackson et al. 2006; Khatoon et al. 2017) despite the barrages at different locations in the Indus River (Braulik 2012; Braulik et al. 2014; Waqas et al. 2012). Similarly, the South of the country hosts the endangered Indus dolphin (Biological name: Platanista gangetica). The Indus wetlands in the South is also critical for the migratory waterfowl population in winter. Moreover, the Indus flyway is recognised worldwide as the fourth major bird migration route (Umar et al. 2018). Similarly among plants, the locally known plant Kut (biological name: Saussurealappa) is common to the alpine regions and is considered a threatened species (Shaheen et al. 2017). Another medicinal plant in the alpine regions is the sea grape (biological name: Ephedra procera) which is more commonly used as a cardiac tonic and a remedy for respiratory asthma and hay fever (IUCN 1997; Qasem 2015). Pakistan has a long history of human settlements starting from Mehergarh to the Mohenjo-Daro to the Gandhara civilisation (Violatti 2013; Yadav et al. 2010). However, these civilisations survived in relative environmental harmony until the 20th century; after which point society negatively impacted the natural environment (Milligan et al. 2009; Vlek and Steg 2007). Human development, industrialisation and urbanisation led to the extinction of many species that were earlier common (McCaffery 2015). Similarly, more forests have been cleared for human settlements and agriculture practices (Piesse 2015).

8.3.2   Supporting Services

Supporting services of freshwater in Pakistan include sediment preservation, soil foundation, nutrient cycling, recycling and nutrients processing (see Table 8.1). As the sediments accumulates, lakes and marshes are naturally filling and drying up leading to the formation of valuable land suitable for agriculture. This process is basically threatened by encroachment or settlements near rivers and lakes. Flood plains near rivers accumulate nutrients and contribute in removing these from building up in both flowing and still waters where they might create problems. This is because high level of nutrients in water can lead to algal blooms which may kill fish and effect water supply. In a country like Pakistan, wetlands play a very useful role in reducing the risk from agricultural runoff with high fertilizer content.

8.3.3   Provisioning Services

The provisioning services of freshwater in Pakistan include production of fish, fruits, fibre, fuelwood, fodder, and water retention for household, industry, and agricultural purposes (see Table 8.1). Fish production in Indus River systems are locally important, promoting aquaculture among local communities. Similarly, Indus Flyway is recognised as one of the primary and important migration route for wildfowl and ducks. Many bird species would not be able to capable to complete their usual migrations without the Indus Flyway. These migratory birds are the important food sources for local people living around the rivers and lakes. Freshwater irrigates 17 million ha of farmland which is the primary source of food production and provides employment to approximately 40% population. Pakistan is an arid country and currently suffering from water shortages, therefore the efficient utilisation and storage of water resources is critical for domestic, industrial, and agriculture usage. Forests surrounding the rivers provide fuelwood and timber for local people. Mangoroes used to be the main source of fuel for river navigation up the Indus in the past.

8.3.4   Regulating

The regulating services include climate regulation (e.g. GHG sinks, variability in regional and local temperature, rainfall and other relevant processes); regulation of water (e.g. hydrological flows); recharge and discharge of groundwater aquifers; waste-water treatment and purification; retrieval and elimination of extra nutrients and other contaminants; erosion regulation, preservation of sediments and soils; flood control; protection against storms; and habitat for pollinators (see Table 8.1). Forests have a significant impact on the local climate as they help shape rainfall patterns, reduce temperatures, and enhance humidity. Glacial melting in the HKH region is of high concern in Pakistan as it provides about 70% of the water source in the Indus River. Since 1930, the glacial area of HKH has fallen by 35% - 50%, and hundreds of smaller glaciers have already disappeared. The Kabul-Indus basin hosts three quarters of Pakistan's populations, providing them food and water security and irrigating 80% percent of the cropland.

Ecosystem degradation in the Indus Delta due to increased abstraction of water upstream may affect downstream communities living in the delta area. The 1991 Water Apportionment Accord among the four provinces indicated a discharge of minimum 10 MAF of freshwater (Rajput 2014) per year below Kotri Barrage. This amount is essential to sustain ESS in the Indus delta. However, it was rarely the case during 1994–2004 where the average annual water release was between 2 MAF (in dry seasons) and 6.8 MAF (in the wet season). Additionally, the waste water treatment, water purification, and removal of surplus nutrients are also some of the important regulating services of freshwater.

Through biological procedures, rivers carry away organic contaminants from households and industries. These biological processes have severe threats from untreated trashes, resulting in degradation of wetlands. Majority of rivers and coastal wetlands of Pakistan are exceptionally polluted due to wastes of major cities. This leads to the loss of biodiversity and efficiency. Similarly freshwater offers erosion regulation which slow down the water flow through a catchment. Furthermore, siltation of dams can reduces the life of dams and reservoirs. The life of dams and reservoirs can only be enhanced through a proper watershed management. Floodplains has the potential to reduce the flood intensity by providing an escape route to floodwater. It is a well-known fact that honeybees provide an important ecosystem service, however its role of pollinators is rarely overlooked.

8.3.5   Cultural Services

The cultural services of freshwater comprise of recreational opportunities, spiritual basis of inspiration; and learning prospects for both informal and formal education (see Table 8.1). Pakistan has quite a few spiritual and religious features of freshwater with tombs and shrines of Saints, Sufis, Pirs, and other sacred personalities in various coastal and wetlands areas. For example, the Kalakahar Lake located in the Salt Range has various archaeological and historical sites that fascinate tourists. There are some fairy-tales associated with lakes and riverine wetlands (e.g. Lake Saiful Muluk). The recreational values and importance of these lakes and rivers have been used to justify the tourism authorities throughout Pakistan, e.g., Tourism Corporation Development of Pakistan (TCDP), Tourism Corporation Khyber Pakhtunkhwa (TCKP), Tourism Development Corporation of Punjab (TDCP), Sindh Tourism Development Corporation (STDC) and The Department of Culture, Archives and Tourism Balochistan. Riverine wetlands, lakes and ponds are being used for research and learning purposes, providing awareness about the importance of water amongst local schools and colleges.

Table 8.1: Major ecosystem services provided by freshwater in Pakistan

Kinds of freshwater	Supporting Services		
	Sediment holding furnish soil formation; organic matter accumulation; storage, recycling, processing, and acquisition of nutrients; freshwater wetlands reduce the risk of pollution from agriculture runoff		
	Provisioning	**Regulating**	**Cultural**
Rainbow water	Huge storage of water on Earth; habitat for birds & insects	Climate regulation, hydrological regulation	Aesthetic (inspiration for art), spiritual (rain Gods/ Gods of thunder), inspiring knowledge)
Blue surface & groundwater	Production of food, fish, fruits, fibre, fuelwood and fodder; preservation of water for household, industrial, and agricultural use; Indus Flyway serve as migration route for ducks and wildfowl which provide local food sources for communities; irrigates 17 million ha of farmland and provides employment to 40% population; Mangroves & riverine forests are source of timber and fuelwood for local population	hydrological flows and groundwater recharge/discharge; water treatment by removing nutrients and other pollutants; natural hazards regulation, protection against storm; pollination habitat for pollinators; Kabul-Indus basin support three quarters of Pakistani population and irrigates 80% cropland; erosion regulation slow down water flow; river beds provide support in flood control and reduce flood intensity	Religious & spiritual aspects of freshwater comprise of shrines in some coastal areas; Kalakahar Lake has archaeological and historical areas and featurs that can attract visitors; fairy-tales attract visitors during festivals; recreational values of lakes, streams and rivers helped in establishing special tourism authorities throughout Pakistan; rivers, lakes & streams used for educational purposes, creating awareness amongst local schools & colleges about the importance of water
Green water	Fodder, food, pastureland, herbs & shrubs	Evaporation (flowing downwind to later fall as precipitation); aquifer recharge	Forests and landscapes for tourism, spiritual needs and education
Grey water	Rice & vegetable production, fodder crops, energy production, mining,	Climate and water regulation, evaporation towards downwind that fals as precipitation	Educational services regarding its potential uses in agriculture (UNEP, 2010)
Black water	Animal fodder, insects & worms as birds' food	Spreads disease unless managed	Educational services regarding its negative effects
White frozen water/glaciers	Habitat for markhor and snow leopard, storage of water	Albedo effect	Preserving data for humans, information about CO_2 in the past, preserving life forms frozen in the past

Source: Modified from Hayat and Gupta 2016

8.4   DRIVERS OF FRESHWATER PROBLEMS IN PAKISTAN

In Pakistan freshwater has both quality and quantity related issues. These issues are further aggravated by several drivers at national and sub-national levels. In this section I discuss national and sub-national level direct (see 8.3.1) and indirect (8.3.2) drivers which also include drivers that are identified in the case studies as well as in the literature whicn might not be case specific.

8.4.1   Direct Drivers

The key direct drivers of water problems in Pakistan are (see Table 8.2): (i) agriculture development (e.g., commercial agricultural practices including animal husbandry, the extractive sector and water use in energy),[78] (ii) industry (including services and infrastructure),[79] (iii) municipal water supply and sanitation services e.g., household usage (drinking water, hygiene and sanitation) and subsistence agriculture,[80] (iv) demographic shifts[81] such as (a) growing population (2% growth rate in 2018) and unsustainable urbanisation (3.19% growth in 2016) where population density has reached a record 260 persons/km^2 as of 2018 in contrast to 60 persons/km^2 in 1961; (b) rapid urbanisation which is further increasing water demand and affecting water quality.

8.4.2   Indirect Drivers

The key indirect drivers of water problems in Pakistan are (see Table 8.2): (a) political dynamics within the state i.e. mistrust and imbalanced power relations among and within provinces where small provinces (Khyber Pakhtunkhwa and Balochistan) blame big provinces (i.e., Punjab and Sindh) for inequitable water distribution despite the 1991 Interprovincial Water Sharing Accord which ensures fixed allocation among provinces, but lacks clearly stated objectives (e.g., to enhance water accounting in the Indus basin and to improve the operating rules) and hence leaves room for interpretation, which is often biased and favours powerful actors,[82] (b) culture and ethnic elements (attitudes regarding access and allocation, wasteful use of resources, etc.),[83] (c) non-water-related policies (agriculture & food security, land tenure, land use, and economic development),[84] (d) economy (economic growth),[85] (e) poverty, (f) technological advances (agriculture intensification),[86] (g) international trade (e.g. _globalisation' or trade in virtual water)[87] and the possible trade route with China, and (h) natural change and variability in weather, droughts, floods, landslides, and tectonic movement.[88]   The impacts of climate change in the Kabul-Indus Basin over the next 25 years are likely to place the Interprovincial Accord under increased scrutiny. Similarly, transboundary water issues with India (in addition to building dams in Afghanistan) are projected to become key challenges in the national and water security context and often mask domestic water governance issues. It is expected that Pakistan may face high temperature rises due to its geographical location. Only 24% of the area in Pakistan receives rainfall between 250-500 mm while the remaining area receives less than 250 mm of rainfall per year. The receding of glaciers in the HKH region due to climate variability may influence major rivers in Pakistan as they are primarily fed by these glaciers due to climate variability (Barnett et al. 2005; Hasson 2016; Stocker 2014). Moreover, Pakistan's climate sensitive agrarian economy has also potential threats from variability in monsoon rains where droughts and floods have caused large scale destruction already.

---

[78] Interviewee 1, 2, 3, 5, 7, 33, 34, 36.
[79] Interviewee 3, 4.
[80] Interviewee 5, 7, 19.
[81] Interviewee 2, 3, 5, 9, 13, 22.
[82] Interviewee 11, 12.
[83] Interviewee 13, 17, 40.
[84] Interviewee 1, 8, 26.
[85] Interviewee, 25, 26, 27.
[86] Interviewee 25, 38, 39.
[87] Interviewee 12, 20, 21, 33.
[88] Interviewee 6, 21, 23, 34.

Table 8.2: Drivers of freshwater challenges in Pakistan

Direct Drivers	Key References
Agriculture development (e.g., commercial agriculture practices including animal husbandry, the extractive sector and water use in energy)	Raza, Ali, and Mehboob 2012; Rehman et al. 2013 Interviewee 1, 2, 3, 5, 7, 33, 34, 36
Industry (including services and infrastructure)	Loayza and Wada 2012; Ahmed, Abbas, and Ahmed 2013; Umer 2018 Interviewee 3, 4
Municipal water supply, sanitation e.g., domestic uses (drinking water, sanitation and hygiene) including sustenance agriculture	Haydar, Arshad, and Aziz 2016; Young et al. 2019 Interviewee 5, 7, 19
Demographic shifts (i.e., migration, population growth, increase in population density, urbanisation, population growth)	Gazdar 2003; Khan, Inamullah, and Shams 2009; Mahmood and Chaudhary 2012; Kugelman 2013 Interviewee 2, 3, 5, 9, 13, 22
Indirect Drivers	
Political dynamics between states	Richter 1979; Malik 1996; Sharif 2010; Anwar and Bhatti 2017 Interviewee 11, 12
Culture and ethnic elements (attitudes about access & allocation, careless resource use)	Briscoe et al. 2005; Lead Pakistan 2018 Interviewee 13, 17, 40
Non-water-related policies (food and agriculture, food security, land tenure, & other economic development activities)	Murgai, Ali, and Byerlee 2001; Ahmad and Farooq 2010; Rehman et al. 2015; Qureshi 2018 Interviewee 1, 8, 26
Economy (economic growth)	Briscoe et al. 2005; Afzal 2009; Yu et al. 2013; Rehman et al. 2015 Interviewee, 25, 26, 27
Poverty	Faruqui 2004; Abbass 2009; Syed Attaullah Shah 2014; Syed A. Shah, Hoag, and Loomis 2017 Interviewee 25, 38, 39
Technological advances (agriculture intensification)	Briscoe et al. 2005; Yu et al. 2013; F. Rehman et al. 2013; A. Rehman et al. 2015 Interviewee 12, 20, 21, 33
International trade (e.g. _globalisation' or trade in virtual water); CPEC	Yu et al. 2013; Ahmed and Mustafa 2014; Ramay 2018 Interviewee 6, 21, 23, 34
Natural change and variability in weather, Droughts; Floods; Landslides, tectonic movement.	Farooqi, Khan, and Mir 2005; Hussain et al. 2005; Ahmed et al. 2016 Interviewee 6, 21, 23, 34

8.5   EVOLUTION OF THE FRESHWATER AND RELATED INSTITUTIONS IN PAKISTAN

8.5.1   Overview of Water Governance Institutions & Practices in Pakistan

Water is a provincial subject according to the 18th amendment in the 1973 Constitution of Pakistan (Constitution 1973: Chap. 3, Art. 155). However, federal government use its authority to ensure equity and access among provinces (National Water Policy of Pakistan 2018). Water sharing among provinces is also regulated and administered by the constitutional and parliamentary bodies e.g. the Council of Common Interests (CCI) and Parliamentary Committee on Water Resources (Ranjan 2012a; Sharif 2010). Water and Power Development Authority (WAPDA) has legitimate authority to carry out water development schemes (UNDP 2016; WAPDA Act 1958: Art. 8). Similarly, the

Indus River System Authority (IRSA) regulate water sharing among provinces under the 1991 Water Apportionment Accord (Anwar and Bhatti 2017; Ranjan 2012b). IRSA contained one representative from each of the four provinces and a federal member. Furthermore, the chairman of IRSA is selected on rotational basis from each province (IRSA 1992: Art. 2/1). At the sub-national levels in Pakistan, provincial irrigation departments and the provincial Environment Protection Agencies (EPA)[89] provide the core regulatory framework (Anwar and Bhatti 2017; Sharif 2010). Similarly, at local level, the local government as well as Water and Sanitation Agencies (WASAs) regulate the provision of clean drinking water and sewage disposal (Khan and Javed 2007).

Water is governed through a mix of informal and formal mechanisms in Pakistan (FoDP 2012; Qureshi 2002). The informal practices date back to over 9000 years, all of which have a unique link with the Indus River (Alam et al. 2007) such as: Baluchistan's Mehergarh (9000-7000 BCE) (Aamir 2015; Notezai 2017); Khyber Pakhtunkhwa's Rehman Dheri (the Pre-Harappan 3300 BCE) (Jan et al. 2008; Khan et al. 2002), Punjab's Harappa (3000 BC) and Sindh's Mohenjo-Daro (2500 BCE) (Angelakis and Rose 2014; Fuller 2001; Possehl 2002). It was discovered that ages ago before the Pharaohs or the Mesopotamians, earlier settlers of Mehergarh were using flood water to grow crops. These settlers had properly trained animals for farming (Grewal 2005; Khan et al. 2014). The remains of the 5000 years old (3300 BCE) large sand and mud dams were also revealed in the Khuzdar district in Baluchistan (Manuel et al. 2018; Shaffer and Thapar 1992). These practices were improved by the Indus Civilisation of Mohenjo-Daro (in Sindh Province) and Harappa (in Punjab's Province) in 3000 BCE which projected that a composite society existed at that time which sustained these sand and mud (Khan et al. 2014; Manuel et al. 2018). The civilisation encouraged water management, water supply, sanitation services, washing platforms and a dedicated waste disposal system (Angelakis and Rose 2014; Cullet and Gupta 2009). Similarly, in the KPK and Punjab Provinces, customary irrigation practices are as old as 330 BCE (Ahmed 2000; Bhutta and Smedema 2007; Mehari et al. 2011). The extensive mysterious _gabarbands[90] that averted water from the dry rivers in that period can still be found in various parts of Sindh and Baluchistan provinces (e.g., Larkana, Dadu and Las Bela districts) (Khan et al. 2014).

The adoption and pattern of formal rules in Pakistan are drawn from the earlier civilisation of the Indus, the Aryans, the Arabs, the Moghuls and more recently, the British (Badruddin 2012; McIntosh 2018). The Islamic Principles are founded on equitable sharing and the recognition of collective control over water (Abderrahman 2000; Cullet and Gupta 2009). The codification of water-related rules and practices in Pakistan were first started by the British Colonial Administration in 1860 from the Pakistan Penal Code to prevent contamination and preserve ecosystems. This was followed by the codification of customary irrigation practices – _warabandi' in 1873 to promote cooperation and ensure equitable and efficient utilisation of shared water resources among farmers. All these rules and practices were recorded in registers called the _Kuliyat-e-Abpashi (set of guidelines for diverting flood flows)' and _Riwajat-e-Abpashi (set of customary practices for

---

[89] Interviewee 5.
[90] The gabarbands are _stone-built dam construction designed to control and store water' (Wright, 2010: 31).

diverting flood water)[91] (Khan et al. 2014). The succeeding British Colonial Rules imposed firm compliance to accomplish management of water shortages (Cullet and Gupta 2009; Mehsud 2015). After the formation of Pakistan, the rules and procedures created by the British Colonial Administration were revised from time to time to meet explicit commitments, but the basic structure of the original rules has persisted (Jurriens et al. 1996; Khan et al. 2014; Subrahmanyam 2006). For evolution of formal/informal multilevel water governance frameworks in Pakistan along with included principles see Figure 8.1 which shows the accumulated inclusion of different categories of principles over different eras. Similarly Figure 8.2(a) presents the number of adopted principles for each category over time; Figure 8.2(b) explains the trends of different categories of principles over time; and Figure 8.2(c) presents the actual progress and regress of different categories of principles over time.

8.5.2   Water Governance before the Common Era (BCE)

The healthy flow of freshwater in rivers gave birth to different civilisations as one civilisation after another rose and fell (Juuti et al. 2016; The DAWN 2012). However, these civilisation died out eventually when the river changed its course or dried up (The DAWN 2012). Obvious examples of such civilisations exist in almost every province of Pakistan. Early cities in these civilisations have died out but left behind traces of earlier development which the modern world follows. Mehergarh (9000-7000 BCE) is located in present day Baluchistan and is seen as the earliest sites of established farming (barley and wheat) (Moulherat et al. 2002; The Express Tribune 2019), metallurgy (Kenoyer 2006; Possehl 2002), and herding (goats, sheeps, and large cattle) in South Asia (Teufel et al. 2010; Zeder 2012). The storing of water and grains for later consumption in that time shows preparedness to cope with food and water insecurities (Angelakis and Zheng 2015; Fardin et al. 2014). The location of the site on the main route between modern day Afghanistan and the Indus Valley illustrates a navigation and trading route between the Indian subcontinent and Near East (Stevens et al. 2016). The structure of the houses on the banks of the Indus River in the flood-resilient locations reveals the skills related to flood protection (Macklin and Lewin 2015)

The remnants of ancient cities Harappa (in the province of Punjab) and Mohenjo-Daro (in the province of Sindh) from 3000-2500 BC discloses that rivers were a significant source of navigation which helped them trade with central and northern India as well as the coastal areas of Mesopotamia and ancient Persia (Majumdar et al. 1978: 34 in Stevens et al. 2016). Water were used mainly for domestic and irrigation purposes (Angelakis and Zheng 2015). The development of earlier irrigation system indicates that agriculture has been the prime source of employment (Weiss and Zohary 2011). However, this too had conseqeunces which caused deforestation and changes in land use (Dellapenna and Gupta 2009). Mohenjo-Daro had over 700 wells while Harappa had as few as 30 (Menon 2018; Ratnagar 2016). The reason for an increased number of wells in Mohenjo-Daro may be due to the fact that it was a bit far from the Indus River and were receiving less winter rain (Petrie

---

[91] _Abpashi' means irrigation; _Haqooq' means rights; _Kuliyat' means comprehensive description; Riwajat' means customs.

2019; Possehl 2002). The Great Bath some 4500 years ago had water channels (Angelakis et al. 2018; Voudouris et al. 2019) and a temple, where priests and rulers bathed in religious ceremonies (Ratnagar 2016; Voudouris et al. 2019). People of Indus Valley had clean water and excellent drains (Angelakis et al. 2018). Almost every home had a separate bathing area and a latrine where the run-off from these designated areas were carefully managed (Antoniou et al. 2016; Khan 2012). The run-off water was reused after removing waste through the sewer into the Indus (Khan 2012). These actions of ancient societies clearly shows that they protected their natural environment to make their resource base stronger.

8.5.3   The Emergence of the Islamic Way of Water Governance

Islam, as a faith and political social order launched in Asia in the 7th century during Muhammad′s lifetime (Khalid 2014; Lewis 2010). It is said that Islam have arrived in 615 AD to Manipur (Northeast India) in the era of silk route trades through Chittagong in Bangladesh (Asimov 1999; Jettmar 1994). Islam's expansion into South Asia caused the development of a flexible legitimate system (Kugle 2001). Islam extended in many directions from Arabia to South and Central Asia in the era of Muhammad – as the first head of the city-state of Medina (622–32) – followed by the Caliphate (i.e., Abu Bakr, Umar, Usman, and Ali) (632–61), later on by the Umayyad (661–751) and then Abbasid (751–1258) (Khan 2015). Islamic law and jurisprudence evolved from the succession of prohibitions enclosed in what followers respect as the exact word of God: the Quran as revealed to the Prophet Muhammad (Coulson 2017; Dutton 2013). After the Quran's revelation, Islamic scholars consistently relied upon the Sunnah[92] (Abdullah and Nadvi 2011). The principles of Sunnah were used to review the diverse Quranic instructions (Khan 2015).

Classical Islamic Law treat water as public property (Abdullah and Nadvi 2011; Naff and Dellapenna 2002; Zahraa and Mahmor 2001) because it is essential for the customary acts of worship and supporting human life (Abdullah and Nadvi 2011). Principles of Islamic Law related to water include a right of thirst to reduce the thirst of human and animals from any available water resources (Faruqui et al. 2001; Koumparou 2018). The Holy Quran identifies significance of water, mentioning its importance in _Surah' 21:30, _we have made every living thing of water' (Abumoghli 2010; Faruqui et al. 2001). Islamic Law usually discourage private ownership of water rights (Faruqui et al. 2001; Sattar 2015) but one can utlize public water resources for as long as they do not damage people's fields or properties (Faruqui et al. 2001). Sahih Bukhari[93] explains that, _it is forbidden to exploit water by polluting or wasting it' (Naff 2009: p. 20). Islamic Law has principles related to dispute resolution. It prescribes techniques for the 148etermining rights by a local expert, who is perceived to have final authority in the matter (Abdalla 2001; Brower and Sharpe 2003; Powell 2015). Islamic scholars and rulers applied principles of Islamic law to the Muslim population

---

[92] Regarded as a second law in Islam or the body of precedents attributed to the Prophet Muhammad (PBUH).
[93] Sahih Bukhari also known as Bukhari Sharif is one of the six major prophetic traditions (or hadith) collections of Sunni Islam. Sunni Muslims view this as one of the two most trusted collections of hadith along with Sahih Muslim. In some circles, it is considered the most authentic book after the Quran. The Arabic word Sahih translates as authentic.

in the past while non-Muslims were encouraged to follow their own systems (Lugo et al. 2013). Table 8.3 summarises the priciples of both customes and sharia.

Table 8.3: Principles of customs and Sharia

Customs	Sharia
Community-based management structures with elected or, more often, selected water masters	In Sharia water is reflected as a community right; Private water ownership rights are forbidden
Water allocation rights are linked with land ownership, the size of the land and the level of the assistance to the construction and maintenance of water infrastructures	Types of water is defined (e.g., lake water, river water, and rainwater)
Traditional structures are not necessarily inclusive and can often be subject to elite capture	Priority of use (i.e., the nearest cultivated plot has first priority) or whose crops are most urgently in need of water may go first
Water can be primarily used for domestic and personal use as well as irrigation	The right to water
Customs encourage reuse of wastewater	No harm principles where Islamic law forbids a person from exploiting or polluting water

8.5.4   Water Governance in the Colonial Era

The history of colonialism in British India (which included Pakistan) dates back to the sixteenth century with the invasion of European colonialists (Stuchtey 2017). During England's industrial revolution period in 1858, the British colonialists transformed the Indian economy from a food production to a commodity-oriented economy (Dellapenna and Gupta 2009; Mukherjee 2010). British Colonialists promoted the market-based economy and transformed customary and social practices (Dellapenna and Gupta 2009; Lange et al. 2006). Subsistence agriculture gave way to commercial agriculture using more natural resources benefitting individuals rather than the community (Banerjee and Iyer 2005). Forests were cleared to gain access to coal, timber and agricultural land (Dangwal 2005; Islam and Hyakumura 2019). In addition, the 1873 Canal and Drainage Act, authorised the colonial administration to control surface water resources and replace community-managed irrigation with State-owned irrigation (Banerjee and Iyer 2005; Mustafa 2001). Similarly, the Indian Forest Act 1927 gave forest ownership to British colonialists (Ghosh and Sinha 2016; Sharma 2017). These colonial activities widened the income disparities and exacerbated poverty in rural areas, and increased water logging and salinity (Angeles 2007; Roy 2018).

The Colonial administration did not interfere with local customs if they did not present a direct threat to the British (Maddison 2013; The DAWN 2010). Infrastructure related to water including irrigation facilities were developed and regulated by the British colonial administration after the

1857 revolution (Dellapenna and Gupta 2009). Colonial water law: 1) protected and regulated water rights through common law principles (e.g., ensuring landowners' access to water resources) (Cullet and Gupta 2009; Smith 2005); and 2) made rules to secure land, safeguard and sustain embankments, and implement these regulations (e.g., see Iyer 2009; Cullet and Gupta 2009). According to Mustafa (2001) and Mosse (2006), the colonial irrigation projects and policies favoured bureaucracy over the needs of farmers and led to the deterioration of customary irrigation institutions.

8.5.5   Water Governance in the Post-Colonial Era

After the partition, Pakistan's water management policies continued the British Colonial legacy[94] (Chuadhry and Chaudhry 2012; Rahman et al. 2018) where they merged and restructured colonial control of water resources rather than aim at public well-being or equity[95] (D'Souza 2006; van Koppen et al. 2007). Currently, freshwater resources are governed by both provincial as well as federal frameworks throughout Pakistan. Overall water management falls under the federal authority i.e. Water and Power Development Authority (WAPDA).[96] The 18th amendment to the 1973 Constitution devolves water to provincial governments[97] and WAPDA only articulates plans for infrastructure development based on data provided by the Indus River System Authority (IRSA).[98] These plans are regulated by WAPDA Act 1958, IRSA Act 1992, Environmental Protection Act 1997 and the Constitution under articles dealing with inter-provincial coordination and conflict resolution by the Council of Common Interests (CCI).[99] The Punjab Canal and Drainage Act of 1873, the Sindh Irrigation Act of 1879, and KP Canal and Drainage Act of 1997 provide the main legal basis at provincial level.[100]

Freshwater resources in present day Pakistan are governed by rules from ancient customs, religion and British Colonisation.[101] Since 1947, two instruments govern freshwater at federal, provincial and local levels: the West Pakistan WAPDA Act 1958 and the 1973 Constitution of the Islamic Republic of Pakistan.[102] These laws target and guide both general and specific uses and users on human rights, access, pollution and conflict prevention and allow women, youth and indigenous peoples to practice their rights. By 1997 another four laws were adopted[103] covering water quality and quantity issues and penalties for violations. The IRSA Accord is one of the successful inter-provincial water sharing agreements of the Indus Basin in this period which is based on the historical use of water by the provinces in Pakistan.[104] Between 2007 and 2017, new policies were adopted: National Drinking

---

[94] Interviewee 10
[95] Interviewee 1, 2
[96] Interviewee 2.
[97] Interviewee 1, 5, 16
[98] Interviewee 8, 9.
[99] Interviewee 1, 2, 3.
[100] Interviewee 12.
[101] Interviewee 3.
[102] Interviewee 1, 2.
[103] Interviewee 10.
[104] Interviewee 10, 11.

Water Policy 2009, Pakistan National Wetlands Policy 2009, National Climate Change Policy 2012;
Biodiversity Action Plan 2015, National Forest Policy 2015 and the Draft National Food Security
Policy 2017. The National Climate Change Policy and the Draft National Food Policy integrates
previous policies and laws with the adoption of modern dimensions related to climate change
andsustainable development.[105]

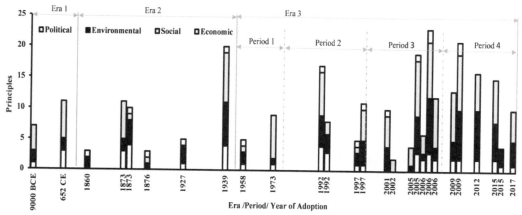

Figure 8.1: Evolution of water governance in Pakistan

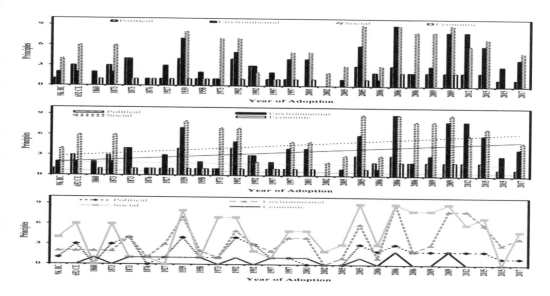

Figure 8.2: (a) Number of included principles (b) overall progress & (c) trend analysis

---

[105] Interviewee 3, 8.

8.6   GOALS, PRINCIPLES AND INSTRUMENTS

8.6.1   Goals of Freshwater Governance Framework at Multiple Level in   Pakistan

Governance frameworks in Pakistan intend to ensure food security and diminish increasing poverty levels by encouraging efficient and sustainable water practices through optimal supply and better management. Furthermore, these policies also aim to accomplish: (1) proper conservation of water resources; (2) water resource development; (3) completion of water related projects in minimum time and cost; (4) equitable distribution of water resources in canal irrigated areas; (5) measuring and reversing fast declining aquifers in low-recharge areas; (6) sustainable groundwater utilisation in high-recharge areas; (7) maximised crop production; (8) improved flood control; (9) ensuring acceptable and safe quality of drinking water; (9) minimisation of salt build-up and other ecological risks in irrigated areas; and (10) institutional reforms in order to make the managing organisations more vibrant and approachable.

8.6.2   Governance Principles in Pakistan's Multilevel Freshwater Governance Frameworks

The governing principles include:

a. **Political Principles:** Political principles in the local level water governance frameworks include notification about emergencies and planned measures; obligation to cooperate; peaceful settlement of disputes; and exchange of information. The provincial level frameworks include data and information exchange; notification about emergency situations and planned measures; peaceful settlement of disputes; obligation to cooperate. Similarly, the federal/national level include notification about emergency situations and planned measures; peaceful settlement of disputes; information exchange; and obligation to cooperate. Moreover, principles, such as notification about emergency situations and planned measures; obligation to cooperate; peaceful settlement of disputes; exchange of information are common at multiple geographic levels.

b. **Social-Relational Principles:** Social principles at the local level include equitable and reasonable use; human right to water and sanitation; ensuring public participation; priority of use; public awareness and education; rights of women, youth, and indigenous peoples; intergenerational equity; and poverty eradication. The provincial level frameworks include prior informed consent; equitable and reasonable use; priority of use. Similarly, the federal/national level principles include rights to water and sanitation; public awareness and education; capacity building; equitable and reasonable use; intergenerational equity; poverty eradication; access of general public to information; participation of users; priority of use; women's rights, including youth and indigenous peoples.

c. **Ecological Principles:** Ecological principles at local level include monitoring; pollution prevention; treating water as a finite resource; and the precautionary principle. Similarly the provincial level frameworks include the river basin as the unit of management; prevention of

152

pollution; precautionary principle; protected areas for water; monitoring; and water as a limited resource. Similarly, the federal/national level includes prevention of pollution; conservation and protection of ecosystems; monitoring; the principle of precautionary use; protected areas for water use; water as a limited resource; EIA; invasive species; protected areas for recharge and zones for discharge; conjunctive use; and subsidiarity.

Annex I presents the evolution of governance principles in the various frameworks. It shows (a) the principles that are currently in use; the principles that have not been included and principles that are no longer being used. Of course, this does not imply that these principles are actually being implemented all over the country.

8.6.3   Governance Instruments in Pakistan's Multilevel Freshwater Governance Frameworks

This section discusses the instruments through which the principles in the national and sub-national level water governance frameworks are operationalised in Pakistan. These instruments are divided into four categories such as: national level water-specific regulatory (fines, licences, metering, mapping & zoning); economic (subsidies, water pricing); suasive (awareness raising); and management (WUAs) instruments as well as non-water related[106] regulatory (fine, punishment, mapping and zoning), economic (water charges), suasive (awareness raising), and no management instrument. Similarly, provincial level instruments consist of regulatory (telemetry system for monitoring, rules formulation, dispute resolution mechanism), economic (water charges, maintenance of drainage infrastructure, cost recovery), suasive (policies and research for addressing salinity and water logging) and management (area water board, farmer organisation, participatory decision making) instruments. Despite the existence of the above-mentioned complete list of instruments in multilevel water governance frameworks, they have not been effective when it comes to operation on the ground. For example, the telemetry system for real-time data collection for monitoring to ensure inter-provincial trust for water sharing has not been effective as it was tampered with One theory is that the system was deliberately tampered with due to an ownership dispute between the irrigation departments and WAPDA so that they could continue with their corrupt practices with the support of some political elites to take advantage of the situation and advance their political agendas on water sharing. Moreover, the other existing instruments have been ineffective due to weak enforcement mechanisms in the form of low amount of fines, fees for permits and licences and imprisonment as described in the Pakistan Penal Code (1860), which are unable to change the behaviour of actors consuming large quantities of water. Furthermore, suasive instruments are scarcely used and as a consequence people are unaware of the need for changed behaviour.

---

[106] This is because non-water related policies can directly impact water governance in the country (e.g. food policies, energy policy, environmental policies etc)

8.7   LEGAL PLURALISM ANALYSIS

The analysis of the historical evolution of water governance indicates that water resources in Pakistan are governed through three parallel governance system: the local customs (mostly at community level), sharia law (mostly at community level) and the state legal codes, laws and policies (also see 7.7). Establishing the true relationship among these governance frameworks is complex and a subject of further investigation. However, these governance systems differentially dominate the different levels in Pakistan; and both water-specific and non-water specific laws and policies are governing freshwater resources (see 8.5). Some of these legal and policy instruments are coherent while others are contradictory.

A legal pluralism analysis reveals, first, that when water governance and many water-related matters were decentralised to the provinces particularly after the 18th amendment to the 1973 Constitution of Pakistan, national laws were not automatically translated to the provincial level. Second, on occasion the State integrates principles and instruments from local customs and Sharia Law in water law (see 7.7). Third, legal pluralism occurs when freshwater governance occurs implicitly and explicitly through multiple non-water policies (e.g. environment, forest, agriculture, pasture, food and mining).

As Table 8.4 shows there are instances where the three types of policies lack an operational relationship (type 1) in addressing water issues. These include: 1) water policy promoting water pricing which is missing in other directly relevant policies (e.g. food security policy 2017); 2) Sharia defines types of water as well as promotes conservation and efficient water use whereas these are not well defined in local customs and practices; 3) mapping and zoning for safe and resilient cities are defined in water policies whereas other relevant policies do not address this; 4) WUAs are empowered to collect water charges but these are so minimal that they hardly cover a small portion of O & M costs and; 5) non-water policies mention drainage charges but water policies at national and provincial levels omit this important aspect.

There are policies and customary practices that contradict each other (type 2). These are as follows: 1) the national policy recommends water pricing for productive sectors but these are insignificant particularly in the agriculture sector which is the largest user of freshwater resources in the country; 2) national policies suggest subsidies for enhancing access to clean drinking water whereas non-water policies recommend charges for water supply and drainage; 3) national water policy goals of sustainable water use can be undermined by the provision of subsidies for groundwater tube wells which can deplete water resources without proper regulation. Local customs and Sharia also have contradictions which can undermine achieving policy goals especially at the sub-national level. These include (type 2): 1) local customs allow private ownership which are forbidden in the Sharia; 2) water is treated as an economic good in state laws and policies whereas it is a human right in the Sharia; 4) Sharia has defined priority of uses for drinking, agriculture and livestock as opposed to local customs which do not prioritise water usage for different purpose.

It is also important that different legal instruments, principles and local customs accommodate (type 3) each other by recognising their relevant elements that apply on addressing similar water issues. One example is the participatory water management practices by formulation of WUAs and FOs at the local level.

There are instruments within the water and non-water policies as well as local customs at the national, sub-national and local levels that are mutually supportive to achieve the national water goals (type 4). These include: 1) fines and licences for regulated use and pollution prevention to address the direct drivers of agriculture and industry in both water and non-water policies; 2) subsidies for increasing water access is supported by metering for water efficiency; 3) mapping and zoning for natural changes in all the policies at national and sub-national levels; 4) information and data sharing as well as real-time monitoring of river flows by IRSA is included to reduce water losses; 5) water charges are introduced to cover the costs of operation and maintenance of water infrastructure from productive sectors; 6) penalties are a salient feature of both Sharia and the national water policy for pollution prevention and; 7) research promotion and awareness raising to address natural changes (floods, droughts) are an important part of the Pakistan water governance framework. Furthermore, Pakistan's water policies encourage metering to increase accessibility by rational water use that can be supported by increasing water quality as envisaged in a number of non-water policies at the national level. Similarly, the National Water Policy recognises the Inter-Provincial Water Apportionment Accord 1991 that ensures pre-defined water sharing among the provinces to address the indirect driver of political dynamics between the provinces.

The analysis of legal pluralistic forms of water governance indicates that the three levels of water governance (national, provincial and local) as well as the three types of legal frameworks (local customs, Sharia and state laws) further complicate water governance and reduce effectiveness of water policies and their implementation. In addition to the issues linked to pluralistic forms of water governance, Pakistan's water governance frameworks are influenced by the national water security narratives that are often linked to the national security paradigm promoted by the security establishment which is a key player in shaping the foreign policy development especially with India and Afghanistan, both of which have transboundary water sharing mechanisms with Pakistan. For example, in India's case, water issues are often linked to India as using Pakistan's water share in the Indus, which overshadows other governance challenges linked to water conservation. Rather than focussing on water conservation issues within the country, the security narrative focuses on water usage by India (and more recently climate change) and advances the construction of large water storage reservoirs[107] to address Pakistan's water challenges.

---

[107] For example, the Supreme Court of Pakistan and the Prime Minister have established a fund for construction of two large water reservoirs (Dimaer-Bhasha and Mohmand): http://www.supremecourt.gov.pk/web/page.asp?id=2757

Table 8.4: Legal pluralism analysis of water governance frameworks in Pakistan

Quality	Weak relations	Strong relations
Contrary	*Type 1: Indifference* - lack operational relationship among principles/instruments - Water pricing (water policy); not addressed in other policies (food policy) - Types of water defined in Sharia, but not specified in local customs (see Table 7.4) - Conservation & efficient water use well defined in Sharia but not in local customs & practices (see Table 7.4) - Mapping & zoning for safe & resilient cities (water policy) - Water charges collection by WUAs can ensure cost recovery for O & M when water pricing is adequate /sufficient - Drainage charges in non-water policies; missing in provincial & national policies	*Type 2: Competition* - contradiction among principles/instruments - Water pricing for productive sectors whereas nominal charges in agriculture sector - Subsidies for increased access (water policy); charges for supply & drainage (non-water policies) - Subsidy for groundwater mining at provincial levels[108] (KP and Balochistan) undermines national water policy goals - Private ownership forbidden in Sharia but allowed in customs (see Table 7.4) - Water as an economic good (state law) & human right (Sharia) (see Table 7.4) - Priority of use in Sharia but not in local customs (see Table 7.4) - Water reuse in customs but prohibited in Sharia (see Table 7.4)
Affirmative	*Type 3: Accommodation* - recognition of principles/instruments - Increasing accessibility by metering (water policy), increasing water quality through fines (non-water policies) - Recognition of Inter-provincial Water Apportionment Accord 1991 (water policy) - Policies & research on salinity & water logging to address agriculture development - Participatory water management through formation of FOs & WUAs including informal conflict resolution mechanisms (e.g. Jirga system) at local level	*Type 4: Mutual support* - principles/ instruments support each other - Fines and licences for regulated use & pollution prevention addressing agriculture & industry - Increasing accessibility by subsidies for groundwater & increasing efficiencies by metering (water policy) - Mapping & zoning for addressing natural changes in water, non-water and provincial policies - Information & data sharing for all water uses - Real-time monitoring of rivers flows by IRSA to address water losses - Water charges recovery from productive for O & M of water infrastructure - Penalties for pollution in both water & Sharia law (see Table 7.4) - Awareness raising and research promotion for addressing natural changes & variability

Source: Modified from Bavinck and Gupta 2014

8.8   PRINCIPLES AND INSTRUMENTS ADDRESSING DRIVERS AND ACHIEVING INCLUSIVE AND SUSTAINABLE DEVELOPMENT

8.8.1   Principles and Instruments Addressing Drivers through the National Water Policy of Pakistan

In this section, I analyse how the four types of instruments at national level (regulatory, economic, suasive and management) are reflected in Pakistan's 2018 National Water Policy (NWP), and to what extent these instruments address the identified direct (agriculture, industry, demographic shifts)

---

[108] These subsidies in under-developed promote commercial use of groundwater, which can affect groundwater levels undermining water sustainability at local levels for subsistence use.

and indirect (political dynamics within states, culture and ethnic elements, non-water policies, economic growth, poverty, technology, international trade and natural changes) drivers that are responsible for freshwater problems at national level (see Table 8.5).[109] The evolution of freshwater governance policies at national level indicates that drivers of freshwater problems are not systematically addressed by the policy instruments in the 2018 NWP. Although normative principles of ESS protection, biodiversity conservation, ensuring access and equitable allocation and distribution of freshwater resources among the key sectors exist in more than half of freshwater and other environment-related policies in Pakistan, they are poorly implemented through instruments (e.g. regulatory, economic, suasive and management). For instance, provision of clean drinking water is a policy objective and a human right in a number of policies in Pakistan. However, 84% of the population does not have access to safe drinking water and only 42% of people have access to sanitation services out of which 65% are in urban areas and 30% are in rural areas (Daud et al. 2017; The DAWN 2017). According to the Global Joint Monitoring Programme Report, 25 million cases of diarrhoea are reported annually, and more than 300 children die every day due to poor sanitation and hygiene services in Pakistan. The high mortality rate for children under the age of five ranks Pakistan second in South Asia (Chao et al. 2018; The Express Tribune 2018). I analyse below whether the instruments adopted address the drivers.

a. **Regulatory instruments:** In terms of regulatory instruments, the National Water Policy relies on fines and licenses for regulated use and pollution prevention to address direct drivers. For example, water use licenses for regulated freshwater use in industries including commercial agriculture, manufacturing and mining (e.g. recent SC regularisation of mineral water companies)[110] is an effective policy measure for addressing scarce freshwater resources in the country. In terms of addressing direct drivers, the policy includes imprisonment for polluting by industries including commercial agriculture, manufacturing and mining. To address demographic drivers (e.g. to cater for the needs of growing population, increasing urbanisation), strict rules on licenses, penalties including imprisonment exists to govern water use sustainably. Moreover, metering of urban water provision services can also enhance access to WSS by reducing water losses and extend piped water to the poor. Penalties can result in enhanced access to improved freshwater quality and reduction in wastewater quantity. Similarly, mapping and zoning for flood protection and urban settlements is an effective instrument for sustainable urbanisation.

b. **Economic Instruments:** The policy includes some economic instruments in the form of subsidies and pricing to address the poverty driver. Although I analyse national level instruments in this section, some instruments also address drivers at subnational level. For example, to enhance the access to water and sanitation services (WSS) for a growing population (provincial level direct driver), provision of subsidies for low-income groups makes freshwater

---

[109] Interviewee 2, 17.
[110] SC summons owners of 11 mineral water companies on Nov 20: https://www.dawn.com/news/1446565.

affordable. The improved access to WSS directly addresses the indirect driver of poverty reduction (by reducing health related costs) and improved health conditions especially for women and children. Water is generally considered as a free and infinite resource by most users which results in inefficient and wasteful usage in a number of sectors. Water pricing for productive sectors can enhance economic and financial sustainability of water services provision and improve sustainable water use practices at the local level. Revenue generation from large productive sectors not only helps in regulating water usage quantities but can be invested in improving water resources. Provision of subsidies for groundwater abstraction can enhance access to WSS, but subsidies for solar pumping can also reduce water levels when these are used for commercial purpose (i.e., large scale agriculture and industries). Moreover, promotion of tube wells through subsidised electricity and technology can result in unsustainable water use due to weak monitoring and governance mechanisms at the local levels. Although subsidies can be effective economic instruments, there are some contradictions in terms of subsidies that can directly affect the objectives of water conservation. The current water charges are minimal and cover a fraction of operation and maintenance costs of irrigation system. Moreover, the recovery of water charges (Abiana) is highly inefficient. Policy documents recommend improving the water charges recovery system but without any enforceable instruments (National Water Policy of Pakistan 2018).

c. **Suasive instruments** also form an important part of the water policy document in the form of awareness raising and research promotion. Public awareness creation has a separate section within the document for sustainable domestic water use. This is implemented through media and inclusion in the curriculum at primary, secondary and tertiary levels of education. It is envisaged that the respective departments of Irrigation and Water Management will develop awareness campaigns within their regular activities. Agriculture is the largest user of freshwater resources with large inefficiencies due to flood irrigation. Improved farming practices in terms of water efficiency and pollution prevention are mentioned as target areas of the policy. Research within the agriculture universities as well as a national research agenda on sustainable water resource use is envisaged with budget allocation for research activities. Although the prescribed suasive instruments can address behaviour change through information dissemination to the general public and curriculum development, they need to be supported by other regulatory and economic instruments for enforcement. In the present form the prescribed suasive instruments resemble normative principles due to lack of effective enforcement measures.

d. **Management Instruments:** The participatory irrigation management system promotion forms a core part of the management instruments for inclusion of all irrigation stakeholders in decision-making processes to address agriculture. Stakeholder participation has a separate section in the 2018 NWP for decision making in the areas of irrigation, drainage, domestic water supply, flood protection and drought mitigation and pollution prevention. Inclusion of women is also a stated objective of NWP, but it does not spell out the specific mechanisms and forums for ensuring participation. Water User Associations (WUAs) are one of the specific measures for involvement of farmers and water users in decision making processes at the local level for equitable water distribution and operation and maintenance of irrigation infrastructure. However, their effectiveness in terms of equitable distribution of water and representation of disadvantaged groups and small famers has been questioned in the Punjab and Sindh provinces especially in the tail end of canals (Qureshi, 2011). Moreover, these participatory organisations lack legitimacy to enforce penalties, fines and punishments due to administrative issues.

Table 8.5: Instruments addressing drivers in Pakistan's national water policy

Direct Drivers	Instruments			
	Regulatory	Economic	Suasive	Management
Agriculture development (e.g., commercial agriculture including animal husbandry, the extractive sector & water use in energy)	Fines & licences for regulated use, pollution prevention	-	Awareness raising	WUAs for O & M, collection of charges
Industry (including services and infrastructure)	Fines & licences for regulated use, pollution prevention	-	-	-
Municipal water supply & sanitation services e.g., household uses (drinking water, sanitation, & hygiene) & subsistence agriculture	Metering for WSS for increasing demand /population	Subsidies for WSS access	-	-
Demographic shifts (i.e., migration, population growth, increase in population density, urbanisation, population growth)	Mapping and zoning for safe & resilient cities	-	-	-
Indirect Drivers				
Political dynamics between/within states	-	-	-	-
Culture & ethnic (attitudes for access & allocation, wasteful use of resources, etc.)	-	-	Awareness raising	-
Non-water-related policies (agriculture & food security, land use, land tenure, economic development)	-	-	-	-
Economy (economic growth)	-	Water pricing productive sector	-	-
Poverty	Metering for WSS Subsidy for groundwater abstraction	Subsidies for access to WSS	Awareness raising	-
Technological advances (agriculture intensification)	-	Subsidy for groundwater abstraction	-	-
International trade (e.g. 'globalisation' or trade in virtual water)	-	-	-	-
Natural change and variability in weather, Droughts; Floods; Landslides, Tectonic movement.	Mapping and zoning	-	Awareness raising & research promotion	-

8.8.2 Principles and Instruments Addressing Drivers through National Level Non- Water Policies/Laws

This section analyses how the four types of instruments are reflected in Pakistan's non-water related policies and whether these instruments address the identified direct and indirect drivers of freshwater problems at national level (see Table 8.6).

a. **Regulatory Instruments:** Except for Pakistan Penal Code 1860 and Cantonment Ordinance 2002, other policies do not contain any effective regulatory instruments to address direct and indirect drivers. Additionally, the punishment and fines in the Penal Code are insufficient to change behaviour of actors when implemented. Similarly, the Cantonment Ordinance only covers the cantonment areas (mostly in cities) and is not applicable to large water bodies, reservoirs and rivers. It also includes fines for pollution prevention, industrial waste disposal, water contamination, damage to water infrastructure, and water diversion in cantonment to address the direct drivers of agriculture, industry and municipal water supply. The rest of the policies consist of normative principles and recommendations and lack effective instruments to change actors' behaviour. For example, mapping and zoning of protected areas is the stated objective of National Wetlands Policy but lacks effective instruments and enforcement measures to address the identified drivers

b. **Economic Instruments:** Except for the Cantonment Ordinance which includes charges, taxes and fees for water supply and drainage but are not applicable to large water bodies including rivers, other policies lack effective economic instruments to address the drivers of freshwater issues. For example, the Drinking Water Policy 2009 mentions water pricing and equitable provision as policy objectives but without proper instruments to support realisation of these policy objectives. Similarly, economic valuation of wetlands and financial incentives for solar water desalination are mentioned as policy measures in the national wetlands and climate change policies respectively, however, these are not backed by proper enforcement mechanisms. Without relevant economic instruments, these goals cannot be achieved.

c. **Suasive Instruments:** Suasive instruments are one of the most common instruments found in all of the analysed non-water policies in my research context. My analysis shows that awareness creation and education are most commonly cited objectives in policy documents covering drinking water, sanitation, wetlands, forestry, environment, biodiversity, climate change and related areas. Many policies also address gender aspects and focus on inclusion of women and disadvantaged groups. These suasive instruments can enhance awareness about water issues linked to a number of drivers, but these instruments alone are insufficient to address the root causes of water governance issues if not supported by regulatory and economic instruments. Some policies recommend removal of subsidies, promotion of clean energy and promotion of water & energy pricing but are not enforceable without supportive instruments to change behaviour of relevant actors. Additionally, the key demographic drivers are difficult to be addressed solely by awareness raising campaigns as envisaged in majority of the policies.

d. **Management Instruments:** Self-regulation and voluntary management mechanisms have been promoted by a growing number of organisations such as large bilateral and multilateral donor organisations and the UN system and are increasingly adopted by government and non-government organisations for management of natural resources including water. These management instruments are reflected in a number of policies linked to water resources. However, when it comes to implementation, the unsupportive administrative rules and legitimacy of these mechanisms hampers their effective implementation. For example, community-managed forestry promotion or community-based water management without the legal powers to sanction and impose fines renders these instruments ineffective for addressing acute water management and related issues.

Table 8.6: Instruments in non-water policies addressing drivers at national level in Pakistan

Direct Drivers	Instruments			
	Regulatory	Economic	Suasive	Management
Agriculture development (e.g., commercial agriculture including animal husbandry, the extractive sector and water use in energy)	Fine & punishment for pollution prevention	-	-	-
Industry (including services and infrastructure)	Fine & punishment for pollution prevention, industrial waste disposal, water diversion, infrastructure damage	-	-	-
Municipal water supply and sanitation services e.g., household uses (drinking water, sanitation, and hygiene) and subsistence agriculture	Fines for pollution from subsistence practices	Charges for supply/drainage	-	-
Demographic shifts (i.e., migration, population growth, increase in population density, urbanisation, population growth)	-	-	-	-
Indirect Drivers				
Political dynamics between/within states	-	-	-	-
Culture and ethnic (attitudes on access/allocation, waste, etc.)	-	-	Awareness raising	-
Non-water-related policies (agriculture & food security, land use, land tenure, economic development)	-	-	-	-
Economy (economic growth)	-	-	-	-
Poverty	-	-	-	-
Technological advances (agriculture intensification)	-	-	-	-
International trade (e.g. _globalisation' or trade in virtual water)	-	-	-	-
Natural change and variability in weather, Droughts; Floods; Earthquakes; Landslides; tectonic movement.	Mapping and zonning	-	-	-

8.8.3   Principles and Instruments Addressing Drivers at Provincial Level in Pakistan

This section analyses how the four types of instruments are reflected in Pakistan's provincial level policies and whether these instruments address the identified direct and indirect drivers of water problems at the provincial level (see Table 8.7).

a. **Regulatory Instrument:** The Indus Water Apportionment Accord is a policy instrument for water distribution among the provinces and does not address any direct drivers of freshwater problems in Pakistan. The Accord permits a minimum water flow of 10 MAF into the sea below Kotri Barage and distributing the remaining among the four provinces based on historical water usage including priorities for industrial and urban usage[111]. In this way, the Accord permits unpredictable water flow in the Indus, and ensures that all provinces gain from surplusses or losswa in supply. The IRSA Accord introduced a telemetry system for monitoring discharge for sharing allocations among the provinces and formed rules for inter-provincial share in water quantities. It set up the Indus Rivers System Authority (IRSA) with administrative powers for inter-provincial dispute resolution mechanisms over water sharing and addresses the indirect driver of political dynamics between and within states. However, it does not contain any penalties for violation of rules. There have been some substantial inter-provincial differences over water despite the IRSA Accord, particularly with Punjab, which is perceived as the larger, upstream province that controls the water infrastructure, by the smaller and downstream provinces.

The Sindh province in particular is worried that the current water flow is inadequate to meet the minimum requirements for inflow into the sea and, therefore, seawater now has reached up to 100 km inland. This saltwater intrusion has caused damages to the agricultural lands in lower Sindh. It has negatively affected ecosystems, the quality of soil, as well as both water quantity and quality for the District Thatta and in lower Sindh, which has resulted in increased diseases and health issues for local people (Mahmood et al. 2014; Qureshi 2011). These political differences and water allocation issues are likely to increase as water shortages in other areas of Pakistan are affecting urban and rural areas. The idea of minimum ecological flows had origins in the inter-provincial water accord where the minimum required flow downstream of Kotri Barrage was agreed at 10 MAF. However, this minimum flow has rarely been the case in practice. There are serious inter-provincial differences about the construction of Kalabagh Dam, which is supported by Punjab and Khyber Pakhtunkhwa (formerly called the North-West-Frontier-Province or NWFP) and opposed by Balochistan and Sindh.

Issues over freshwater resources among the provinces in Pakistan are very similar. For example, the Province of Khyber Pakhtunkhwa (KP) is anxious about the communities' displacement and land

---

[111] Apportionment of the waters of the Indus River System between the provinces of Pakistan: http://pakirsa.gov.pk/WAA.aspx

loss in Peshawar, Nowshera, and Charsadda due to flooding in the case of earthquakes. Balochistan is worried about the available freshwater supplies as the dam might strengthen Punjab's position to control the water resources. Sindh as the tailender is worried about the decreasing water quantity reaching the province. In April 2006, the foundation stone for the Diamer-Bhasha dam was laid by the then President General Pervez Musharraf in Northern Pakistan to divert people's attention away from the controversial Kalabagh Dam. It is said that the Diamer-Bhasha reservoir will not affect the KP Province but can have consequences for the Chilas valley in the Northern Areas which might be affected. Furthermore, it is not likely to affect the water shares of other provinces since water rights are clearly safeguarded in the Indus Water Accord-1991. For the improvement of the Accord, it has been suggested that: i) water audits prepared by IRSA should be made publicly available, ii) the terminology should be properly defined to avoid differing interpretations, iii) financial penalties for violation of the Principles of Accord should be adopted, and iv) a third party independent water auditor should be engaged to audit the national water resources, with the broader objective of increasing trust and credibility in the data and information released by IRSA to stakeholders (Anwar and Bhatti 2017).

The Canal and Drainage Act 1997 also refers to the Water Apportionment Accord (1991) for inter-provincial allocations as the basis and aims to ensure equitable and reliable irrigation water distribution, effective drainage and flood control.

a. **Economic Instruments:** The Canal and Drainage Act uses economic instruments where the Farmers' Organisations are accountable to control, accomplish and enhance irrigation and drainage infrastructure, collect water charges and other fees to be used for operation and maintenance and institutional sustainability of the FOs. These two economic instruments address the direct driver of agriculture development and the indirect driver of natural changes (flood protection).

b. **Suasive Instruments:** The Act promotes policies and research for tackling the severe issues of water logging and salinity for addressing the most important driver of agriculture development, which is the largest consumer of freshwater resources in Pakistan.

c. **Management Instruments:** The Provincial Act also relies on management instruments by formation of FOs for self-regulation and management of the irrigation systems at the distributary level and addresses the direct driver of agriculture development. The Act aims to improve water use efficiency through formation of the Area Water Boards (AWBs) and the Farmers Organisations (FOs) for management and development of infrastructure related to irrigation, drainage and iflood control. The Act aims to authorise water users by employing regulatory instruments where the AWBs are to articulate and implement policies with a view to accomplish and uninterruptedly enhance effective, economical and well-organized operation of irrigation water at its disposal and promote development, growth and development of FOs including pilot projects for FOs.

Table 8.7: Instruments addressing drivers at provincial level in Pakistan

Direct Drivers	Instruments			
	Regulatory	Economic	Suasive	Management
Agriculture development (e.g., commercial agriculture including animal husbandry, the extractive sector and water use in energy)	-	-	Policies & research for salinity/ water logging (CDA 1997)	Area Water Board, FOs for participatory decision making O & M through FOs
Industry (including services and infrastructure)	-	-	-	-
Municipal water supply and sanitation services e.g., household uses (drinking water, sanitation, and hygiene) and subsistence agriculture	-	-	-	-
Demographic shifts (i.e., migration, population growth, increase in population density, urbanisation, population growth)	-	-	-	-
Indirect Drivers				
Political dynamics between/within states	Telemetry systems for water share monitoring Rules formulation for inter-provincial water share Formulation of IRSA for inter-provincial dispute resolution	-	-	-
Culture and ethnic (attitudes for access and allocation, wasteful use of resources, etc.)	-	-	-	-
Non-water-related policies (agriculture & food security, land use, land tenure, economic development)	-	-	-	-
Economy (economic growth)	-	-	-	-
Poverty	-	-	-	-
Technological advances (agriculture intensification)	-	-	-	-
International trade (e.g. globalisation' or trade in virtual water)	-	-	-	-
Natural change and variability in weather, Droughts; Floods; Earthquakes; Landslides, tectonic movement.	-	Maintenance of drainage infrastructure for flood protection; O & M cost recovery for flood protection	-	-

8.8.4   Instruments Contribution in Achieving Inclusive and Sustainable Development

The inclusion of instruments in the governance frameworks through the three periods (i.e., pre-colonial, colonial and modern) are different. Remarkable progress is recorded in the last two decades where many new and diverse frameworks were developed with the inclusion of regulatory, economic, suasive and management instruments to promote inclusive and sustainable development. To protect the environment there are twelve normative environmental and social principles in the governance frameworks but there is no strong regulatory or economic instrument to endorse these normative principles. The analysis of the inclusion and distribution of principles and instruments over three major eras and across all water governance frameworks reveals weak linkages between instruments and dimensions of inclusive and sustainable development.

The inclusion of economic principles in considerable numbers specifies the economic direction of the policy to consider water as an economic good. Similarly, normative environmental principles are adopted in considerable numbers because Pakistan has ratified various Multilateral Environmental Agreements (including Kyoto Protocol, the United Nations Framework Convention on Climate Change, the Ramsar Convention on Wetlands, United Nations Convention on Biological Diversity, etc.),[112] but in terms of addressing environmental issues there is no single instrument to protect and preserve ecosystems by endorsing penalties or developing strict regulations. Pakistan has very obvious environmental challenges including threats to freshwater resources, receding glaciers and low rainfall which can become worse with the growing challenges of climate change. Additionally, industrial activities including large scale mining as well as population growth have degraded water quality and reduced water availability due to weak enforcement mechanisms and low penalties in the Pakistan Penal Code. Addressing these challenges requires strict regulatory instruments such as mapping and zoning which can prevent pollution and may also protect water recharge and discharge zones.[113] These instruments can effectively counter the power politics challenges if it comes in conjunction with an information sharing mechanism among actors and stakeholders involved.

Similarly, subsidies for social uses and solar tube wells are two contradicting instruments which can further reduce access for the poor to WSS. Despite the inclusion of dozens of normative principle, various governance frameworks are still unable to cope with severe pollution-related issues throughout Pakistan.[114] Key drivers, such as water supply for food production and human activities, political dynamics among provinces to share water resources equitably and efficiently and non-water related policies which contradict many water conservation-related efforts play a key role in making the water scarcity situation worse. Further, instruments to deal with social drivers such as mounting poverty and access to clean and potable drinking water and sanitation principles are commonly designed to address the freshwater role in providing supplies in emergencies and dealing with

---

[112] Interviewee 2.
[113] Interviewee 1, 10.
[114] Interviewee 3.

changing geopolitics, demographic shifts, as well as to support equitable or rights-based allocation. In terms of the scopes of inclusive and sustainable development the lack of interest of the Government of Pakistan towards social-ecological development reveals less investment in empowering communities through providing education. On the other hand, it also indicates that the lack of social-ecological development can lead to lack of coordination among different levels and stakeholders. As such, there is less potential for inclusive development and little prospect for sustainable development.

8.9   HOW DO POWER AND INSTITUTIONS INFLUENCE WATER SHARING IN PAKISTAN?

Similar to Afghanistan, water governance in the pre-colonial era in today's Pakistan was also based on local customs and principles of Sharia where the community collectively conserved and managed water resources. With colonialism and the introduction of formal water laws, the institutional architecture in Pakistan has resulted in a legal pluralistic form of governance. The devolution of water governance from federal to provincial level in 2010 and the, contradiction between customs, sharia and modern water laws and policies and the disaggregation of water governance in non-water laws and policies has created an incoherent water governance system that does not address the drivers of water problems and hence is unable to change the behaviour of water users.

Pakistan's water laws continue to reflect colonial era priorities (e.g. Canal & Drainage Act 1873) such as market-based principles that promote individual water rights, unilateral control and regulation, which result in income disparities between small and large farmers as well as consolidated centralised water management system. Historically, water infrastructure development including construction of large water reservoirs and dams in the upstream area and water sharing among the provinces continues to be a highly politicised issue despite the 1991 Inter-provincial Water Apportionment Accord, which ensures fixed allocation among provinces through a monitoring mechanism. Due to some ambiguities and room for interpretation in the Accord, it favours the powerful upstream province of Punjab. Inter-provincial water issues have created mistrust and imbalanced power relations among provinces allowing for the inclusion of some groups and the exclusion of others, where socio-economically weak provinces (Khyber Pakhtunkhwa and Balochistan) and the downstream province of Sindh blame the stronger upstream province of Punjab for inequitable water distribution. As a result of the centralised water bureaucracy, colonial legacies and corrupt practices, water governance reforms have been slow where water policies continue to promote conventional irrigation and agricultural practices that favour large farmers as opposed to small land-holders. Similarly, domestic water governance issues are masked by highlighting transboundary water issues with India and Afghanistan (due to security-based foreign policy towards these countries) which further tightens the grip of national government over water policies. The resulting centralised policy orientation – highly influenced by the water security narrative in the context of transboundary issues with Afghanistan and India – promotes construction of large water reservoirs as opposed to addressing traditional water conservation and agricultural practices to tackle inter-provincial water sharing and diffuse power struggles among the provinces.

8.10   INFERENCES

This chapter has described and analysed multilevel freshwater governance in Pakistan. It has done so by looking at (1) how the various characteristics including ESS and drivers of freshwater problems are taken into account at multiple geographic levels in Pakistan; (2) how governance frameworks have evolved at multiple geographic levels in Pakistan; (3) which governance instruments address the drivers of freshwater problems at multiple geographic levels in Pakistan; (4) how legal pluralism can be observed at multiple geographic levels in Pakistan; and (5) how power and institutions influence freshwater governance frameworks at multiple geographic levels in Pakistan. Through these sub-questions, this chapter draws five conclusions.

First, water sharing among provinces in Pakistan is a highly politicised issue and is often employed as an instrument to promote the political interests of different national and local actors. The devolution of power from the federal government to the provinces after the 18th Constitutional Amendment has complicated the governance of natural resources including cases where small provinces (e.g. KP and Balochistan) blame large provinces (Punjab and Sindh) for water theft. This has created mistrust among the provinces despite the 1991 Interprovincial Water Apportionment Accord among the provinces. Historical water infrastructure development and controversies surrounding the construction of large dams contribute to further tensions between local and national actors. Furthermore, with the disclosure of CPEC's Master Plan, the Chinese objectives and concerns in Pakistan are very obvious where, for example, there are plans to lease out thousands of acres of productive agricultural land to a number of Chinese enterprises which may cause displacement and exclusion of small local farmers leading to social and political unrest and further water allocation issues.

Second, since a large population directly and indirectly depends on freshwater related BESS, the severity of water scarcity, poor water governance and climatic changes and variability can directly affect food security, livelihoods and the economy of the country. The diverse climate zones of Pakistan can be significantly altered affecting its flora and fauna. Human development, industrialisation and urbanisation in Pakistan has led to the extinction of many species that were still common at the turn of the century. The Indus River System is vital for Pakistan's water and food security since more than three quarters of Pakistan's population lives in the Kabul-Indus basin. Furthermore, water from the Kabul-Indus Basin irrigates more than 80% of the nation's cropland. Pakistan has recently hit the _water scarce mark' (with current annual per capita water availability of 1017 m3) and according to some estimates Pakistan could _run dry' by 2025.

Third, identification of key direct and indirect drivers of freshwater challenges is the first step towards addressing water challenges in the country. Key direct drivers include: a) increasing water demand for agriculture and industrial practices including commercial agriculture, manufacturing and mining, as well as other water-intensive activities under the China-Pakistan-Economic-Corridor (CPEC); b) water and sanitation needs of the growing population (2% growth rate in 2018) and unsustainable rapid urbanisation. Key indirect drivers of the water problems are related to political dynamics within states, including mistrust and unbalanced power relations among provinces, despite the 1991 Interprovincial Water Apportionment Accord which ensures fixed allocation with a monitoring mechanism among provinces. Transboundary water issues with India (in addition to building dams in Afghanistan) are considered key issues in the national and water security context, which often mask domestic water governance issues. Negligence about the domestic water crisis including climatic and environmental changes and legacies of the colonial laws which are still in practice, support conventional irrigation and water management practices. Demographic, socioeconomic changes, and the impacts of climate change in the Kabul-Indus Basin over the next 25 years are likely to place the Interprovincial Accord under increased scrutiny.

Fourth, the institutional architecture is characterised by legal pluralism because of the continued application of colonial laws, local customs and Sharia, as well as three levels of often different water governance frameworks (national, provincial and local). This has reduced the effectiveness of water policies and their implementation in Pakistan. Moreover, Pakistan's water governance frameworks are influenced by the national water security narratives that are often linked to the national security paradigm promoted by the security establishment. Due to the contradictions and lack of coherence in these plural governance frameworks various direct and indirect drivers are unlikely to influence important actors' behaviour towards achieving the goal of equitable water allocations among provinces and key sectors. Water governance in the pre-colonial era in today's Pakistan was based on local customs and principles of Sharia where the community collectively conserved and managed water resources. Subsequently, some of the principles of local customs (such as _Warabandi') were codified by the colonial administration. In local customs, access to water resources is attached to ownership and size of irrigated land as well as to contributions in the construction and maintenance of infrastructure. The British Colonial Administration introduced market-based principles and instruments based on individual rights for natural resource utilisation including water, which resulted in income disparities between rural and urban populations as well as between small and large farmers. Water laws and other policies created during the colonial administration were mainly to control and regulate water unilaterally and do not serve the needs of current day Pakistan. In addition, the security establishment is a key player in shaping foreign policy development especially with India and Afghanistan, both of which have transboundary water sharing mechanisms with Pakistan.

Fifth and finally, the analysis of the inclusion and distribution of principles and instruments over three major eras and across all water governance frameworks reveals weak linkages between instruments and dimensions of inclusive and sustainable development. The inclusion of economic principles in national water policies in considerable numbers specifies that water should be treated as an economic good. Similarly, normative environmental principles have been adopted in considerable numbers because Pakistan has ratified various Multilateral Environmental Agreements (including the Kyoto Protocol, the United Nations Framework Convention on Climate Change, the Ramsar Convention on Wetlands, the United Nations Convention on Biological Diversity, etc.). However, in terms of addressing environmental issues there is hardly any effective instrument for preservation of ecosystems by endorsing penalties or developing strict regulations or adopting subsides.

9

MULTI-LEVEL INTEGRATED ANALYSIS FOCUSING ON ISSUES FOR RE-DESIGN

9.1   INTRDUCTION

This chapter integrates elements of freshwater governance from the different geographic levels of the KRB and aims to answer the question: How do power and institutions influence multilevel freshwater governance in the KRB and the achievement of inclusive and sustainable development? This chapter also answers some sub-questions: (1) How are various characteristics including biodiversity, ESS and drivers of freshwater problems taken into account at multiple levels of governance in the KRB? (2) How have freshwater governance frameworks evolved at multiple levels of governance in the KRB? (3) Which governance instruments address the drivers of freshwater problems at multiple levels of governance in the KRB? (4) How does legal pluralism occur at multiple levels of governance in the KRB? (5) How do power and institutions influence water sharing at multiple level of governance in the KRB? And (6) How can the current designs of the KRB multilevel institutional architecture become consistent with the key global institutions to achieve inclusive and sustainable development?

To answer these questions, this chapter applies the methodology explained in Chapter 2 and continues as follows. First, it describes the political organisation of multilevel governance in the KRB (9.2), multilevel biodiversity and ESS in the KRB (9.3) and multilevel drivers of freshwater problems (9.4). Second, it assesses the evolution of the multilevel institutional context (9.5) and identifies the relevant goals, principles and instruments (9.6) within these institutions. Third, it explains the instances of legal pluralism (9.7). Fourth, it explores the relationship between drivers and principles/instruments to achieve inclusive and sustainable development (9.8). Fifth, it explains the linkages between power and intuitions as a hybrid approach (9.9). At the end this chapter draws inferences (9.10).

9.2   THE CONTEXT OF MULTILEVEL WATER GOVERNANCE IN THE KRB

General Stanley McChrystal in *The Atlantic Monthly,* 2010 stated: ‚The insurgency is only fundamentally effective in the Pashtun belt - The critical part of the population is where the water and the roads are. People near water are more important economically along the Helmand and Kabul rivers. You secure these areas, and you take the oxygen out of the insurgency'. In Chapters 6, 7 and 8, I have discussed the context of water governance at transboundary level (Chap 6), and national and subnational levels (Chap 7 and 8) in the Kabul River Basin. The context specifies that: (1) the growing water issues between Afghanistan and Pakistan needs to be evaluated in the context of increasing population, urbanisation, industrialisation and climate change rather than only in the context of security and strategic discussions; (2) shortages and poor management of freshwater resources in this region contributes to geopolitical turmoil; and (3) there is no formal bilateral cooperation between Afghanistan and Pakistan on technical information exchange, flow monitoring and water planning, nor on rights and equitable sharing of freshwater. Afghanistan's plans for constructing reservoirs, hydro-power, irrigation, and fishing could ultimately start tensions, particularly given the decades-long unresolved Durand Line issue between Afghanistan and Pakistan (see Chap 6). The border area between Afghanistan and Pakistan is famous for ideology-based extremism, terrorism, and historical tribal conflict. Nevertheless, the growing water issues between Afghanistan and Pakistan is hardly noted, let alone factored into the reasons of local conflict. As the demand for freshwater will grow in the next decades and as climate changes affect precipitation patterns, the pressure on the Kabul River water sharing will increase. This is because the Kabul River supports over seven million people in Afghanistan in addition to more than two million on Pakistan and contributes approximately 26% of Afghanistan's total annual flow (Bokhari et al. 2018; Iqbal 2017; Tariq et al. 2014; Yousaf 2017).

9.3   MULTILEVEL BIODIVERSITY AND ECOSYSTEM SERVICES IN THE KRB

This section analyses the biodiversity and ESS in the KRB in a multilevel institutional context. There is similar ESS and biodiversity which can lead to an enabling environment for transboundary cooperation. In the post-colonial period of Afghanistan, there are seven different multilevel legal frameworks that address different aspects of freshwater-related biodiversity and ESS. However, in Pakistan conservation and protection of biodiversity and different ESS appear in three different legal texts from both the colonial and post-colonial periods. At the transboundary level, the legal texts that address biodiversity and ESS only appeared in the colonial era and no longer apply. The multilevel biodiversity and ESS are elaborated in Table 9.1 and Annex K.

9.3.1   Multilevel Biodiversity in the KRB

In this section I discuss multilevel biodiversity in the KRB. By identifying similarities and differences in biodiversity at multiple levels, it is possible to find common ground for both riparians to collaborate on protection activities that can enhance social and environmental sustainability. For example, the snow leopard is a unique and endangered species for which international efforts are underway for its protection and conservation. Since both Afghanistan and Pakistan are parties to the CBD, joint conservation efforts can be planned and implemented for its protection. Conservation efforts in one country may not be effective if these species travel to the bordering country where they are not protected. Similarly, both countries can work to protect migratory birds as their route crosses both the countries where joint efforts can provide conducive habitats for migratory birds. A number of similar flora species at transboundary level can be protected and promoted by adopting joint efforts in terms of pesticide usage and introduction of suitable crop varieties for the region's ecosystem and environment. Fish species are one of the most important aspects of transboundary rivers since activities in the upper riparian country can significantly affect these species where joint efforts can protect and promote fish species that benefit a large population and economy in both countries.

Table 9.1: Multilevel biodiversity in the KRB

	Transboundary	Afghanistan	Pakistan
Fauna	35 fish species including the endangered Masheer (king of river fish)	Nine local sheep breeds in Afghanistan;eight cattle breeds and seven goat breeds	Snow leopard (Panthera uncia)
	Pintail, shoveller, widgeon, mallard, garganey, tufted and ruddy shelduck, lapwings, herons, egrets, gulls and terns		Indus dolphin (Platanista gangetica)
	Common cranes are occasionally sighted		The migratory and guest waterfowl population
	A number of turtle species along many parts of the river		
Flora		Fuel wood	Plant Kut (Saussurealappa)
		Pistachio/ juniper forests	The coniferous and other rain forests
		Crop plants	Sea grape (Ephedra procera)
		Wheat and other local crops	
Habitat Migratory Routes	Habitat of migratory bird species (e.g. wtarefowl, cranes, and waders		The Indus flyway is considered as the fourth major bird migration route in the world

9.3.2   Multilevel Ecosystem Services in the KRB

As described in Annex H, various types of ESS are relevant at different geographic levels in both countries. These services are dealt with by different institutions (in some cases similar while different in others) and can have implications for transboundary level interactions. For example, in the category of provisioning services, the governance of hydro-energy in Afghanistan is a federal subject whereas in KP Province of Pakistan hydro-energy can be produced and sold by the local and provincial governments to the State (see Annex H). The dissimilarity in this category can have negative consequences for transboundary level interaction because interests at local, provincial and national levels can undermine the institutional cooperation between the two countries. Furthermore, dialogues for a joint hydro-power project between the WAPDA (federal level authority in Pakistan) and Afghanistan's Khost Province in 2006 did not materialise primarily due to the disparities in interests between the two administrative levels of both the countries. Additionally, the multilevel institutional analysis indicates that provisioning services gets much attention in the policy arena since these services are materially more visible and politically charged issues in both countries. In Pakistan's case, the multilevel water governance arrangements also pose a difficult challenge in terms of accommodating the contextual and local issues of different provinces in national level policymaking. When provincial level priorities and ESS (in the case of KP province) are not fully acknowledged and included at national level policymaking, the transboundary cooperation between the two countries might not foreground the importance of various ESS and their benefits.

My analysis (see Chapter 6, 7, and 8) indicates that regulating services usually get less priority and institutional support at national level and fewer financial resources are allocated to these issues. Since these regulating services are generally perceived as regional and international issues, both countries seek and depend on external financial resources as well as technical expertise to address these issues.

Similarities in terms of cultural services are high due to the customary Pakhtun practices as Pakhtun communities follow the Pakhtun Code of Conduct called _Pakhtunwali' whether living in Afghanistan or Pakistan. Since transboundary cooperation occurs at national level, similarities at provincial and local level tend to get ignored, undermining transboundary water cooperation. People to people linkages are stronger between both sides despite dissimilarities at the formal institutional level. In Afghanistan where the Pakhtun are in the majority and politically powerful, their local customs are recognised in national level policies processes and institutional building, while in Pakistan the Pakhtun community does not form the majority and hence the local Pakhtun code of conduct does not inform the national level policy processes.

Supporting services (solid formation, habitat provision and nutrient cycling) are typically natural processes which are seriously affected by human interaction but are generally not prioritised in the policy processes. However, they are the same at the local level, so they can contribute to transboundary level cooperation. In terms of provisioning, cultural and supporting ESS most of the elements are relevant at different geographic levels but are dealt with by similar level institutions. However, since the regulating service (e.g., climate and water regulations) is typically perceived as natural and _international', national governments depend on foreign assistance.

Due to donor interest or funding availability, regulating services are typically discussed at transboundary level (e.g., GLOFs and climate change related events) while politically sensitive issues (e.g., water governance) are seen as important state secrets and issues of sovereignty and national security. In the context of inclusive development, it is important that local level institutional mechanisms (Pakhtun customs, such as on equitable water sharing, water reuse and conflict resolution) inform those national level policy and institutional development processes that are important in transboundary level cooperation. However, this is not the case in Afghanistan where local level Pakhtun customs are an integral part of the constitutional as well as formal institutional mechanism dealing with water and other related issues. This dissimilarity can undermine the transboundary cooperation. Therefore, it is important to ensure participation at local level to meaningfully inform national policy processes.

9.4   THE MULTILEVEL DRIVERS OF CONFLICT

Based on the literature review, I have already identified a range of direct and indirect drivers influencing freshwater governance. Additionally, the analysis of various national and subnational as well as transboundary policy documents and laws and fieldwork in both countries have resulted in the identification of ten drivers within Afghanistan, eight in Pakistan and eight at the transboundary level (see 6.4, 7.4, and 8.4). Multilevel analysis shows that there are four direct and six indirect drivers which are relevant for analysis. These drivers can be linked to both natural and anthropogenic activities. For example, flash floods caused by GLOFs in the HKH region can affect the freshwater quality in the KRB at multiple levels. Similarly, population growth and economic development through industries and agriculture development can depreciate the quality and quantity of freshwater.

Table 9.3 shows that most of the direct and indirect drivers are similar at multiple levels except for the municipal level water supply and sanitation services. This shows that highlighting these drivers and linking them to similar issues of both countries (e.g. agriculture development, industry, economic growth, environmental degradation, unemployment and militancy) can result in common problem framing at the transboundary level where solutions can be discussed based on a shared understanding of the issues, and ultimately feed into policy making processes.

Table 9.2: Multilevel drivers of conflict in the KRB

Direct Drivers
Agriculture development (e.g., commercial agriculture practices including animal husbandry, the extractive sector and water use in energy)
Industry (including services and infrastructure)
Municipal water supply and sanitation services for household uses (drinking water, sanitation, hygiene) and subsistence agriculture
Demographic shifts (i.e., migration, population growth, increase in population density, urbanisation, population growth)
Indirect Drivers
Political dynamics between/within states (e.g. on Durand line)
Culture and ethnic elements (attitudes about access and allocation, wasteful use of resources, etc.)
Non-water-related policies (agriculture & food security, land use, land tenure, economic development; China-Pakistan-Afghanistan economic corridor related projects)
Economy (economic growth)
Poverty
Technological advances (agriculture intensification)
International trade (e.g. ‚globalisation‘ or trade in virtual water)
Natural change and variability in weather, Droughts; Floods; Earthquakes; Landslides, tectonic movement

Bold: Non-common Drivers; **Non-Bold:** Common Drivers

9.5   GOALS, PRINCIPLES AND INSTRUMENTS IN MULTILEVEL GOVERNANCE FRAMEWORKS

9.5.1   Goal of the Multilevel Governance Framework

The multilevel goals in the KRB include: (1) the goals at transboundary level; (2) goals of the Afghan governance frameworks; (3) and goals of the Pakistan governance framework. As no formal regulatory framework exists at transboundary level in the KRB, there are no goals on social and ecological inclusion. However, both countries have accepted the Sustainable Development Goals and hence have agreed to transboundary water collaboration. The Afghan water governance framework imposes the principles enshrined in Article Nine of the Afghanistan's Constitution for conservation, equitable distribution and sustainable use of freshwater resources, support for the national economy and securing water users' rights, in accordance with the principles of Islamic Law and the local customs. Likewise, the objective of Pakistan's water governance framework is to contribute to food security and diminish rising poverty levels by promoting sustainable productivity of freshwater through better management. This indicates that there are some differences in goal setting between the two countries. Pakistan's water goal emphasises increased productivity through better management while Afghan water laws (apparently) foster human rights, equitable distribution and conservation accordingly by incorporating local customs and Sharia. Despite some differences, the goals of both the countries' water governance frameworks are comprehensive and have elements that can support steps for transboundary water cooperation as explained in 9.5.2.

9.5.2   Principles in the Multilevel Governance Framework

Multilevel principles in the freshwater governance frameworks at the transboundary, national, and sub-national levels include Political, Social-relational, and Ecological Principles (see 6.6.2; 7.6.2; 8.6.2).

Table 9.3: Multilevel principles inclusion (denoted by X)

Categories	Principles	T/boundary	Afghanistan	Pakistan
Political Principles	**Exchange of Information**	X	X	X
	Notification of Emergency Situations	-	X	X
	Notification of Planned Measures	-	X	X
	Obligation to Cooperate	X	X	X
	Peaceful Resolution of Disputes	X	X	X
	Limited Territorial Sovereignty/ No Harm	X	X	-
Environmental Principles	Aquifer/basin as the Unit of Management	-	X	X
	BATT	-	-	
	Conjunctive Use	-	X	X
	EIA	-	X	X
	Invasive Species	-	-	X
	Monitoring	-	X	X
	Pollution Prevention	-	X	X
	Precautionary Principle	-	X	X
	Protected Areas for water	-	X	X
	Protected Recharge & Discharge Zones	-	X	X
	Ecosystem Protection & Preservation	X	X	X
	Polluter Pays Principle	-	X	X
Social Principles	**Capacity Building**	X	X	X
	Equitable & Reasonable Use	-	X	X
	Human Right to Water & Sanitation	X		X
	Intergenerational Equity	-	X	X
	Poverty Eradication	X	-	-
	Prior Informed Consent	-	X	X
	Priority of Use	-	X	X
	Public Access to Information	-	X	X
	Public Awareness & Education	-	X	X
	Public Participation	-	X	X
	Rights of Women, Youth, & Indigenous Peoples	-	X	X
	Food Security	-	-	-
	Human Well-being	-	-	-
	Quality Education	-	-	-
	Clean Energy	-	-	-
	Economic Growth	-	-	-
	Infrastructure	-	-	-
	Reduced Inequality	-	-	-
	Sustainable Urbanisation	-	-	-
	Responsible Consumption & Production	-	-	-

Source: Modified from Conti 2017; **Bold:** Common Principles; **Non-Bold:** Non-Common Principles

Although a number of principles are missing at the transboundary level, there are important elements at the national level that can be useful for transboundary level cooperation such as information exchange, peaceful resolution of disputes, obligation to cooperate,, limited territorial sovereignty/ no harm, and ecosystem protection and preservation, among others. Pakistan's water frameworks include almost all of the four categories of principles while Afghan water laws and policies do not cover some important principles such as BATT, invasive species, the human right to water and sanitation and a number of social principles (food security, human well-being, education, economic growth and inequality etc.). As can be observed in Table 9.4, most of the principles are not present at all levels of governance except for the exchange of information, obligation to cooperate, ecosystem protection and capacity building. In this direction, the UN Watercourses Convention can offer support by addressing legal weaknesses, providing guidance for policy coherence, facilitate the work of bilateral and multilateral institutions in promoting transboundary cooperation by creating an impartial level playing ground among riparian states, and integrate social and ecological concerns into the management and development of transboundary watercourses. I would recommend ratifying the Watercourses Convention as a first step to resolving water sharing issues and considering the ratification of the UNECE Water Law – as that may enable a common understanding of the ecological principles and instruments.

9.5.3   Instruments in the Multilevel Governance Framework

Instruments in the multilevel governance frameworks include instruments from the transboundary normative frameworks; instruments of the 2009 Water Law in Afghanistan; and the 2018 National Water Policy in Pakistan. As there is no regulatory mechanism at transboundary level (see 6.6.3), there is no instrument in the existing framework. However, the 2009 Afghan Water law has some regulatory, economic, suasive and management instruments which can prevent pollution from agriculture and industries as well as meet the growing demand of population and cities (see 7.6.2). Similarly, the recently approved National Water Policy of Pakistan also has instruments in all four categories which can address the drivers and change the behaviour of some relevant and non-relevant actors towards the sustainable use of freshwater resources (see 8.6.2).

Table 9.4: Multilevel instruments inclusion/exclusion

Categories	Instruments	Transboundary	Afghanistan	Pakistan	Multilevel
Regulatory	Permit		x		
	Procedures				
	Penalties		x		
	EIA			x	
	Fines		x	x	
	Licences			x	
	Metering			x	
	Mapping and zoning			x	
Economic	Property rights				
	Taxes				
	Tradable quotas				
	Tariffs		x		
	Subsidies			x	
	Grants				
Suasive	Education				
	Awareness trainings			x	
	Award schemes				
	Disclosure requirements				
Management	Self-regulation		x	x	x
	Voluntary management processes		x	x	x

Source: Modified from Conti 2017

My analysis shows that none of the instruments are present at the transboundary level due to the absence of a treaty. Afghanistan's water frameworks comprise only a few regulatory instruments (permits, penalties and fines), one economic instrument and two management instruments without any suasive mechanisms for awareness creation. However, Pakistan's water policies and legal frameworks are more comprehensive and a cover a range of all the four types of instruments. At the multilevel only two management instruments are present.

9.6   THE MULTILEVEL LEGAL PLURALISM ANALYSIS

The legal pluralism analysis of the transboundary and national governance framework is discussed in detail in Sections 6.7, 7.7 and 8.7). In Pakistan's case, there are three levels of water governance (national, provincial and local) as well as three types of legal frameworks (local customs, Sharia and state laws) that further complicate water governance and reduce the effectiveness of water policies and their implementation. Similarly, there are different laws for water, energy, land, food and agriculture that treat water differently, which in turn reduces their effectiveness and implementation. Despite the presence of formal legal frameworks for water, the local customs and Sharia laws are historically grounded in local contexts which undermines some elements within the formal laws that contradict the local customs. At the transboundary level, there are only three political, three socio-relational and one ecologically normative principles based on colonial and existing practices which are unlikely to address the direct and indirect drivers of freshwater problems within the KRB. It can be concluded that there are only a few normative principles and colonial era treaties which may not be sufficient for transboundary cooperation at multiple levels, however, global institutions (e.g., 1997 UN WCC) include a number of effective instruments that can provide guidance where both countries can start dialogues for cooperation.

9.7   APPLICABILITY OF ARTICLE 5 AND 6 FOR ENHANCING MULTILEVEL WATER GOVERNANCE IN THE KRB

In line with addressing the issues of quality, quantity and climate change in the KRB, the UNWCC, especially Articles 5 & 6 on equitable and reasonable utilisation, have the potential to enhance cooperation in transboundary water issues. For example, adapting similar principles can bring harmony in the national level legal and policy frameworks that can improve future cooperation. Both countries will also benefit when their sub-national and national level legal and policy frameworks are harmonised on similar principles of equitable and reasonable use. Similarly, Article 6 on considering relevant factors is helpful in identifying the similarity of drivers of water issues in both countries of the KRB. When similar drivers of water issues (e.g. agriculture development, industry, economic growth, environmental degradation, unemployment and militancy) are identified, it can help in common problem framing and pave the way for mutual strategies and policies in addressing them through, for example, cooperation and information exchange to address these drivers.

Article 6 on factors relevant for equitable and reasonable utilisation is of particular relevance given the similarity of identified drivers of freshwater problems in the KRB. For example, due to lack of cooperation, no information is shared between Afghanistan and Pakistan in terms of population growth, urbanisation, the areas under cultivation, nor other changes along the river and catchment areas, especially in times of crisis and disasters caused by climate change and environmental variabilities such as the floods of 2010 and GLOFs in the region. The 2010-2013 monsoon floods were massive and unprecedented, killing hundreds of people, affecting land area and millions of people, that caused losses of billions of dollars because of damages to infrastructure, agriculture and livestock, housing, and other family assets in both Afghanistan and Pakistan. Human lives, infrastructure, economy and livelihoods could have been protected on both sides of the border if there had been an effective information sharing mechanism in the basin. The floods and GLOFs also damage the already weak irrigation and other related infrastructure due to the non-exchange of information.

Currently, donor organisations in both countries work individually on irrigation improvement where the lack of information exchange and cooperation can cause damage due to disasters (such as floods and GLOFs). These losses can be minimised and it will save their investment by enhancing transboundary cooperation and information exchange especially in times of disasters. Moreover, cooperation on the KRB can enhance the bargaining power of both countries with donor organisations to invest in beneficial infrastructure projects for enhancing water use efficiency and utilisation. Since climate change has regional implications, including for the KRB, transboundary cooperation is urgently needed. The formation of a River Basin Organisation (RBO) can directly contribute towards institutional strengthening that can last beyond the short-term political priorities of different parties that come into power in both countries. Despite some problems, the Indus Water Treaty and the Indus Water Commission is one such example that has withstood some serious and longstanding conflicts between India and Pakistan for over five decades now. Independent transboundary institution on the KRB may be able to endure the political pressure of the ruling governments to contribute towards long-lasting water cooperation in the region. These kinds of institutions can be strengthened through sustained support for capacity building by the donor countries that have strategic interest in the region for reducing militancy and promoting peace and stability through dialogues and cooperation.

As highlighted in chapters 7 (Afghanistan) and chapter 8 (Pakistan), water related biodiversity and ecosystem services (ESS) are vital for survival of millions of people on both sides of the border in the KRB region. In this direction, the principles of equitable and reasonable water utilisation can be translated into allocating sufficient water for protection and sustainability of these ESS. The importance of the ESS can be highlighted by application of reasonable water use principle for ESS, which sometimes is considered as waste of water. For example, a large number of people depend on forests and related biodiversity for their livelihoods which are directly dependent on sustained water supplies in the KRB, which can be enhanced by applying the principles of equitable and reasonable utilisation.

9.8   INFERENCES

This chapter has integrated elements of freshwater governance at various geographic levels of the KRB in order to answer the question of how power and institutions influence multilevel freshwater governance in the KRB to facilitate the achievement of inclusive and sustainable development. It has done so by looking at (1) how various characteristics including biodiversity, ESS and drivers of freshwater problems are taken into account at multiple governance levels in the KRB; (2) how freshwater governance frameworks have evolved at multiple levels of governance in the KRB; (3) which governance instruments address the drivers of freshwater problems at multiple levels of governance in the KRB; (4) how legal pluralism can be observed at multiple levels of governance in the KRB; (5) how power and institutions influence water sharing at multiple governance levels in the KRB; and (6) how the current designs of the KRB multilevel institutional architecture can become consistent with the key global institutions to achieve inclusive and sustainable development. Through answering these sub-questions, the chapter draws four conclusions.

First, due to four decades of conflict in the KRB, the ideological-based insurgencies have seriously influenced the foreign policies of Afghanistan and Pakistan. These long-standing border disputes – such as rejection of Durand Line by Afghanistan as an internationally recognised border, Taliban proxies supported by Pakistan, and use of extremist ideologies by both the countries to destabilise each other – restrict both countries in initiating dialogues and solving various bilateral issues including transboundary water issues. Currently cooperation over transboundary water in the region is minimal due to power asymmetries between Afghanistan and Pakistan. Water issues are seen through the lens of territorial sovereignty where water data is treated as state secrets prohibiting information sharing. Pakistan, being a hydro-hegemon in this case can use its powerful position to initiate dialogue for transboundary water cooperation, also by involving international players.

Second, since both Afghanistan and Pakistan are signatories of many international environmental conventions and treaties (e.g. SDGs, CBD, Ramsar, HRWS), the BESS based approaches can provide an enabling environment and common ground for cost-effective transboundary cooperation including water. My analysis shows that the hydro-energy (provisioning service) is governed at different levels in Afghanistan (federal) and Pakistan (provincial & local) can have negative consequences for transboundary level interaction since interests and administrative issues at different levels can undermine transboundary water cooperation. Therefore, new knowledge and evidence by applying the valuation of ESS can also inform the policy narrative of transboundary water cooperation by highlighting the win-win scenarios.

Third, highlighting the anthropogenic and natural drivers and linking them to similar issues of both the countries (e.g. agriculture development, industry, economic growth, environmental degradation, unemployment and militancy) can result in common problem framing at the transboundary level where solutions can be discussed at a similar understanding of issues, and ultimately feed into policy making processes. Moreover, other large regional projects (e.g. CPEC, TAPI)[115] can potentially create an opportunity for powerful actors and donor countries to play their role in bringing stability and cooperation in the KRB which can protect their long-term investments in the region. This can ultimately lead to creating an enabling environment for cooperation including transboundary water issues.

Fourth, as no formal regulatory framework exists at transboundary level in the KRB, there are no goals on social and ecological inclusion. Pakistan's water goals, principles and instruments are mostly based on local priorities while Afghanistan's are heavily influenced by the donors and have some common elements with the global instruments. Pakistan's water goals emphasise increased productivity through better management while Afghan water laws (apparently) foster human rights, equitable distribution and conservation accordingly by incorporating local customs and Sharia. In this scenario, the UNWC can offer support by addressing the weak legal aspects, provide guidance for policy coherence, and facilitate the work of bilateral and multilateral institutions to foster transboundary cooperation by establishing a level playing field among riparian states, and incorporate social and environmental aspects for the management and development of international water resources.

---

[115] Turkmenistan–Afghanistan–Pakistan–India (TAPI) gas pipeline project is a natural gas pipeline project connecting the four countries. The project is co-funded and jointly developed by the Asian Development Bank (ADB) and the Galkynysh Pipeline Company Limited (Joshi 2011).

10
CONCLUSION

10.1   INTRODUCTION

This chapter brings together findings and analysis of chapters 1-9 in terms of addressing the main research question and sub questions. Overall, this thesis presents a multilevel institutional analysis of transboundary river basins. It has taken an integrated approach to assimilate the hydro-geological characteristics and ESS knowledge into current understandings of freshwater governance. It has identified the important role of both water and non-water related relevant actors as well as patterns in the freshwater governance frameworks. It positions inclusive and sustainable development as the guiding norm and highlighting the importance of multilevel institutional frameworks in order to deal with power politics and draws conclusions about how existing freshwater governance frameworks may be further improved using the case study of the transboundary Kabul River Basin.

10.2   RECALLING THE QUESTIONS

Based on the numerous existing and potential challenges concerning transboundary freshwater resources and their linkages to the inclusive and sustainable development of the populations, it addresses the question: _How can regional hydro-politics and institutions be transformed at multiple levels of governance through inclusive development objectives and incorporate the relationships with non-water sectors in addressing issues of water quality, quantity and climate change?' In order to answer this question, four sub-questions were developed: 1) How can the concept of biodiversity and ESS be incorporated in a framework to analyse the effectiveness of institutions, and the role of power, in governing transboundary water resources? 2) Which principles and instruments address the causes/drivers of freshwater problems in transboundary river basins at multiple geographic levels? 3) How does legal pluralism affect transboundary water cooperation? 4) How do power politics and institutions influence water governance in transboundary river basins at multiple geographic levels? These questions are explored with special reference to the Kabul River that flows through Afghanistan and Pakistan.

To respond to these questions, I adapted Oran Young's institutional analysis model to accommodate the key concepts of my thesis (see 2.4). In this framework, it is first important to comprehend the context and the driving forces which lead to lack of cooperation in the Kabul River Basin. I looked at how power has shaped the existing transboundary, national and local institutions in the KRB. I then identified the key instruments that aim to change the behaviour of actors. Then I analysed whether these instruments have the potential to change actors' behaviour, given the context and drivers in such a way as to ensure social and ecological inclusion and alter relational issues. Finally, based on an assessment of which principles and instruments work and which do not (in terms of addressing contextual challenges and the drivers; and in mobilising changed behaviour in actors), I suggest some recommendations for redesign of institutions regarding how the institutional approach can be improved and discuss whether this can change the win-set which might influence the underlying power politics and therefore lead to the development of mutually satisfactory conclusions. The entire conceptual framework is applied at multiple levels of governance focusing on the transboundary, national, provincial, and local levels as well as the relationship among these levels (see 2.4.1). The four sub-questions are designed to explain the above-mentioned transboundary water issues in the KRB based on my theoretical framework.

10.3   CONCLUSIONS AND RECOMMENDATIONS

The overarching question is: how can regional hydro politics and institutions be transformed through inclusive development approaches? This is now answered through an integrated set of seven conclusions as listed and explained in Table 10.1.

Table 10.1: Integrated conclusions and recommendations

	No.	Conclusions	Recommendations
Differences	1	Current cooperation: This is frozen as both countries use sovereignty approaches on water. This is also because both countries define their relationship in terms of security and strategic issues and ignore the water related issues. As a consequence, data and information on the river is also securitised and secret.	Water collaboration could provide gains to both countries. Pakistan, as the regional hegemon, could invest in and promote win-win collaboration with Afghanistan (see recommendations below). Establishing a river basin organisation is critical as a first step.
	2	Border dispute: The contested Durand line prevents water collaboration. However, the Pakhtun population living on both sides of the Durand line have similar water related customs. Discussing water sharing in line with the Watercourses Convention may be counter-productive as the border itself is disputed.	However, one could perhaps address the water problems without waiting for the Durand line problem to be solved by using Pakhtun customs to develop common water strategies in the contested border areas. Pakistan needs to take the initiative.
Similarities	3	Biodiversity and Ecosystem Services: Although there are differences (see Table 9.1 and Annex K), there are more similarities in recognising the huge social and economic value of protecting these services. In particular, if the water level falls too much, salt water intrusion can destroy agricultural land in coastal Pakistan.	Recognise the social and economic value of BESS and see if a joint collaborative approach can be more cost-effective for both.
	4	Drivers: Both countries have similar drivers. At municipal level the drivers are also similar (see Table 9.3). Furthermore, both countries face the problem that non-water related policies dictate water use and pollution. The role of China as an investor in trade routes to Pakistan and Afghanistan can also be a major driver of water use and pollution.	Pooling knowledge and resources to address common drivers can be cost-effective. Developing an agricultural, industrial and trade policy that takes water limits into account is critical for the long-term sustainability of development policy.
	5	Address contradictions: The principles (see Table 9.4) and instruments (see Table 9.5) at national level show that there is sometimes considerable consistency. Differences exist in power as water governance is centralised in Afghanistan and is devolved to state level in Pakistan making e.g. collaboration on dams difficult.	Tables 9.4 and 9.5 shows the common principles and instruments at national level that could be included in a transboundary collaborative instrument. Both countries should address domestic policy contradictions; and develop appropriate policy mixes.
	6	Resource limitation: The research reveals that resource limitations seriously hamper the operationalisation and implementation of policy, its monitoring and enforcement. At the same time, instruments that limit the potential for corruption are often deliberately sabotaged by political actors. This leads to a focus on a short-term focus on economic growth and dependence on aid agencies.	A socially and ecologically inclusive system will be sustainable in the long-term. A focus on short-term economic growth will lead to externalisation of social and ecological impacts with long-term impact on security and livelihoods. Seeking out collaborative, locally developed, cost-effective solutions is critical for enhancing livelihoods and wellbeing.
	7	Knowledge and Dialogue: There is an absence of collective and integrated knowledge based on data and experiences as well as dialogue at different levels of cross-border governance to be able to craft and refine each of the above recommendations in more detail.	The need for mobilising cross-border knowledge generation in schools, universities and life-long learning institutions and dialogue between civil society and governments is critical to address the long-term problems of water sharing.

10.3.1 Defrost frozen collaboration for inclusive development

On the basis of research in this thesis, I conclude, first, that current cooperation is frozen as both countries see their own interests as distinct and adopt absolute sovereignty/territorial integrity approaches which prevents transboundary collaboration. This is also because both countries define their relationship in terms of security and strategic issues and ignore the water related issues. As a consequence, data and information on the river is also securitised and secret. Regional hydro-politics is mostly ignored and over-shadowed by the security and strategic discussions between Pakistan and Afghanistan. This leads to secrecy with respect to knowledge about water resources, uses and users.

However, there is enough evidence to recommend that water collaboration could provide the basis for security and strategic collaboration. There is more to gain from transboundary collaboration than conflict. As Pakistan is downstream, more powerful, and has more to gain from collaboration, it is thus up to Pakistan to draw inspiration from hegemonic stability theory to invest in and promote win-win collaboration with Afghanistan, based on mutual trust. Pakistan is the regional hegemon because of its political strength (despite some issues, it is a democratic country with successive elected governments); its military strength (being the 17th largest military strength in the world and nuclear capability), its active participation in the international arena (i.e. in the UN, South Asia Association for Regional Cooperation, and Shanghai Cooperation); its ability to make laws and policies; its growing economic power and enormous growth; its superior geography, technological innovation, ideology, abundant resources, and other factors (Yilmaz 2010; Liu 2011). Being a land-locked country, all the trade and supplies for Afghanistan as well as international forces and NATO are routed through Pakistan, which gives it enormous power to bargain and negotiate trade terms.

Such win-win collaboration can build upon their similarities (see below) and facilitate establishing a river basin organisation to identify and refine such similarities, as a first step to a transboundary agreement, is critical. Clearly Pakistan as the regional hegemon will have to invest more in this relationship, both financially and intellectually, to make it attractive to Afghanistan to join. It will have to demonstrate that such collaboration will lead to social, ecological and political inclusion.

10.3.2 Bypass border dispute for inclusive development

Second, the border dispute hampers collaboration. Water issues between Afghanistan and Pakistan are politically charged issues as Afghanistan rejects the Durand Line as an internationally recognised border which allows Pakistan to claim the water flows from the Hindukush Mountains of Pakistan. Afghanistan argues that the source of the Kunar River which originates in Pakistan actually belongs to them. The question is whether cooperation on water issues should be postponed until after the dispute over the border is settled. However, one could perhaps address the water problems without waiting for the Durand line problem to be solved. On both sides of the contested border area, Pakhtun people live and use the water in accordance with their customs on equitable water sharing, water reuse and conflict resolution. However, while on the Afghanistan side these customs are recognised in the Constitution and dominant, on the Pakistani sides these customs are not formally included in national water policy as these people have less representation. This dissimilarity can undermine transboundary cooperation. It may be wise of the Pakistani government to see to what extent it can use Pakhtun codes of conduct in the collaboration in this contested region as a way of pacifying its own people and seeking common ground with Afghanistan in resolving transboundary issues. Principles of social and relational inclusion including participation could be of key relevance here.

10.3.3 Use biodiversity and ecosystem services (BES) for inclusive development

Third, the scholarly literature on biodiversity and ecosystem services (BES) shows that it is worth many times the global GDP. Although it is difficult to calculate these services, and may not always be sensible, it does convey the message that it is extremely short-sighted to ignore these services. The research in this thesis shows that through centuries, the Kabul-Indus Basin (KIB) has provided ecosystem services which the people of these two countries have benefitted from. My analysis of BES at sub-national, national and transboundary levels shows the centrality of these services for a large population as _GDP of the poor'. Degradation of water related ESS in the KRB can directly affect these marginalised groups of people. On the other hand, conservation of these services can empower them by allowing the needed environmental flows agreed by the two riparian countries (such as in the case of IRSA which allows minimum flow of 10 MAF into the Indus Delta). However, the thesis shows that currently the water quality and quantity in the Kabul River Basin is being badly affected, modifying the quality of the ecosystem services which therefore affects human lives, livelihoods and biodiversity. These have been analysed in great detail in the foregoing chapters. However, although there are differences (see Table 9.1 and Annex K), there are more similarities in recognising the enormous social and economic value of protecting these services. Furthermore, protection and conservation of the snow leopard, markhor, migratory birds (Syberian Ducks), and fish species (particularly the indigionious Mahasheer) can only be planned through joint efforts under the Convention on Biological Diversity and using efforts at glacier protection. Reducing pollution of the river in upstream areas in both countries can protect and promote animal and fish species that indirectly or directly benefit a large population and economy in both countries. While both countries are exploiting the provisioning uses of water since these services are materially more visible and politically charged issues, they do so at the cost of the regulating, supportive and cultural uses of water, if the latter three are not protected the provisioning services will be reduced. In particular, if the water level falls too much, salt water intrusion can destroy agricultural land in coastal Pakistan. If the water is polluted, fish will be affected. There is enough evidence in the scholarly literature to show that protecting biodiversity and ecosystem services (BES) can enhance the livelihood prospects of people along the basin while increasing economic development. Given that both countries are sometimes upstream and sometimes downstream, they could draw inspiration from the experience of the Boundary Waters Agreement between the US and Canada and develop a common agenda on biodiversity and ecosystem services (BES). In order to implement this, they could adopt the ecological principles that emerge from, in particular the UNECE Water Law (see 5.3.2) and ensure that the regulating, economic and management instruments are in line with these principles. In order to influence policy makers, further detailed socio-ecological information (economic valuation) of biodiversity and ecosystems services (BES) is needed in future research projects to support local to transboundary water negotiations, ecological inclusiveness and enable a more cost-effective and sustainable approach for both countries.

10.3.4 Address Drivers for inclusive development

Fourth, the above-mentioned ESS and biodiversity of freshwater are affected by a number of direct and indirect drivers. Both countries have similar drivers although there are differences in nuance. Direct drivers of poor water quality include agriculture, industry, domestic water use and demographic shifts. Indirect drivers are political issues between and within states, culture and ethnic elements, non-water-related policies including economic growth and poverty reduction policies, the economic motivation of local industries, poverty of the local people, technological advances, international trade, climate variability and change and other natural factors.

In Afghanistan's case, the drivers of water problems are agriculture; industrial development including mining and manufacturing industries; demographic shifts; increasing demand for clean drinking water and improved sanitation; and natural changes due to climate and weather variability such as flooding caused by rivers overflowing and glacial lake outbursts (GLOFs) in the western region and central belt, while drought in the southwest and northern regions have put farmers out of work and degraded water quality. The indirect drivers of water conflict in Afghanistan are: (a) political drivers where (i) the state's weak regulation encourages privately owned water pumping stations and other actors to control the flow and consumption and provide safe drinking water to approximately 80% of the population, (ii) four decades of war has wrecked the water-carrying infrastructure while the government has also held back plans to build new infrastructure, as security challenges are disincentive to building infrastructure; (iii) solving Afghanistan's water crisis could cost as much as USD $11 Billion – money that is scarcely available; (b) social drivers including poverty where about half of the population has access to clean drinking water, and just about 35% use improved sanitation facilities; (c) economic drivers including national economic growth, consumption and pollution from agriculture and industry including mining and manufacturing, and the interest of private sector; (d) cultural drivers including wasteful behaviour towards water consumption and pollution, low technical knowledge, and low education which puts Afghanistan at the 169 position on the Human Development Index (HDI).

Direct drivers of freshwater issues in Pakistan's case include: increasing water demand for agriculture and industrial practices including commercial agriculture, manufacturing and mining, as well as other intensive activities under the indirect driver of China-Pakistan-Economic-Corridor (CPEC); water and sanitation needs of the growing population (2% growth rate in 2018) and unsustainable rapid urbanisation which is further increasing water demand and affecting water quality. Similarly, the key indirect drivers of the water problems in Pakistan include political dynamics within states, i.e. mistrust and imbalanced power relations among provinces despite the 1991 interprovincial water apportionment treaty which ensures fixed allocation with a monitoring mechanism among provinces. In the same vein, transboundary water issues with India (in addition to building dams in Afghanistan) are considered key issues in the national and water security context which often mask domestic water governance issues such as: negligence about the domestic water crisis including climatic and environmental changes; legacies of the colonial laws which are still in practice and support conventional irrigation and water management practices; plural legal systems governing water through informal and formal laws including weak regulation for water allocation and quality control; and growing poverty.

The key direct drivers of KRB conflict are: agriculture development; industry; municipal water supply and sanitation services and demographic shifts. Similarly, the indirect drivers include: political dynamics between states; culture and ethnic elements, non-water-related policies, economy (economic growth, poverty, technological advances, international trade and natural change and variability in weather), droughts; floods; earthquakes; landslides, and tectonic movement. The political context of extremism, Taliban proxies, and Pakhtun nationalism inhibit collaboration. Moreover, the Hindukush region is prone to earthquakes and climate variability which can also indirectly influence freshwater resources. Furthermore, both countries face the problem that non-water related policies dictate water use and pollution. The role of China as an investor in trade routes to Pakistan and Afghanistan can also be a major driver of water use and pollution.

Given the similarities in (a) the need to address drivers; and (b) the nature of the drivers, both countries could pool knowledge and resources to address common drivers as this can be cost-effective. Addressing multilevel drivers can lead to effective transboundary water governance. For example, continuous flooding (a common phenomenon in both countries) has recently resulted in meaningful dialogues, since both the countries have equally suffered. Similarly Afghanistan and Pakistan have been facing acute energy deficiencies over the last two decades which has resulted in increased poverty and unemployment levels. Due to these similar issues, both countries have prioritised hydro-energy development in negotiations at all levels. Furthermore, increased poverty, lack of economic opportunities and education, and a weak institutional setup has contributed to various social issues such as radicalisation, extremism, and militancy (especially in conflict areas such as Afghanistan and Pakistan). Both Afghanistan and Pakistan may face an increased level of conflict, enhanced transaction costs, lack of cooperation, deteriorated natural environment, and lack of foreign investment, if relevant drivers at multiple levels are not addressed through the institutional context in order to achieve inclusive and sustainable development. This shows that highlighting these drivers and linking them to similar issues of both the countries (e.g. agriculture development, industry, economic growth, environmental degradation, unemployment and militancy) can result in common problem framing at the transboundary level where solutions can be discussed with a shared understanding of issues, and ultimately feed into policy making processes. Finally, a key problem in both Afghanistan and Pakistan are agricultural, industrial and mining policies as these are seen as critical to economic growth. However, if such policies are to be sustainably implemented they need to draw inspiration from the Sustainable Development Goals. They should be managed within a clear water budget and should climate-proof the water use strategy for these sectors. This is another common substantive issue that both countries could benefit by engaging in multi-level dialogue.

10.3.5 Remove contradictions in the policy environment to promote inclusive development

Fifth, the principles (see Table 9.4) and instruments (see Table 9.5) at national level show that there is sometimes considerable consistency which can be scaled up to the transboundary level. However, there are four key policy contradictions which stand out in this thesis. (a) While, on the one hand, the Sharia recognises equitable use of water and the need to respect water and treat it as a gift of God, local practices often violate these principles. (b) Furthermore, while the vast majority of people are quite poor in both countries, the principle of cost recovery is pushed by donors in relation to meeting the human right to water and sanitation. For example, in Afghanistan, although water is described as a human right, it is clearly considered as an economic good through the development of a number of policy instruments. This is problematic. (c) Third, differences in governance of provisioning services (e.g. hydropower at federal level, see Annex H) can have negative consequences for transboundary level interaction because interests at local, provincial, and national levels can undermine the institutional cooperation between the two countries. (d) Finally, policies in both countries are sometimes not supported by other policies which undermines their effectiveness. For example, Afghanistan's water governance policies (e.g. Water Law 2009) includes some functions (construction, development, rehabilitation, protection and monitoring of irrigation infrastructure) including agricultural activities to mitigate flood risks (indirect driver) and protect and maintain the environment. However, these measures are not supported by any regulatory or economic instruments such as subsidies. Although the policy recommends establishing River Basin Councils that should include representatives of water users, and relevant federal and local departments of the line ministries in the river basin, the provinces are not empowered and all matters fall directly under the administrative authority of the Federal Ministry of Energy and Water (MEW). Furthermore, in terms of addressing the indirect drivers of natural change and variability (e.g. flood and drought forecasting through data collection and analysis), normative management principles are applied without any effective policy instruments. Although it can improve equitable distribution at the canal level, the permit system is likely to provide benefits to the existing water bureaucracy by providing avenues for financial benefits and control over water recourses decision making (Lee 2006). In terms of non-water related laws, the other laws provide principles and instruments but many have not been updated and calibrated (e.g. the fines are set too low to be effective) to address modern problems or are simply not implemented. Moreover, post-colonial economic instruments often subsidise water extraction for agriculture and industry rather than promoting their sustainable use; and end up supporting the larger users at the cost of the smaller users exacerbating existing inequalities. Finally, instruments to share water between provinces such as the Indus Water Apportionment Accord in Pakistan are not fully implemented: e.g. the objective telemetry system which would help to monitor the sharing is the subject of controversy as many political actors do not wish it to become successful as it would affect their role in influencing water policy. There are serious issues among the four provinces over the Kalabagh Dam, where the Punjab province favours its construction and Khyber Pakhtunkhwa (formerly known as North-West-Frontier-Province or

NWFP), and Balochistan and Sindh are against the propoject for various reasons. The province of Khyber Pakhtunkhwa (KP) is concerned about the loss of fertile land and displacement of local population in the fertile Peshawar, Nowshera, and Charsadda Valley, and the potential damages due to earthquakes. Moreover, Balochistan is concerned about its share of available water resources and fears that the dam might strengthen historic levels of control by the Punjab province. Sindh's concerns are about the diminished water supplies reaching the province as it is the last downstream country with a fragile ecosystem of Indus delta. For the improvement of the Accord, it has been suggested that: i) water audits prepared by the IRSA should be made publicly available, ii) the terminology should be properly defined to avoid differing interpretations, iii) financial penalties for violation of the Principles of Accord should be adopted, and iv) a third party independent water auditor should be engaged to audit the national water resources, with the broader objective of increasing trust and credibility in the data and information released by the IRSA to stakeholders (Anwar and Bhatti 2017).

However, the analysis of global institutions (Chapter 5) shows that there are a number of relevant instruments (e.g. in UNWC and UNECE) to address a majority of the identified drivers, and can provide building blocks and guidelines for working towards a cooperation mechanism within the KRB. Moreover, the unaddressed principles are covered by the SDGs, which are universal, if both the countries are committed towards achieving these goals based on their national development priorities. The SDGs and other global institutions can provide an effective basis for working towards a transboundary water sharing mechanism since the SDGs have a specific water related goal (Goal 6) with a target on transboundary water sharing (target 6.5). Although the SDGs are voluntary and not binding, less developed countries have incentives to achieve these targets with the support from international cooperation. The SDGs and global water law instruments (e.g. in UNWC and UNECE) can inspire the design of a treaty for transboundary cooperation to achieve the goals of inclusive and sustainable development. Article 5 & 6 of the UNWC on equitable and reasonable water use is specifically relevant for water cooperation in the KRB (see 10.7 on redesign). Since most of the existing policy instruments do not address the identified drivers, my analysis shows that these policies were based on colonial legacies, ignored realities on the ground (non-participatory policy formulation mechanisms) and are supported by vested interests that support the continuation of the status quo.

10.3.6 Combat resource limits and dependence by promoting collaboration on long-term cost effective solutions for inclusive development

The research reveals that resource limititations seriously hamper the operationalisation and implementation of policy, its monitoring and enforcement (e.g. the use of cost-recovery on water and sanitation, and subsidies for water withdrawal for productive purposes). At the same time, instruments that limit the potential for corruption are often deliberately sabotaged by political actors (e.g. the discussions around the telemetry system). This leads to a focus on a short-term political focus on economic growth which is also promoted by, and results with, a dependence on aid agencies. However, instead of an attempt at rapid economic growth, a socially and ecologically inclusive system aiming at well-being will be more sustainable in the long-term. A focus on short-term economic growth will lead to externalisation of social and ecological impacts with long-term negative impacts on security and livelihoods. Seeking out collaborative, transboundary and locally developed, cost-effective solutions is critical for enhancing livelihoods and wellbeing.

10.3.7 Knowledge and dialogue on inclusive development

Finally, the research reveals that there is inadequate information about the transboundary water system, which is shrouded in secrecy, and inadequate local assessments of the kinds of knowledge needed to address the transboundary water system. This thesis has made an effort to integrate the secondary information and to generate primary information to establish a base line of information and pathways to inclusive development and the achievement of the Sustainable Development Goals. However, this needs to be supplemented by much more data collection and integrated knowledge as well as dialogue at different levels of cross-border governance to be able to craft and refine each of the above recommendations in more detail. The need for mobilising cross-border knowledge generation in schools, universities and life-long learning institutions and dialogue between civil society and governments is critical to address the long-term problems of water sharing. Such knowledge is also necessary to create the necessary arguments to convince the politicians and decision-makers to take the water challenge seriously and to find indigenous ways of addressing it. Such knowledge could be used to empower the hydro hegemon on the Kabul River to promote an equitable and sustainable knowledge based solution to Afghanistan and to promote an equitable solution within Pakistan building on the best of Sharia law (i.e. its focus on equity and community) and the latest scholarly information. After all, equitable and reasonable use has the potential to reconcile conflicting interests in multilevel and transboundary issues (ILA 2001).

10.4   CAN PAKISTAN USE HEGEMONIC STABILITY THEORY TO PROMOTE WATER RELATED PEACE?

Hegemonic stability theory argues that a hegemon can use its power and resources to shape institutional design in such a way that it leads to a win-win situation for the countries involved. I argue that by using its hegemonic character, Pakistan has the ability to promote peace in the region by trading in some of its existing advantages for long-term cooperation and sustainable water resources development. I have argued above that Pakistan could consider the Pakhtun customs in shaping transboundary cooperation which could meet some of Afghanistan's concerns; that it could promote collaboration on understanding how biodiversity and ecosystem services can be protected for the benefit of both countries; that agricultural, industrial and mining policy could be made climate-proof and be undertaken with a water budget and related constraints in order to prevent future problems; that multi-level policy contradictions such as between Islamic, customary and international law principles could perhaps be jointly resolved; that a better understanding of the common elements of Islamic Law and the principles of equitable and optimal utilisation of water resources may enable both countries to actually begin a discussion of water sharing; that existing and new river basin organisation and dispute resolution mechanisms could be encouraged as a way of managing river basin issues on a continuous basis. All of these could be part of the ideas that the Pakistani government uses to promote greater trust and collaboration with Afghanistan. But it would require Pakistan to hold out enough incentives for Afghanistan to find its worth while joining such a discussion. This would require Pakistan to distance itself from the absolute territorial integrity doctrine and convince Afghanistan to waive it as well. It would require Pakistan and Afghanistan to draw inspiration from the Sustainable Development Goals and to see if they can manage their water resources within ecosystemic limits while meeting the needs of their society and promoting human well-being. It might require that both countries focus more on inclusive development than economic growth as the latter may have long-term social and ecological costs for society.

REFERENCES

References

Aamir, A.: Mehrgarh: The Hidden Archeological Treasure – Balochistan Point, 24th July [online] Available from: http://thebalochistanpoint.com/mehrgarh-the-hidden-archeological-treasure/ (Accessed 17 August 2018), 2015.

Abbas, H., Hussain, A., Shafique, S. and Hassan, W. U.: Inland Navigation - An Instrument of Peace and Prosperity in Shared Basins, 32, 2018.

Abbass, Z.: Climate change, poverty and environmental crisis in the disaster prone areas of Pakistan, Oxfam Policy Pract. Clim. Change Resil. 5(3), 1–96, 2009.

Abdalla, A.: Principles of Islamic interpersonal conflict intervention: a search within Islam and western literature, J. Law Relig., 15, 151–184, 2001.

Abdallah, S., Thompson, S. and Marks, N.: Estimating worldwide life satisfaction, Ecol. Econ., 65(1), 35–47, 2008.

Abdelzaher, D. M., Kotb, A. and Helfaya, A.: Eco-Islam: Beyond the Principles of Why and What, and Into the Principles of How, J. Bus. Ethics, 1–21, 2017.

Abderrahman, W. A.: Application of Islamic Legal Principles for Advanced Water Management, Water Int., 25(4), 513–518, doi:10.1080/02508060008686865, 2000.

Abdolvand, B., Mez, L., Winter, K., Mirsaeedi-Gloßner, S., Schütt, B., Rost, K. T. and Bar, J.: The dimension of water in Central Asia: security concerns and the long road of capacity building, Environ. Earth Sci., 73(2), 897–912, 2015.

Abdullah, M. and Nadvi, M. J.: Understanding the Principles of Islamic World-View., Dialogue Pak., 6(3), 2011.

Abumoghli, I.: Sustainable development in Islam, Retreived Httpeboo Browse Comsustainable-Dev.--Islam.-Law-Iyadabumoghli-Doc-D30254102, 2010.

Acevedo-Whitehouse, K. and Duffus, A. L.: Effects of environmental change on wildlife health, Philos. Trans. R. Soc. B Biol. Sci., 364(1534), 3429–3438, 2009.

Adams-Jack, U.: Rethinking the South African Crisis. Nationalism, Populism, Hegemony, Taylor & Francis., 2015.

Afzal, M.: Population growth and economic development in Pakistan, Open Demogr. J., 2(1), 2009.

Agardy, T., Alder, J., Dayton, P., Curran, S., Kitchingman, A., Wilson, M., Catenazzi, A., Restrepo, J., Birkeland, C. and Blaber, S. J. M.: Coastal systems, 2005.

Agyenim, J. B.: Investigating institutional arrangements for integrated water resource management in developing countries: The case of White Volta Basin, Ghana, 2011.

Ahmad, M. and Farooq, U.: The state of food security in Pakistan: Future challenges and coping strategies, Pak. Dev. Rev., 903–923, 2010.

Ahmad, M. and Wasiq, M.: Water resource development in Northern Afghanistan and its implications for Amu Darya Basin, The World Bank., 2004.

Ahmad, M., Latif, Z., Tariq, J. A., Akram, W. and Rafique, M.: INVESTIGATION OF ISOTOPES AND HYDROLOGICAL PROCESSES IN INDUS RIVER SYSTEM, PAKISTAN, , 30, 2009.

Ahmad, O.: Afghanistan's coming water crisis, Third Pole [online] Available from: https://www.thethirdpole.net/en/2016/10/24/afghanistans-coming-water-crisis/ (Accessed 13 January 2019), 2016.

Ahmad, S.: Towards Kabul Water Treaty: Managing Shared Water Resources–Policy Issues and Options, IUCN Karachi Pak., 15 [online] Available from: https://cmsdata.iucn.org/downloads/pk_ulr_d3_1.pdf (Accessed 4 January 2016), 2010.

Ahmadullah, R. and Dongshik, K.: Assessment of Potential Dam Sites in the Kabul River Basin Using GIS, Int. J. Adv. Comput. Sci. Appl., 6(2), doi:10.14569/IJACSA.2015.060213, 2015.

Ahmadzai, S. and McKinna, A.: Afghanistan electrical energy and trans-boundary water systems analyses: Challenges and opportunities, Energy Rep., 4, 435–469, 2018.

Ahmed, R. and Mustafa, U.: Impact of CPEC projects on agriculture sector of Pakistan: Infrastructure and agricultural output linkages, Pak. Inst. Dev. Econ. PIDE Islamabad Pak., 2014.

Ahmed, S.: Indigenous water harvesting systems in Pakistan, Water Resour. Res. Inst. WRRI Natl. Agric. Res. Cent. NARC Islamabad Pak., 2000.

Ahmed, T., Scholz, M., Al-Faraj, F. and Niaz, W.: Water-related impacts of climate change on agriculture and subsequently on public health: A review for generalists with particular reference to Pakistan, Int. J. Environ. Res. Public. Health, 13(11), 1051, 2016.

Ahmed, V., Abbas, A. and Ahmed, S.: Public Infrastructure and Economic Growth in Pakistan: A Dynamic CGE-Micro Simulation Analysis, Infrastruct. Econ. Growth Asia, 117, 2013.

Aich, V., Akhundzadah, N., Knuerr, A., Khoshbeen, A., Hattermann, F., Paeth, H., Scanlon, A. and Paton, E.: Climate Change in Afghanistan Deduced from Reanalysis and Coordinated Regional Climate Downscaling Experiment (CORDEX)—South Asia Simulations, Climate, 5(2), 38, doi:10.3390/cli5020038, 2017.

Akbari, A. M., Chornack, M. P., Coplen, T. B., Emerson, D. G., Litke, D. W., Mack, T. J., Plummer, N., Verdin, J. P. and Verstraeten, I. M.: Water Resources Availability in Kabul, Afghanistan, in AGU Fall Meeting Abstracts., 2008.

Akhtar, F.: Water availability and demand analysis in the Kabul River Basin, Afghanistan, 2017.

Akhtar, F., Awan, U. K., Tischbein, B. and Liaqat, U. W.: Assessment of Irrigation Performance in Large River Basins under Data Scarce Environment—A Case of Kabul River Basin, Afghanistan, Remote Sens., 10(6), 972, doi:10.3390/rs10060972, 2018.

Akhtar, N., Saeed, K. and Khan, S.: Current status of Mammals in District Buner Khyber Pakhtunkhwa, Pakistan, Int. J. Mol. Evol. Biodivers., 4(4), 2014.

Akhtar, S. M. and Iqbal, J.: Assessment of emerging hydrological, water quality issues and policy discussion on water sharing of transboundary Kabul River, Water Policy, wp2017119, 2017.

Alaerts, G. J.: Institutional arrangements, Water Pollut. Control, 219–244, 1997.

Alahuhta, J., Joensuu, I., Matero, J., Vuori, K.-M. and Saastamoinen, O.: Freshwater ecosystem services in Finland, 2013.

Alam, U., Sahota, P. and Jeffrey, P.: Irrigation in the Indus basin: A history of unsustainability?, Water Sci. Technol. Water Supply, 7(1), 211–218, doi:10.2166/ws.2007.024, 2007.

Alberti, M., Asbjornsen, H., Baker, L. A., Brozovic, N., Drinkwater, L. E., Drzyzga, S. A., Jantz, C. A., Fragoso, J., Holland, D. S. and Kohler, T. T. A.: Research on coupled human and natural systems (CHANS): approach, challenges, and strategies, Bull. Ecol. Soc. Am., 92(2), 218–228, 2011.

Alcamo, J., Henrichs, T. and Rösch, T.: World water in 2025: Global modeling and scenario analysis for the world commission on water for the 21st century, 2017.

ALEP: Afghanistan Legal Education Project (ALEP). [online] Available from: https://www-cdn.law.stanford.edu/wp-content/uploads/2016/03/ALEP-Law-of-Afghanistan-3d-Ed_English.pdf (Accessed 20 November 2018), 2011.

Alexy, R.: On the structure of legal principles, Ratio Juris, 13(3), 294–304, 2000.

Ali, A. and Shaoliang, Y.: Highland rangelands of Afghanistan: Significance, management issues, and strategies, High-Alt. Rangel. Their Interfaces Hindu Kush Himalayas, 15, 2013.

Ali, Y. A.: Understanding Pashtunwali, The Nation, 8th June [online] Available from: https://nation.com.pk/06-Aug-2013/understanding-pashtunwali (Accessed 10 August 2017), 2013.

Al-Jayyousi, O. R.: Greywater reuse: towards sustainable water management, Desalination, 156(1–3), 181–192, 2003.

Allan, J. A.: Virtual water-the water, food, and trade nexus. Useful concept or misleading metaphor?, Water Int., 28(1), 106–113, 2003.

Allen, L., Christian-Smith, J. and Palaniappan, M.: Overview of greywater reuse: the potential of greywater systems to aid sustainable water management, Pac. Inst., 654, 19–21, 2010.

Alpa, G.: General principles of law, Ann Surv Intl Comp L, 1, 1, 1994.

Andersson, K.: Sanitation, wastewater management and sustainability: from waste disposal to resource recovery, UN Environmental Programme Global Programme of Action for the protection of …., 2016.

Andonova, L. B., Betsill, M. M. and Bulkeley, H.: Transnational climate governance, Glob. Environ. Polit., 9(2), 52–73, 2009.

Andreatta, F. and Koenig-Archibugi, M.: Which synthesis? Strategies of theoretical integration and the neorealist-neoliberal debate, Int. Polit. Sci. Rev., 31(2), 207–227, 2010.

Angelakis, A. and Zheng, X.: Evolution of water supply, sanitation, wastewater, and stormwater technologies globally, Multidisciplinary Digital Publishing Institute., 2015.

Angelakis, A. N. and Rose, J. B.: Evolution of sanitation and wastewater technologies through the centuries, IWA Publishing., 2014.

Angelakis, A. N., Chiotis, E., Eslamian, S. and Weingartner, H.: Underground Aqueducts Handbook, CRC Press., 2016.

Angelakis, A. N., Asano, T., Bahri, A., Jimenez, B. E. and Tchobanoglous, G.: Water reuse: from ancient to modern times and the future, Front. Environ. Sci., 6, 26, 2018.

Angeles, L.: Income inequality and colonialism, Eur. Econ. Rev., 51(5), 1155–1176, 2007.

Angoua, E. L. E., Dongo, K., Templeton, M. R., Zinsstag, J. and Bonfoh, B.: Barriers to access improved water and sanitation in poor peri-urban settlements of Abidjan, Côte d'Ivoire, PloS One, 13(8), e0202928, 2018.

Antoniou, G., De Feo, G., Fardin, F., Tamburrino, A., Khan, S., Tie, F., Reklaityte, I., Kanetaki, E., Zheng, X. and Mays, L.: Evolution of toilets worldwide through the millennia, Sustainability, 8(8), 779, 2016.

Antunes, S. and Camisão, I.: Realism, E-International Relations Publishing., 2017.

Anwar, A. A. and Bhatti, M. T.: Pakistan's water apportionment Accord of 1991: 25 years and beyond, J. Water Resour. Plan. Manag., 144(1), 05017015, 2017.

Arfanuzzaman, M. and Syed, M. A.: Water demand and ecosystem nexus in the transboundary river basin: a zero-sum game, Environ. Dev. Sustain., 20(2), 963–974, 2018.

Armah, F. A., Yawson, D. O., Yengoh, G. T., Odoi, J. O. and Afrifa, E. K.: Impact of floods on livelihoods and vulnerability of natural resource dependent communities in Northern Ghana, Water, 2(2), 120–139, 2010.

Ashbolt, N. J.: Microbial contamination of drinking water and disease outcomes in developing regions, Toxicology, 198(1–3), 229–238, 2004.

Asimov, M. S.: History of Civilizations of Central Asia, Motilal Banarsidass Publ., 1999.

Awesti, A.: The European Union, New institutionalism and types of multi-level governance, Polit. Perspect., 1(2), 1–23, 2007.

Azam, S.: KABUL RIVER TREATY: A NECESSITY FOR PEACE-N-SECURITY BETWEEN AFGHANISTAN AND PAKISTAN, AND PEACE IN SOUTH ASIA, Gomal Univ. J. Res., 31(2), 2015.

Aziz, K.: Need for a Pak-Afghan Treaty on Management of Joint Water Courses, Criterion Q., 2(4), 18, 2013.

Azizi, P. M.: Special Lecture on Water Resources in Afghanistan, , 10, 2007.

Bache, I. and Flinders, M.: Multi-level governance and the study of the British state, Public Policy Adm., 19(1), 31–51, 2004.

Badie, B., Berg-Schlosser, D. and Morlino, L.: International encyclopedia of political science, Sage., 2011.

Badruddin, A.: A Muslim majority Indus Valley Civilization?, DAWN.COM, 22nd June [online] Available from: http://www.dawn.com/news/728611 (Accessed 15 November 2018), 2012.

Bakker, M. H. and Duncan, J. A.: Future bottlenecks in international river basins: where transboundary institutions, population growth and hydrological variability intersect, Water Int., 42(4), 400–424, 2017.

Ball, A. and Craig, R.: Using neo-institutionalism to advance social and environmental accounting, Crit. Perspect. Account., 21(4), 283–293, 2010.

Bandaragoda, D. J.: A framework for institutional analysis for water resources management in a river basin context, IWMI., 2000.

Banerjee, A. and Iyer, L.: History, institutions, and economic performance: The legacy of colonial land tenure systems in India, Am. Econ. Rev., 95(4), 1190–1213, 2005.

Banting, E.: Afghanistan: The People, Crabtree Publishing Company., 2003.

Barandat, J. and Kaplan, A.: International water law: regulations for cooperation and the discussion of the international water convention, in Water in the Middle East, pp. 11–30, Springer., 1998.

Barfield, T.: 1 Afghan Customary Law and Its Relationship to Formal Judicial Institutions, 2003.

Barfield, T.: Culture and custom in nation-building: law in Afghanistan, Me Rev, 60, 347, 2008.

Barlow, M.: Blue covenant: The global water crisis and the coming battle for the right to water, McClelland & Stewart., 2009.

Barlow, M. and Clarke, T.: Blue gold: the battle against corporate theft of the world's water, Routledge., 2017.

Barnett, T. P., Adam, J. C. and Lettenmaier, D. P.: Potential impacts of a warming climate on water availability in snow-dominated regions, Nature, 438(7066), 303, 2005.

Barraqué, B.: Urban Water Conflicts: UNESCO-IHP, CRC Press., 2011a.

Barraqué, B.: Urban Water Conflicts: UNESCO-IHP, CRC Press., 2011b.

Barrett, C. B., Lee, D. R. and McPeak, J. G.: Institutional arrangements for rural poverty reduction and resource conservation, World Dev., 33(2), 193–197, 2005.

Bassiouni, M. C. and Rothenberg, D.: An assessment of justice sector and rule of law reform in Afghanistan and the need for a comprehensive plan, in Conference on the Rule of Law in Afghanistan, Rome, vol. 2, pp. 1101–78., 2007.

Basu, A. and Shankar, U.: Balancing of competing rights through sustainable development: role of Indian judiciary, Jindal Glob. Law Rev., 6(1), 61–72, 2015.

Bates, B., Kundzewicz, Z. and Wu, S.: Climate change and water, Intergovernmental Panel on Climate Change Secretariat., 2008.

Bates, R. H. and Block, S. A.: Revisiting African agriculture: institutional change and productivity growth, J. Polit., 75(2), 372–384, 2013.

Bavinck, M. and Gupta, J.: Legal pluralism in aquatic regimes: a challenge for governance, Curr. Opin. Environ. Sustain., 11, 78–85, 2014.

Baxter, P. and Jack, S.: Qualitative case study methodology: Study design and implementation for novice researchers, Qual. Rep., 13(4), 544–559, 2008.

BBC: The Afghans forced to go home, , 28th August [online] Available from: https://www.bbc.com/news/world-asia-37163857 (Accessed 25 January 2019), 2016.

Benda- Beckmann, F. von: Legal pluralism and social justice in economic and political development, IdS Bull., 32(1), 46–56, 2001.

Bergdolt, J., Sharvelle, S. and Roesner, L.: Estimation of graywater constituent removal rates in outdoor free-water-surface wetland in temperate climate, J. Environ. Eng., 139(5), 766–771, 2012.

Berkes, F.: Rethinking community- based conservation, Conserv. Biol., 18(3), 621–630, 2004.

Bernard, N.: Multilevel governance in the European Union, Springer., 2002.

Bhaduri, A., Manna, U., Barbier, E. and Liebe, J.: Climate change and cooperation in transboundary water sharing: an application of stochastic Stackelberg differential games in Volta river basin, Nat. Resour. Model., 24(4), 409–444, 2011.

Bhutta, M. N. and Smedema, L. K.: One hundred years of waterlogging and salinity control in the Indus valley, Pakistan: a historical review, Irrig. Drain. J. Int. Comm. Irrig. Drain. 56(S1), S81–S90, 2007.

Bibi, S., Khan, R. U., Nazir, R., Khan, P., Rehman, H. U., Shakir, S. K., Naz, S., Waheed, M. A. and Jan, R.: Heavy Metals Analysis in Drinking Water of Lakki Marwat District, KPK, Pakistan, World Appl. Sci. J., 34(1), 15–19, 2016.

Birnie, P. W. and Boyle, A. E.: International law and the environment. 1994.

Biswas, A.: Durand Line: History, Legality & Future, , 56, 2013.

Biswas, A. K.: Cooperation or conflict in transboundary water management: case study of South Asia, Hydrol. Sci. J., 56(4), 662–670, 2011.

Bjelica, J. and Ruttig, T.: The State of Aid and Poverty in 2018 in Afghanistan | Afghanistan Analysts Network, [online] Available from: https://www.afghanistan-analysts.org/the-state-of-aid-and-poverty-in-2018-a-new-look-at-aid-effectiveness-in-afghanistan/ (Accessed 27 January 2019), 2018.

Boelens, R.: The politics of disciplining water rights, Dev. Change, 40(2), 307–331, 2009.

Bokhari, S. A. A., Ahmad, B., Ali, J., Ahmad, S., Mushtaq, H. and Rasul, G.: Future Climate Change Projections of the Kabul River Basin Using a Multi-model Ensemble of High-Resolution Statistically Downscaled Data, Earth Syst. Environ., 2(3), 477–497, 2018.

Booth, K.: Realism and world politics, Routledge London., 2011.

Bortolotti, L. and Antrobus, M.: Costs and benefits of realism and optimism, Curr. Opin. Psychiatry, 28(2), 194, 2015.

Bos, K. and Gupta, J.: Inclusive development, oil extraction and climate change: a multilevel analysis of Kenya, Int. J. Sustain. Dev. World Ecol., 23(6), 482–492, 2016.

Boyd, J. and Banzhaf, S.: What are ecosystem services? The need for standardized environmental accounting units, Ecol. Econ., 63(2–3), 616–626, 2007.

Brady, D., Blome, A. and Kleider, H.: How politics and institutions shape poverty and inequality, Oxf. Handb. Soc. Sci. Poverty, 117, 2016.

Brasseur, B. L.: Recognizing the Durand Line - A Way Forward for Afghanistan and Pakistan? 24, 2011.

Braulik, G. T.: Conservation ecology and phylogenetics of the Indus River dolphin (Platanista gangetica minor), University of St Andrews., 2012.

Braulik, G. T., Arshad, M., Noureen, U. and Northridge, S. P.: Habitat fragmentation and species extirpation in freshwater ecosystems; causes of range decline of the Indus River Dolphin (Platanista gangetica minor), PloS One, 9(7), e101657, 2014.

Braun, C. L. and Smirnov, S. N.: Why is water blue? J. Chem. Educ., 70, 612–612, 1993.

Braune, E. and Adams, S.: Regional diagnosis for the sub-Saharan Africa region. Groundwater governance–A global framework for action, GEF, UNESCO-IHP, FAO, World Bank and IAH. www.groundwatergovernance.org, 2013.

Briscoe, W. J., Qamar, U., Contijoch, M., Amir, P. and Blackmore, D.: Pakistan's Water Economy: Running Dry, , 140, 2005.

Brochmann, M. and Gleditsch, N. P.: Shared rivers and conflict–A reconsideration, Polit. Geogr., 31(8), 519–527, 2012.

Brochmann, M. and Hensel, P. R.: The effectiveness of negotiations over international river claims, Int. Stud. Q., 55(3), 859–882, 2011.

Brohi, N.: Women, Violence and Jirgas, 47, 2016.

Brower, C. N. and Sharpe, J. K.: International arbitration and the Islamic world: The third phase, Am. J. Int. Law, 97(3), 643–656, 2003.

Brown, C.: Greywater Recycling: Risks, benefits, costs and policy, in 49th NZWWA annual conference: Whose water. 2007.

Brown, C.: Structural realism, classical realism and human nature, Int. Relat., 23(2), 257–270, 2009.

Bulmer, S. J.: New institutionalism and the governance of the Single European Market, J. Eur. Public Policy, 5(3), 365–386, 1998.

Burby, R. J. and May, P. J.: IntergovernmentalEnvironmental Planning: Addressing the Commitment Conundrum, J. Environ. Plan. Manag. 41(1), 95–110, 1998.

Burki, S. J.: Economics of Pak-Afghan Relations, , 15, 2013.

Campbell, J.: A dry and ravaged land: Investigating water resources in Afghanistan, EARTH Mag. [online] Available from: https://www.earthmagazine.org/article/dry-and-ravaged-land-investigating-water-resources-afghanistan (Accessed 27 January 2019), 2015.

Capoccia, G.: When do institutions ―bite"? Historical institutionalism and the politics of institutional change, Comp. Polit. Stud., 49(8), 1095–1127, 2016.

Caponera, D. A.: Patterns of cooperation in international water law: Principles and institutions, Nat. Resour. J., 25(3), 563–587, 1985.

Caponera, D. A.: Ownership and transfer of water and land in Islam, Water Manag. Islam, 94–102, 2001.

Caravella, K. D.: Mimetic, coercive, and normative influences in institutionalization of organizational practices: The case of distance learning in higher education, Florida Atlantic University., 2011.

Carberry, S. and Faizy, S.: Afghanistan's Forests A Casualty Of Timber Smuggling, NPR.org [online] Available from: https://www.npr.org/2013/03/18/174200911/afghanistans-forests-a-casualty-of-timber-smuggling (Accessed 25 February 2018), 2013.

Cardoso, P. G., Raffaelli, D., Lillebø, A. I., Verdelhos, T. and Pardal, M. A.: The impact of extreme flooding events and anthropogenic stressors on the macrobenthic communities' dynamics, Estuar. Coast. Shelf Sci., 76(3), 553–565, 2008.

Carr, G.: Stakeholder and public participation in river basin management—an introduction, Wiley Interdiscip. Rev. Water, 2(4), 393–405, 2015.

Carr, G. M. and Neary, J. P.: Water quality for ecosystem and human health, UNEP/Earthprint., 2008.

Cascão, A. E.: Ethiopia–challenges to Egyptian hegemony in the Nile Basin, Water Policy, 10(S2), 13–28, 2008.

Catafago, S.: Restructuring water sector in Lebanon Litani River authority-facing the challenges of good water governance, Food Secur. Water Scarcity Middle East Probl. Solut. 2005.

Chao, F., You, D., Pedersen, J., Hug, L. and Alkema, L.: National and regional under-5 mortality rate by economic status for low-income and middle-income countries: a systematic assessment, Lancet Glob. Health, 6(5), e535–e547, 2018.

Chapagain, A. and Orr, S.: UK Water Footprint: the impact of the UK's food and fibre consumption on global water resources Volume two: appendices, WWF-UK Godalming, 31–33, 2008.

China Daily: Economic corridor links China, Pakistan dreams - Business - Chinadaily.com.cn, [online] Available from: http://www.chinadaily.com.cn/business/2013-09/02/content_16937071.htm (Accessed 29 January 2019), 2013.

Christensen, M.: Judicial Reform in Afghanistan: Towards a Holistic Understanding of Legitimacy in Post-Conflict Societies, Berkeley J Middle E Islam. L, 4, 111, 2011.

Chuadhry, A. G. and Chaudhry, H. U. R.: Development chronicle of Pakistan: A case of colonial legacy, FWU J. Soc. Sci., 6(1), 48, 2012.

Ciervo, M.: Geopolitica dell'Acqua., Carocci. 2009.

Collier, A. C.: Pharmaceutical contaminants in potable water: potential concerns for pregnant women and children, EcoHealth, 4(2), 164–171, 2007.

Conti, K. I.: Norms in multilevel groundwater governance and sustainable development, Unpubl. Ph Thesis Univ. Amst. Neth., 2017.

Conti, K. I. and Gupta, J.: Protected by pluralism? Grappling with multiple legal frameworks in groundwater governance, Curr. Opin. Environ. Sustain. 11, 39–47, 2014.

Corcoran, E.: Sick water?: the central role of wastewater management in sustainable development: a rapid response assessment, UNEP/Earthprint., 2010.

Cordesman, A. H.: Agriculture, Food, and Poverty in Afghanistan, Agric. Food Poverty Afghan. [online] Available from: https://www.csis.org/analysis/agriculture-food-and-poverty-afghanistan (Accessed 5 February 2018), 2010.

Correia, F. N. and da Silva, J. E.: International framework for the management of transboundary water resources, Water Int., 24(2), 86–94, 1999.

Corvalan, C., Hales, S., McMichael, A. J., Butler, C. and McMichael, A.: Ecosystems and human well-being: health synthesis, World Health Organization., 2005.

Cosgrove, W. J. and Loucks, D. P.: Water management: Current and future challenges and research directions, Water Resour. Res., 51(6), 4823–4839, 2015a.

Cosgrove, W. J. and Loucks, D. P.: Water management: Current and future challenges and research directions, Water Resour. Res., 51(6), 4823–4839, 2015b.

Coulson, N.: A history of Islamic law, Routledge. 2017.

Crawford, S. E. and Ostrom, E.: A grammar of institutions, Am. Polit. Sci. Rev., 89(3), 582–600, 1995.

Cullet, P.: Governing the Environment without CoPs-The Case of Water, Intl Comm Rev, 15, 123, 2013.

Cullet, P. and Gupta, J.: India: evolution of water law and policy, in The evolution of the law and politics of water, pp. 157–173, Springer. 2009.

Dallas, H. F. and Rivers-Moore, N.: Ecological consequences of global climate change for freshwater ecosystems in South Africa, South Afr. J. Sci., 110(5–6), 01–11, 2014.

Damkjaer, S. and Taylor, R.: The measurement of water scarcity: Defining a meaningful indicator, Ambio, 46(5), 513–531, 2017.

Dangwal, D. D.: Commercialisation of forests, timber extraction and deforestation in Uttaranchal, 1815-1947, Conserv. Soc., 3(1), 110, 2005.

Daniel, T. C., Muhar, A., Arnberger, A., Aznar, O., Boyd, J. W., Chan, K. M., Costanza, R., Elmqvist, T., Flint, C. G. and Gobster, P. H.: Contributions of cultural services to the ecosystem services agenda, Proc. Natl. Acad. Sci., 109(23), 8812–8819, 2012.

Daniell, K. A.: Co-engineering and participatory water management: organisational challenges for water governance, Cambridge University Press., 2012.

Daoudy, M.: Asymmetric power: Negotiating water in the Euphrates and Tigris, Int. Negot., 14(2), 361–391, 2009.

Daud, M. K., Nafees, M., Ali, S., Rizwan, M., Bajwa, R. A., Shakoor, M. B., Arshad, M. U., Chatha, S. A. S., Deeba, F. and Murad, W.: Drinking water quality status and contamination in Pakistan, BioMed Res. Int., 2017, 2017.

De Bruyne, C. and Fischhendler, I.: Negotiating conflict resolution mechanisms for transboundary water treaties: A transaction cost approach, Glob. Environ. Change, 23(6), 1841–1851, 2013.

De Sadeleer, N.: Environmental principles: from political slogans to legal rules, Oxford University Press on Demand., 2002.

Deephouse, D. L. and Suchman, M.: Legitimacy in organizational institutionalism, Sage Handb. Organ. Institutionalism, 49, 77, 2008.

Delgado, J. V. and Zwarteveen, M.: Modernity, exclusion and resistance: Water and indigenous struggles in Peru, Development, 51(1), 114–120, 2008.

Dellapenna, J. and Gupta, J.: Toward global law on water, Glob. Gov. Rev. Multilateralism Int. Organ., 14(4), 437–453, 2008.

Dellapenna, J. W.: The evolution of riparianism in the United States, Marq Rev, 95, 53, 2011.

Dellapenna, J. W. and Gupta, J.: The evolution of global water law, in The evolution of the law and politics of water, pp. 3–19, Springer., 2009.

Dent, D.: Green Water Credits, presentation at the FAO, in Netherlands Conference on Water for Food and Ecosystems, The Hague, Netherlands, vol. 31., 2005.

Díaz, S., Demissew, S., Carabias, J., Joly, C., Lonsdale, M., Ash, N., Larigauderie, A., Adhikari, J. R., Arico, S. and Báldi, A.: The IPBES Conceptual Framework—connecting nature and people, Curr. Opin. Environ. Sustain. 14, 1–16, 2015.

Dimitrov, R. S.: Knowledge, power, and interests in environmental regime formation, Int. Stud. Q., 47(1), 123–150, 2003.

Dinar, A., Dinar, S., McCaffrey, S. and McKinney, D.: Bridges Over Water: Understanding Transboundary Water Conflict, Negotiation and Cooperation Second, World Scientific Publishing Company. 2013.

Dinar, S.: Negotiations and international relations: a framework for hydropolitics, Int. Negot., 5(2), 375–407, 2000.

Dinar, S.: Assessing side-payment and cost-sharing patterns in international water agreements: The geographic and economic connection, Polit. Geogr., 25(4), 412–437, 2006.

Dinar, S.: Scarcity and cooperation along international rivers, Glob. Environ. Polit., 9(1), 109–135, 2009.

Dirzauskaite, G. and Ilinca, N.: Understanding Hegemony in International Relations Theories, Aalborg: Aalborg University. 2017.

Dıaz, S., Tilman, D., Fargione, J., Chapin III, F. S., Dirzo, R. and Ktzberber, T.: Biodiversity regulation of ecosystem services, Trends Cond., 279–329, 2005.

Dokhani, S. and Ramezani, M.: Study floodwater spreading systems and solutions to increase their efficiency (Case study watershed Defyh Rafsanjan Iran), 2017.

Dolatyar, M. and Gray, T. S.: The politics of water scarcity in the Middle East, Environ. Polit., 9(3), 65–88, 2000.

Döll, P., Douville, H., Güntner, A., Schmied, H. M. and Wada, Y.: Modelling freshwater resources at the global scale: challenges and prospects, Surv. Geophys. 37(2), 195–221, 2016.

Dore, J., Lebel, L. and Molle, F.: A framework for analysing transboundary water governance complexes, illustrated in the Mekong Region, J. Hydrol., 466, 23–36, 2012.

Doris, F., Edward, W., Caroline, T. and Mary, A.: Assessment of the Coping Strategies of Flood Victims in the Builsa District, Am. J. Environ. Sci. Eng., 2(1), 17–25, 2018.

Douglass, C.: North, Institutions, institutional change and economic performance, Cambridge: Cambridge university press. 1990.

Drieschova, A. and Eckstein, G.: Cooperative transboundary mechanism, 2014.

D'souza, A. and Jolliffe, D.: Food insecurity in vulnerable populations: coping with food price shocks in Afghanistan, Am. J. Agric. Econ., 96(3), 790–812, 2013.

D'Souza, R.: Water in British India: the making of a _Colonial Hydrology,' Hist. Compass, 4(4), 621–628, 2006.

Duffield, J.: What are international institutions? Int. Stud. Rev., 9(1), 1–22, 2007.

Dupree, N. H.: An Historical Guide to Afghanistan, Rev. and enl edition. Afghan Air Authority, Afghan Tourist Organization., 1977.

Duratovic, D.: Spatial hydropolitics between the upstream and the downstream states-A case study of the state, territory and identity formation of the Rogun HPP, 2016.

Dutton, Y.: The Origins of Islamic Law: The Qur'an, the Muwatta'and Madinan Amal, Routledge. 2013.

Earle, A. and Neal, M. J.: Inclusive transboundary water governance, in Freshwater governance for the 21st century, pp. 145–158, Springer. 2017.

Easter, K. W.: Introduction to the special section on Institutional Arrangements and the Effects on Water Resources, Water Resour. Res., 40(12), 2004.

Easton, G.: Critical realism in case study research, Ind. Mark. Manag. 39(1), 118–128, 2010.

Eckerberg, K. and Joas, M.: Multi-level Environmental Governance: a concept under stress? Local Environ., 9(5), 405–412, 2004.

Eckstein, G. and Sindico, F.: The law of transboundary aquifers: many ways of going forward, but only one way of standing still, Rev. Eur. Comp. Int. Environ. Law, 23(1), 32–42, 2014.

Elhance, A. P.: Hydropolitics in the Third World: Conflict and cooperation in international river basins, US Institute of Peace Press. 1999.

Emerton, L. and Bos, E.: Value: Counting ecosystems as water infrastructure, Iucn. 2004.

Enderlein, H., Walti, S. and Zurn, M.: Handbook on multi-level governance, Edward Elgar Publishing., 2010.

Erfurt, P. J.: An assessment of the role of natural hot and mineral springs in health, wellness and recreational tourism. 367, 2011.

European Commission: EUROPEAN UNION - AFGHANISTAN, , 17, 2012.

European Union: The positive impact of EU support to Afghanistan's agriculture sector, EEAS - Eur. Comm. [online] Available from: https://eeas.europa.eu/headquarters/headquarters-homepage/60117/positive-impact-eu-support-afghanistans-agriculture-sector_fr (Accessed 28 March 2019), 2019.

Eynon, K.: Complex Interdependence: Watering Relations Between Ethiopia and Egypt, 2016.

Falkenmark, M.: Freshwater as shared between society and ecosystems: from divided approaches to integrated challenges, Philos. Trans. R. Soc. Lond. B Biol. Sci., 358(1440), 2037–2049, 2003.

Falkenmark, M. and Rockström, J.: The new blue and green water paradigm: Breaking new ground for water resources planning and management, American Society of Civil Engineers., 2006.

Fardin, H. F., Hollé, A., Gautier-Costard, E. and Haury, J.: Sanitation and water management in ancient South Asia, Evol. Sanit. Wastewater Technol. Centuries, 43–53, 2014.

Farooqi, A. B., Khan, A. H. and Mir, H.: Climate change perspective in Pakistan, Pak. J Meteorol, 2(3), 2005.

Faruqui, N. I.: Responding to the water crisis in Pakistan, Int. J. Water Resour. Dev., 20(2), 177–192, 2004.

Faruqui, N. I., Biswas, A. K. and Bino, M. J.: Water management in Islam, United Nations University., 2001.

Favre, R. and Kamal, G. M.: Watershed atlas of Afghanistan, working document for planners, parts I and II 1st edn. Kabul: Government of Afghanistan, Ministry of Irrigation, Water Resour. Environ., 60, 2004.

Finnemore, M. and Sikkink, K.: International norm dynamics and political change, Int. Organ., 52(4), 887–917, 1998.

Fitzgerald, P. and Gould, E.: Invisible history: Afghanistan's untold story, City Lights Books., 2009.

Fitzmaurice, M.: Convention on the Law of the Non-Navigational Uses of International Watercourses, Leiden J. Int. Law, 10(3), 501–508, doi:10.1017/S0922156597000368, 1997.

FoDP: A Productive and Water-Secure Pakistan: Infrastructure, Institutions, Strategy, , 196, 2012.

Foster, S. and Garduño, H.: Groundwater-resource governance: Are governments and stakeholders responding to the challenge?, Hydrogeol. J., 21(2), 317–320, 2013.

Frey, F. W. and Naff, T.: Water: an emerging issue in the Middle East?, Ann. Am. Acad. Pol. Soc. Sci., 482(1), 65–84, 1985.

Friend, R. and Thinphanga, P.: Urban water crises under future uncertainties: the case of institutional and infrastructure complexity in Khon Kaen, Thailand, Sustainability, 10(11), 3921, 2018.

Frischmann, P.: Afghanistan Resource Corridor Development: Water Strategy Final Kabul River Basin Report, [online] Available from: http://documents.worldbank.org/curated/en/599831468196445996/pdf/797210WP0v205-0Box0379789B00PUBLIC0.pdf (Accessed 25 January 2019), 2012.

Fritz, D., Miller, U., Gude, A., Pruisken, A. and Rischewski, D.: Making poverty reduction inclusive: Experiences from Cambodia, Tanzania and Vietnam, J. Int. Dev. J. Dev. Stud. Assoc., 21(5), 673–684, 2009.

Fuller, D. Q.: Harappan seeds and agriculture: some considerations, Antiquity, 75(288), 410–414, 2001.

Furlong, K.: Hidden theories, troubled waters: International relations, the ‚territorial trap‘, and the Southern African Development Community's transboundary waters, Polit. Geogr., 25(4), 438–458, 2006.

Gada, M. Y.: Environmental Ethics in Islam: Principles and Perspectives, , 9, 2014.

Gadgil, A.: Afghanistan's Water Crisis, HydrateLife [online] Available from: https://www.hydratelife.org/afghanistans-water-crisis/ (Accessed 27 January 2019), 2012.

Galassi, D. M., Lombardo, P., Fiasca, B., Di Cioccio, A., Di Lorenzo, T., Petitta, M. and Di Carlo, P.: Earthquakes trigger the loss of groundwater biodiversity, Sci. Rep., 4, 6273, 2014.

Gall, C.: The wrong enemy: America in Afghanistan, 2001–2014, HMH., 2014.

Gasparotti, C.: A modern approach of water management in the Danube River Basin, Manag. Mark., 7(4), 777, 2012.

Gaston, E. and Dang, L.: Addressing Land Conflict in Afghanistan, , 16, 2015.

Gazdar, H.: A review of migration issues in Pakistan, Refug. Migr. Mov. Res. Unit Bangladesh DFID, 2003.

Gerlak, A. K. and Schmeier, S.: River basin organizations and the governance of transboundary watercourses, in The Oxford Handbook of Water Politics and Policy., 2016.

Ghosh, M. and Sinha, B.: Impact of forest policies on timber production in I ndia: a review, in Natural Resources Forum, vol. 40, pp. 62–76, Wiley Online Library., 2016.

Ghulami, M.: Assessment of climate change impacts on water resources and agriculture in data-scarce Kabul basin, Afghanistan, Université Côte d'Azur., 2017.

Gilli, F.: Islam, water conservation and public awareness campaigns, in 2nd Israeli-Palestinian-international academic conference on water for life, Antalya, pp. 10–14., 2004.

Gilpin, R.: War is too important to be left to ideological amateurs, Int. Relat., 19(1), 5–18, 2005.

Gilpin, R. and Palan, R.: International Political Economy, Princeton University Press, Princeton., 1987.

Ginsburg, T. and Huq, A.: What Can Constitutions Do?: The Afghan Case, J. Democr., 25(1), 116–130, 2014.

Giordano, M. A. and Wolf, A. T.: Sharing waters: Post- Rio international water management, in Natural Resources Forum, vol. 27, pp. 163–171, Wiley Online Library., 2003.

GIZ: Financial Sustainability of International River Basin Organizations. [online] Available from: https://transboundarywaters.science.oregonstate.edu/sites/transboundarywaters.science.oreg onstate.edu/files/Publications/GIZ%202014%20Financing%20International%20River%20B asin%20Organizations.pdf, 2014.

Glasbergen, P.: Mechanisms of private meta-governance: an analysis of global private governance for sustainable development, Int. J. Strateg. Bus. Alliances, 2(3), 189–206, 2011.

Gleick, P. H.: Basic water requirements for human activities: Meeting basic needs, Water Int., 21(2), 83–92, 1996.

Glover, J. L., Champion, D., Daniels, K. J. and Dainty, A. J. D.: An Institutional Theory perspective on sustainable practices across the dairy supply chain, Int. J. Prod. Econ., 152, 102–111, 2014.

Goddard, S. E. and Nexon, D. H.: The dynamics of global power politics: A framework for analysis, J. Glob. Secur. Stud., 1(1), 4–18, 2015.

Gohar, A. A., Ward, F. A. and Amer, S. A.: Economic performance of water storage capacity expansion for food security, J. Hydrol., 484, 16–25, 2013.

Goldstein, A.: Rising to the challenge: China's grand strategy and international security, Stanford University Press., 2005.

Gordon, L. J., Finlayson, C. M. and Falkenmark, M.: Managing water in agriculture for food production and other ecosystem services, Agric. Water Manag., 97(4), 512–519, 2010.

Gorissen, M.: The establishment of UNTAC. Power or persuasion?, 2016.

Gough, I. and McGregor, J. A.: Wellbeing in developing countries, Cambridge University Press., 2007.

Government of Afghanistan: Islamic Republic of Afghanistan- National Biodiversity Strategy & Action Plan: Framework for Implementation 2014 - 2017. [online] Available from: https://www.cbd.int/doc/world/af/af-nbsap-01-en.pdf (Accessed 3 March 2018), 2014.

Grewal, R.: Five Thousand Years of Urbanization: The Punjab Region, Manohar Publishers., 2005.

Griffiths, J. and Lambert, R.: Free Flow: Reaching Water Security through Cooperation, Unesco., 2013.

Griffiths-Sattenspiel, B. and Wilson, W.: The carbon footprint of water, River Netw. Portland, 2009.

Guirdham, M.: Communicating across cultures at work, Macmillan International Higher Education., 2011.

Guo, L., Zhou, H., Xia, Z. and Huang, F.: Evolution, opportunity and challenges of transboundary water and energy problems in Central Asia, SpringerPlus, 5(1), 1918, 2016.

Gupta, J.: 11 Toward Sharing Our Ecospace, New Earth Polit. Essays Anthr., 271, 2016a.

Gupta, J.: The watercourses convention, hydro-hegemony and transboundary water issues, Int. Spect., 51(3), 118–131, 2016b.

Gupta, J. and Conti, K.: Global climate change and global groundwater law: their independent and pluralistic evolution and potential challenges, Water Int., 42(6), 741–756, 2017.

Gupta, J. and Lebel, L.: Access and allocation in earth system governance: Water and climate change compared, Int. Environ. Agreem. Polit. Law Econ., 10(4), 377–395, 2010.

Gupta, J. and Pahl-Wostl, C.: Global water governance in the context of global and multilevel governance: its need, form, and challenges, Ecol. Soc., 18(4), 2013a.

Gupta, J. and Pahl-Wostl, C.: Global water governance in the context of global and multilevel governance: its need, form, and challenges, Ecol. Soc., 18(4), 2013b.

Gupta, J. and Pouw, N.: Towards a trans-disciplinary conceptualization of inclusive development, Curr. Opin. Environ. Sustain., 24, 96–103, 2017.

Gupta, J. and Vegelin, C.: Sustainable development goals and inclusive development, Int. Environ. Agreem. Polit. Law Econ., 16(3), 433–448, 2016.

Gupta, J., Van Der Grijp, N. and Kuik, O.: Climate change, forests and REDD: lessons for institutional design, Routledge., 2013.

Gupta, J., Pouw, N. R. and Ros-Tonen, M. A.: Towards an elaborated theory of inclusive development, Eur. J. Dev. Res., 27(4), 541–559, 2015.

Guzmán-Arias, I. and Calvo-Alvarado, J.: Water Resource Planning in Latin America and the Caribbean, Rev. Tecnol. En Marcha, 29, 14–32, 2016.

Habib, H.: Water related problems in Afghanistan, Int. J. Educ. Stud., 1(3), 137–144, 2014.

Hajer, M. A.: The politics of environmental discourse: ecological modernization and the policy process, Clarendon Press Oxford., 1995.

Halder, J. N. and Islam, M. N.: Water pollution and its impact on the human health, J. Environ. Hum., 2(1), 36–46, 2015.

Haleem, M. A.: Water in the Qur'an, Islam. Q., 33(1), 34, 1989.

Hall, P. A. and Taylor, R. C.: Political science and the three new institutionalisms, Polit. Stud., 44(5), 936–957, 1996.

Hamed, S. E.-D.: Seeing the environment through islamic eyes: Application ofShariah to natural resources planning and management, J. Agric. Environ. Ethics, 6(2), 145–164, 1993.

Hamid, S. S.: Influence of Western Jurisprudence over Islamic Jurisprudence: A Comparative Study, North. Univ. J. Law, 4, 13–26, 2013.

Hamner, J. H. and Wolf, A. T.: Patterns in international water resource treaties: the transboundary freshwater dispute database, Colo J Intl Envtl Pol, 9, 157, 1998.

Hanasz, P.: The politics of water security in the Kabul river basin, Future Dir., 10, 2011a.

Hanasz, P.: The Politics of Water Security in the Kabul River Basin, Future Dir. Int. [online] Available from: http://www.futuredirections.org.au/publication/the-politics-of-water-security-in-the-kabul-river-basin/ (Accessed 26 January 2019b), 2011.

Hanasz, P.: Muddy Waters: International Actors and Transboundary Water Cooperation in the Ganges-Brahmaputra Problemshed., Water Altern., 10(2), 2017.

Hanauer, L. and Chalk, P.: India's and Pakistan's Strategies in Afghanistan, RAND corporation., 2012.

Hanna, E. and Braithwaite, R.: The Greenland ice sheet: A global warming signal?, Weather, 58(9), 351–357, 2003.

Harrison, N. E.: Complexity in world politics: Concepts and methods of a new paradigm, SUNY Press., 2012.

Haseena, M., Malik, M. F., Javed, A., Arshad, S., Asif, N., Zulfiqar, S. and Hanif, J.: Water pollution and human health., Environ. Risk Assess. Remediat., 1(3), 2017.

Hashemi, S. M. S., Barari, M. and Bagheri, M., A.: TOWARDS AN INTEGRATED LAND AND WATER RESOURCES GOVERNANCE SYSTEM IN LAKE ZRÊBAR BASIN, IRAN: A COMMUNITY-BASED APPROACH, 2015.

Hashimi, G.: Helping Afghanistan's Informal Dispute Resolution System Follow Afghan Law in Criminal Matters: What Afghanistan Can Learn from Native American Peacemaking Program, Mich St Intl Rev, 25, 77, 2017.

Hassan, A., Suhid, A., Abiddin, N. Z., Ismail, H. and Hussin, H.: The role of Islamic philosophy of education in aspiring holistic learning, Procedia-Soc. Behav. Sci., 5, 2113–2118, 2010.

Hassenforder, E. and Barone, S.: Institutional arrangements for water governance, Int. J. Water Resour. Dev., 1–25, 2018.

ul Hasson, S.: Future water availability from Hindukush-Karakoram-Himalaya Upper Indus Basin under conflicting climate change scenarios, Climate, 4(3), 40, 2016.

Hay, C.: Political analysis: a critical introduction, Macmillan International Higher Education., 2002.

Hayat, S. and Gupta, J.: Kinds of freshwater and their relations to ecosystem services and human well-being, Water Policy, wp2016182, 2016.

Haydar, S., Arshad, M. and Aziz, J. A.: Evaluation of drinking water quality in urban areas of Pakistan: A case study of Southern Lahore, Pak. J. Eng. Appl. Sci., 2016.

Hellin, J., Ratner, B. D., Meinzen-Dick, R. and Lopez-Ridaura, S.: Increasing social-ecological resilience within small-scale agriculture in conflict-affected Guatemala, 2018.

Helmke, G. and Levitsky, S.: Informal institutions and comparative politics: A research agenda, Perspect. Polit., 2(4), 725–740, 2004.

Hendrickson III, K. E.: The encyclopedia of the industrial revolution in world history, Rowman & Littlefield., 2014.

Hensengerth, O., Dombrowski, I. and Scheumann, W.: Benefit-sharing on dams on shared rivers, 2012.

Héritier, A.: Explaining institutional change in Europe, Oxford University Press., 2007.

Hessami, E.: Afghanistan's Rivers Could Be India's Next Weapon Against Pakistan, Foreign Policy, 13th November [online] Available from: https://foreignpolicy.com/2018/11/13/afghanistans-rivers-could-be-indias-next-weapon-against-pakistan-water-wars-hydropower-hydrodiplomacy/ (Accessed 27 January 2019), 2018.

Hessami, E. B.: Afghanistan's Water Plans Complicated by Worried Neighbors, New Secur. Beat [online] Available from: https://www.newsecuritybeat.org/2017/03/afghanistans-water-plans-complicated-worried-neighbors/ (Accessed 26 January 2019), 2017.

Hoekstra, A. Y., Mekonnen, M. M., Chapagain, A. K., Mathews, R. E. and Richter, B. D.: Global monthly water scarcity: blue water footprints versus blue water availability, PLoS One, 7(2), e32688, 2012.

Hoff, H.: Global water resources and their management, Curr. Opin. Environ. Sustain., 1(2), 141–147, 2009.

Holman, I. P., Hollis, J. M., Bramley, M. E. and Thompson, T. R. E.: The contribution of soil structural degradation to catchment flooding: a preliminary investigation of the 2000 floods in England and Wales, Hydrol. Earth Syst. Sci. Discuss., 7(5), 755–766, 2003.

Hooghe, L., Marks, G. and Wilson, C. J.: Does left/right structure party positions on European integration?, Comp. Polit. Stud., 35(8), 965–989, 2002.

Hooper, B.: Integrated river basin governance, IWA publishing., 2017.

Hooper, B. P.: Integrated water resources management and river basin governance, J. Contemp. Water Res. Educ., 126(1), 3, 2003.

Hopwood, B., Mellor, M. and O'Brien, G.: Sustainable development: mapping different approaches, Sustain. Dev., 13(1), 38–52, 2005.

Horton, J.: What Might it Mean for Political Theory to Be More _Realistic'?, Philosophia, 45(2), 487–501, 2017.

Howlett, M.: Managing the ‑hollow state": Procedural policy instruments and modern governance, Can. Public Adm., 43(4), 412–431, 2000.

Huitema, D. and Meijerink, S.: The politics of river basin organizations, Ecol. Soc., 22(2) [online] Available from: http://www.jstor.org/stable/26270064, 2017.

Humphreys, A. R. C.: Kenneth Waltz and the limits of explanatory theory in international relations, University of Oxford., 2007.

Hunter, P. R., MacDonald, A. M. and Carter, R. C.: Water supply and health, PLoS Med., 7(11), e1000361, 2010.

Hurlbert, M. and Gupta, J.: The split ladder of participation: a diagnostic, strategic, and evaluation tool to assess when participation is necessary, Environ. Sci. Policy, 50, 100–113, 2015.

Husain, K.: Exclusive: CPEC master plan revealed, DAWN.COM, 14th May [online] Available from: https://www.dawn.com/news/1333101 (Accessed 29 January 2019), 2017.

Hussain, S. S., Mudasser, M., Sheikh, M. M. and Manzoor, N.: Climate change and variability in mountain regions of Pakistan implications for water and agriculture, Pak. J. Meteorol., 2(4), 2005.

Hussain, Z.: Sources of Tension in Afghanistan and Pakistan: A Regional Perspective, CIDOB Policy Res. Proj., 2011.

ICG: China-Pakistan Economic Corridor: Opportunities and Risks, Int. Crisis Group [online] Available from: https://www.crisisgroup.org/asia/south-asia/pakistan/297-china-pakistan-economic-corridor-opportunities-and-risks (Accessed 29 January 2019), 2018.

Ikenberry, G.: John and Michael Mastanduno. 2003, Int. Relat. Theory Asia-Pac., 2005.

Ikenberry, G. J.: The rise of China and the future of the West-Can the liberal system survive, Foreign Aff, 87, 23, 2008.

Ikenberry, G. J.: Power, order, and change in world politics, Cambridge University Press., 2014.

International Law Association: Helsinki rules on the uses of the waters of international rivers, International Law Association., 1967.

International Law Association: Berlin rules on water resources, in International law association, report of the seventy-first conference, Berlin, pp. 334–480., 2004.

International Law Commission: Draft articles on the Law of Transboundary Aquifers, [online] Available from: http://legal.un.org/ilc/texts/instruments/english/draft_articles/8_5_2008.pdf (Accessed 25 August 2018), 2008.

Iqbal, M. S., Dahri, Z. H., Querner, E. P., Khan, A. and Hofstra, N.: Impact of Climate Change on Flood Frequency and Intensity in the Kabul River Basin, Geosciences, 8(4), 114, 2018.

Iqbal, S.: Quantifying the impact of socioeconomic development and climate change on Escherichia coli concentrations in the Pakistani Kabul River, Wageningen University., 2017.

Islam and Tanaka: Impacts of pollution on coastal and marine ecosystems including coastal and marine fisheries and approach for management: a review and synthesis, Mar. Pollut. Bull., 48(7–8), 624–649, 2004.

Islam, K. K. and Hyakumura, K.: Forestland Concession, Land Rights, and Livelihood Changes of Ethnic Minorities: The Case of the Madhupur Sal Forest, Bangladesh, Forests, 10(3), 288, 2019.

IUCN: 1994-057.pdf, [online] Available from: https://portals.iucn.org/library/efiles/documents/1994-057.pdf (Accessed 25 January 2019), 1994.

IUCN: Biological Diversity in Pakistan. [online] Available from: https://portals.iucn.org/library/sites/library/files/documents/1997-073.pdf (Accessed 13 December 2018), 1997.

IUCN: towards_kabul_water_treaty.pdf. [online] Available from: http://www.cawater-info.net/afghanistan/pdf/towards_kabul_water_treaty.pdf (Accessed 26 January 2019), 2013.

IWMI: Kabul hosts Afghanistan's first-ever National Media dialogue on the Indus basin :: IWMI, Int. Water Manag. Inst. IWMI [online] Available from:

http://www.iwmi.cgiar.org/2016/10/kabul-hosts-the-indus-basin-national-media-dialogue/ (Accessed 27 January 2019), 2016.

Iyer, R. R.: Water and the Laws in India, SAGE Publishing India., 2009.

Jabeen, S., Mahmood, Q., Tariq, S., Nawab, B. and Elahi, N.: Health impact caused by poor water and sanitation in district Abbottabad, J. Ayub Med. Coll. Abbottabad, 23(1), 47–50, 2011.

Jackson, R. M., Roe, J. D., Wangchuk, R. and Hunter, D. O.: Estimating snow leopard population abundance using photography and capture- recapture techniques, Wildl. Soc. Bull., 34(3), 772–781, 2006.

Jacobs, I. M.: The politics of water in Africa: Norms, environmental regions and transboundary cooperation in the Orange-Senqu and Nile Rivers, A&C Black., 2012.

Jacobs, J. W.: The Mekong River Commission: transboundary water resources planning and regional security, Geogr. J., 168(4), 354–364, 2002.

Jager, N. W.: Transboundary Cooperation in European Water Governance–A set- theoretic analysis of International River Basins, Environ. Policy Gov., 26(4), 278–291, 2016.

Jägerskog, A.: The Sanctioned Discourse œ a Crucial Factor for Understanding Water Policy in the Jordan River Basin, SOAS, University of London, Occas. Pap., (41), 2001.

Jain, R.: In Parched Afghanistan, Drought Sharpens Water Dispute With Iran, US News World Rep. [online] Available from: https://www.usnews.com/news/world/articles/2018-07-16/in-parched-afghanistan-drought-sharpens-water-dispute-with-iran (Accessed 26 January 2019), 2018.

Jamil, M. A. and Haddad, L.: A Study on Environmental Issues with Reference to the Qur'an and the Sunna, ISESCO., 1999.

Jan, Z., Ali, I. and Khan, S. N.: SOME NEWLY DISCOVERED BRONZE AGE SITES IN THE GOMAL VALLEY, NWFP, PAKISTAN, , 16, 2008.

Janjua, M. Q.: In the Shadow of the Durand Line: Security, Stability, and the Future of Pakistan and Afghanistan. [online] Available from: https://apps.dtic.mil/dtic/tr/fulltext/u2/a501684.pdf, 2009.

Jarvis, T. and Wolf, A.: Managing water negotiations and conflicts in concept and in practice, Transbound. Water Manag. Princ. Pract., 125–141, 2010.

Jasanoff, S., Martello, M. L. and Haas, P. M.: Earthly politics: local and global in environmental governance, MIT press., 2004.

Jettmar, K.: History of civilizations of Central Asia. Volume 1, JSTOR., 1994.

Jiménez Cisneros, B. E., Oki, T., Arnell, N. W., Benito, G., Cogley, J. G., Doll, P., Jiang, T. and Mwakalila, S. S.: Freshwater resources, 2014.

Joshi, M.: Turkmenistan-Afghanisthan-Pakistan-India Pipeline, Possibility Pipe Dream, 2011.

Juneja, T. and Chaudhary, A.: Assessment of water quality and its effects on the health of residents of Jhunjhunu district, Rajasthan: A cross sectional study, J. Public Health Epidemiol., 5(4), 186–191, 2013.

Jurriens, M., Mollinga, P. P. and Wester, P.: Scarcity by design: protective irrigation in India and Pakistan, ILRI., 1996.

Juuti, P. S., Katko, T. S. and Vuorinen, H. S.: A brief history of water and health from ancient civilizations to modern times, Retrieved Dec, 29, 2016, 2016.

Kagan, R.: America and Europe in the New World Order, , 53, 2003.

Kakakhel, S.: Afghanistan-Pakistan Treaty on the Kabul River Basin?, Third Pole [online] Available from: https://www.thethirdpole.net/en/2017/03/02/afghanistan-pakistan-treaty-on-the-kabul-river-basin/ (Accessed 25 January 2019), 2017.

Kakakhel, S.: Agreements and Institutions Related to Shared Rivers within South Asia and Beyond, in South Asian Rivers, pp. 45–67, Springer., 2018.

Kakoyannis, C. and Stankey, G. H.: Assessing and evaluating recreational uses of water resources: implications for an integrated management framework., Gen Tech Rep PNW-GTR-536 Portland US Dep. Agric. For. Serv. Pac. Northwest Res. Stn. 59 P, 536, 2002.

Kamal, D., Khan, A. N., Rahman, M. A. and Ahamed, F.: Study on the physico chemical properties of water of Mouri River, Khulna, Bangladesh, Pak. J. Biol. Sci., 10(5), 710–717, 2007.

Kamble, S. M.: Water pollution and public health issues in Kolhapur city in Maharashtra, Int. J. Sci. Res. Publ., 4(1), 1–6, 2014.

Kassim, H. and Le Galès, P.: Exploring governance in a multi-level polity: A policy instruments approach, West Eur. Polit., 33(1), 1–21, 2010.

Kasymov, S.: Water resource disputes: conflict and cooperation in drainage basins, Int. J. World Peace, 81–110, 2011.

Kaura, V.: The Durand Line: A British Legacy Plaguing Afghan-Pakistani Relations, Middle East Inst. [online] Available from: https://www.mei.edu/publications/durand-line-british-legacy-plaguing-afghan-pakistani-relations (Accessed 26 January 2019), 2017.

Kawasaki, S., Watanabe, F., Suzuki, S., Nishimaki, R. and Takahashi, S.: Current situation and issues on agriculture of Afghanistan, J. Arid Land Stud., 22(1), 345–348, 2012.

Kayathwal, M. K. and Kayathwal, M. K.: PAK-AFGHAN RELATIONS : DURAND LINE ISSUE, Indian J. Asian Aff., 7(2), 37–46, 1994.

Keck, M. E. and Sikkink, K.: Transnational advocacy networks in international and regional politics, Int. Soc. Sci. J., 51(159), 89–101, 1999.

Kehl, J. R.: Water Security in Transboundary Systems: Cooperation in Intractable Conflicts and the Nile System, Water Secur. Middle East Essays Sci. Soc. Coop., 1, 39, 2017.

Keithly, D. M.: The USA and The World 2013, Rowman & Littlefield., 2013.

Kenoyer, J. M.: Ancient cities of the Indus valley civilization, American Institute of Pakistan Studies., 1998.

Kenoyer, J. M.: Cultures and societies of the Indus tradition, Hist. Roots Mak. _ Aryan, 21–49, 2006.

Keohane, R. O.: The demand for international regimes, Int. Organ., 36(2), 325–355, 1982.

Keohane, R. O.: International institutions: Two approaches, Int. Stud. Q., 32(4), 379–396, 1988.

Keohane, R. O.: After hegemony: Cooperation and discord in the world political economy, Princeton University Press., 2005.

Keohane, R. O. and Martin, L. L.: The promise of institutionalist theory, Int. Secur., 20(1), 39–51, 1995.

Kerry, J. F., Boxer, B., Menendez, R., Cardin, B. L., Casey, R. P., Webb, J., Shaheen, J., Coons, C. A., Durbin, R. J., Udall, T., Lugar, R. G., Corker, B., Risch, J. E., Rubio, M., Inhofe, J. M., Demint, J., Carolina, S., Isakson, J., Barrasso, J., Lee, M. and Lowenstein, F. G.: COMMITTEE ON FOREIGN RELATIONS. [online] Available from: https://www.foreign.senate.gov/imo/media/doc/Senate%20Print%20112-

10%20Avoiding%20Water%20Wars%20Water%20Scarcity%20and%20Central%20Asia%20Afgahnistan%20and%20Pakistan.pdf, 2011.

Keys, P., Barron, J. and Lannerstad, M.: Releasing the pressure: water resource efficiencies and gains for ecosystem services, UNEP. 2012.

Khalid, A.: Islam after communism: religion and politics in Central Asia, Univ of California Press., 2014.

Khalid, S., Qasim, M. and Farhan, D.: Hydro-meteorological characteristics of Indus River Basin at extreme north of Pakistan, J Earth Sci Clim Change, 5(1), 2013.

Khan, A.: A review of the wetlands of Afghanistan, , 7, 2006.

Khan, F., Knox, R. and Thomas, K.: West of the Indus: the chronology of settlement in the protohistoric culture phases, with special reference to the Bannu region, , XV [online] Available from: http://journals.uop.edu.pk/papers/AP_v15_119to125.pdf (Accessed 28 July 2019), 2002.

Khan, F. J. and Javed, Y.: Delivering Access to Safe Drinking Water and Adequate Sanitation in Pakistan Working Paper Series 2007: 30. Retrieved 2009., 2007.

Khan, H., Inamullah, E. and Shams, K.: Population, environment and poverty in Pakistan: linkages and empirical evidence, Environ. Dev. Sustain., 11(2), 375–392, 2009.

Khan, H. M.: Islamic Law, Customary Law, and Afghan Informal Justice. 14, 2015.

Khan, M. and Ghouri, A. M.: Environmental pollution: Its effects on life and its remedies, 2011.

Khan, M. and Khan, M.: Water quality characteristics of the Kabul river in Pakistan under high flow conditions, J.-Chem. Soc. Pak., 19, 205–209, 1997.

Khan, M. A.: Introduction to tourism, Anmol. 2005.

Khan, R. S., Nawaz, K., van Steenbergen, F., Nizami, A. and Ahmad, S.: THE DRY SIDE OF THE INDUS, [online] Available from: http://metameta.nl/wp-content/uploads/2014/09/Book_Dry_Side_Indus_SF.pdf, 2014.

Khan, S.: Chapter 2: Sanitation and wastewater technologies in Harappa/Indus valley civilization (ca. 2600-1900 BC), Khan Sanit. Wastewater Technol. HarappaIndus Val. Civiliz., 2012.

Khan, S., Shahnaz, M., Jehan, N., Rehman, S., Shah, M. T. and Din, I.: Drinking water quality and human health risk in Charsadda district, Pakistan, J. Clean. Prod., 60, 93–101, 2013.

Khatoon, R., Hussain, I., Anwar, M. and Nawaz, M. A.: Diet selection of snow leopard (Panthera uncia) in Chitral, Pakistan, Turk. J. Zool., 41(5), 914–923, 2017.

Khuram, I., Barinova, S., Ahmad, N., Ullah, A., Din, S. U., Jan, S. and Hamayun, M.: Ecological assessment of water quality in the Kabul River, Pakistan, using statistical methods, Oceanol. Hydrobiol. Stud., 46(2), 140–153, 2017.

Kiani, K.: Pakistan, Afghanistan mull over power project on Kunar River, DAWN.COM, 26th August [online] Available from: http://www.dawn.com/news/1038435 (Accessed 7 May 2016), 2013.

King, M. and Sturtewagen, B.: Making the Most of Afghanistan's River Basins, East-West Inst. N. Y., 2010.

Kipping, M.: Water security in the senegal river basin: water cooperation and water conflicts, in Facing Global Environmental Change, pp. 675–684, Springer., 2009.

Kohler- Koch, B. and Rittberger, B.: The _governance turn'in EU studies, JCMS J. Common Mark. Stud., 44, 27–49, 2006.

van Koppen, B., Giordano, M., Butterworth, J. and Mapedza, E.: Community-based water law and water resource management reform in developing countries: rationale, contents and key messages., 2007.

Koremenos, B., Lipson, C. and Snidal, D.: The rational design of international institutions, Int. Organ., 55(4), 761–799, 2001.

Koskenniemi, M. and Leino, P.: Fragmentation of international law? Postmodern anxieties, Leiden J. Int. Law, 15(3), 553–579, 2002.

Koumparou, D.: The right of thirst: water as a human right and as a commons, Glob. NEST J., 20(3), 637–645, 2018.

Kraft, M. E. and Furlong, S. R.: Public policy: Politics, analysis, and alternatives, Cq Press., 2012.

Krishna, R. and Salman, S. M.: International groundwater law and the World Bank policy for projects on transboundary groundwater, World Bank Tech. Pap., 163–190, 1999.

Krishnan, S. and Indu, R.: Groundwater contamination in India: discussing physical processes, health and socio-behavioral dimensions, International Water Management Institute., 2006.

Kromah, L. M.: The institutional nature of US hegemony: post 9/11., 2009.

Kugelman, M.: Urbanisation in Pakistan: causes and consequences, Wash. DC NOREF Nor. Peacebuilding Resour. Cent., 2013.

Kugle, S. A.: Framed, blamed and renamed: the recasting of Islamic jurisprudence in colonial South Asia, Mod. Asian Stud., 35(2), 257–313, 2001.

Lacey, A.: The research process, Res. Process Nurs., 13–26, 2010.

Lahiri, S. K. and Sinha, R.: Tectonic controls on the morphodynamics of the Brahmaputra River system in the upper Assam valley, India, Geomorphology, 169, 74–85, 2012.

Lange, M., Mahoney, J. and Vom Hau, M.: Colonialism and development: a comparative analysis of Spanish and British colonies, Am. J. Sociol., 111(5), 1412–1462, 2006.

Larigauderie, A. and Mooney, H. A.: The Intergovernmental science-policy Platform on Biodiversity and Ecosystem Services: moving a step closer to an IPCC-like mechanism for biodiversity, Curr. Opin. Environ. Sustain., 2(1–2), 9–14, 2010.

Laruelle, G. G.: Quantifying nutrient cycling and retention in coastal waters at the global scale. Geologica Ultraiectina (312), Departement Aardwetenschappen., 2009.

Lashkaripour, G. R. and Hussaini, S. A.: Water resource management in Kabul river basin, eastern Afghanistan, The Environmentalist, 28(3), 253–260, 2008.

Lau, M.: Afghanistan's Legal System and its Compatibility with International Human Rights Standards, [online] Available from: https://www.refworld.org/pdfid/48a3f02c0.pdf (Accessed 23 November 2018), 2003.

Lawson, V.: Reshaping economic geography? Producing spaces of inclusive development, Econ. Geogr. 86(4), 351–360, 2010.

Lead Pakistan: Prospects for Benefit Sharing in the Transboundary Kabul River Basin.pdf, [online] Available from: http://www.lead.org.pk/lead/Publications/Prospects%20for%20Benefit%20Sharing%20in%20the%20Transboundary%20Kabul%20River%20Basin.pdf (Accessed 26 January 2019), 2017.

Lead Pakistan: Shifting From Water Scarcity to Surplus in Pakistan, [online] Available from: http://www.lead.org.pk/talks/talk20.htm (Accessed 1 May 2019), 2018.

Leao, I., Ahmed, M. and Kar, A.: Unlocking the Potential of Agriculture for Afghanistan's Growth. [online] Available from:

https://www.worldbank.org/en/country/afghanistan/publication/unlocking-potential-of-agriculture-for-afghanistan-growth (Accessed 13 June 2018), 2018.

Lechner, S.: Why anarchy still matters for International Relations: On theories and things, J. Int. Polit. Theory, 13(3), 341–359, 2017.

Lecours, A.: New institutionalism, New Institutionalism Theory Pract., 3–27, 2005.

Lee, J. L.: Water management, livestock and the opium economy: social water management, Afghanistan Research and Evaluation Unit., 2006.

Leeds, B. A.: The Design and Performance of Regional Institutions, Blackwell Publishing Ltd Oxford, UK. 2009.

Leicht, K. T. and Jenkins, J. C.: Handbook of politics: State and society in global perspective, Springer. 2009.

LeMarquand, D. G.: International rivers: The politics of cooperation, Westwater Research Centre, University of British Columbia., 1977.

Levi, M.: A logic of institutional change'in Cook, KS and Levi, M. (eds) The Limits of Rationality, The University of Chicago Press, Chicago., 1990.

Levy, B. S. and Sidel, V. W.: Water rights and water fights: preventing and resolving conflicts before they boil over, American Public Health Association., 2011.

Lewis, B.: Faith and power: Religion and politics in the Middle East, Oxford University Press., 2010.

Lewis, J.: The Council of the European Union and the European Council, Routledge Handb. Eur. Polit., 2015.

Lewis, J. and Wiser, R.: A review of international experience with policies to promote wind power industry development, Prep. Energy Found. China Sustain. Energy Program, 2005.

Liao, K.-H.: A theory on urban resilience to floods—a basis for alternative planning practices, Ecol. Soc., 17(4), 2012.

Link, M. P., Piontek, F., Scheffran, J. and Schilling, J.: Impact of climate change on water conflict and cooperation in the Nile River Basin, Clim. Change Secur. Trondheim, 21–24, 2010.

Liu, J., Dietz, T., Carpenter, S. R., Alberti, M., Folke, C., Moran, E., Pell, A. N., Deadman, P., Kratz, T. and Lubchenco, J.: Complexity of coupled human and natural systems, science, 317(5844), 1513–1516, 2007.

Liu, J., Zehnder, A. J. and Yang, H.: Global consumptive water use for crop production: The importance of green water and virtual water, Water Resour. Res., 45(5), 2009.

Liu, T. T.-T. and Ming-Te, H.: Hegemonic Stability and Northeast Asia: What Hegemon? What Stability?, J. Altern. Perspect. Soc. Sci., 3(2), 404–418, 2011.

Loayza, N. and Wada, T.: Public infrastructure trends and gaps in Pakistan, 2012.

Lopes, P. D.: Governing Iberian Rivers: from bilateral management to common basin governance?, Int. Environ. Agreem. Polit. Law Econ., 12(3), 251–268, 2012.

López Zavala, M. Á., Castillo Vega, R. and López Miranda, R. A.: Potential of Rainwater Harvesting and Greywater Reuse for Water Consumption Reduction and Wastewater Minimization, Water, 8(6), 264, 2016.

López-Hoffman, L.: Transboundary ecosystem services: a new vision for managing the shared environment of the US and Mexico, Udall Cent. Stud. Public Policy Univ. Ariz. Environ. Policy Work. Pap. (2), 2010.

Loschmann, C., Parsons, C. R. and Siegel, M.: Does Shelter Assistance Reduce Poverty in Afghanistan? World Dev., 74, 305–322, 2015.

Lugo, L., Cooperman, A., Bell, J., O'Connell, E. and Stencel, S.: The world's Muslims: Religion, politics and society, World, 2013.

Lukes, S. and Haglund, L.: Power and luck, Arch. Eur. Sociol. J. Sociol. Arch. Für Soziol., 45–66, 2005.

Lustick, I. S.: Hegemony and the riddle of nationalism: The dialectics of nationalism and religion in the Middle East, 2002.

Luzi, S.: International River Basins: Management and Conflict Perspectives, CSS Environ. 2007.

Mack, T. J.: Afghanistan's Kabul Basin Faces Major Water Challenges. USGS Newsroom, Httpwww Usgs Govnewsroomarticle Asp ID 2521, 2010.

Macklin, M. G. and Lewin, J.: The rivers of civilization, Quat. Sci. Rev., 114, 228–244, 2015.

Maddison, A.: Class structure and economic growth: India and Pakistan since the Moghuls, Routledge. 2013.

Mahmood, H. and Chaudhary, A. R.: FDI, population density and carbon dioxide emissions: A case study of Pakistan, Iran. J. Energy Environ. 3(4), 354–360, 2012.

Mahmood, K., Alamgir, A., Khan, M. A., Shaukat, S. S., Anwar, M. and SHERWANI, S. K.: Seasonal variation in water quality of lower Sindh, Pakistan, FUUAST J. Biol., 4(2), 147–156, 2014.

Majidyar, W.: Afghanistan and Pakistan's Looming Water Conflict, The Diplomat [online] Available from: https://thediplomat.com/2018/12/afghanistan-and-pakistans-looming-water-conflict/ (Accessed 25 January 2019), 2018.

Majoor, S. and Schwartz, K.: Instruments of urban governance, in Geographies of Urban Governance, pp. 109–126, Springer. 2015.

Malakootian, M. and Nouri, J.: Chemical Variations of Ground Water Affected by the Earthquake in bam region Malakootian, M, Int. J. Environ. Res., 4(3), 443–454, 2010.

Malik, I.: State and civil society in Pakistan: Politics of authority, ideology and ethnicity, Springer. 1996.

Malyar, I.: Transboundary water institutions in developing countries: a case study in Afghanistan, 2016.

Malyar, I. and Hearns, G.: A Review of Current and Possible Future Relations in Kabul River Basin, Duran Res. Anal. [online] Available from: http://duran.af/a-review-of-current-and-possible-future-relations-in-kabul-river-basin/ (Accessed 26 January 2019), 2014.

Mangi, H. O.: Tide Management in the Elbe River and Changes in Ecosystem Services, Adv. Ecol., 2016, 2016.

Manuel, M., Lightfoot, D. and Fattahi, M.: The sustainability of ancient water control techniques in Iran: an overview, Water Hist., 10(1), 13–30, 2018.

Marks, G. and Hooghe, L.: National identity and support for European integration, WZB Discussion Paper., 2003.

Mashal, M.: What Iran and Pakistan Want from the Afghans: Water, Time [online] Available from: http://world.time.com/2012/12/02/what-iran-and-pakistan-want-from-the-afghans-water/ (Accessed 27 January 2019), 2012.

Masood, A. and Mushtaq, H.: Spatio-Temporal Analysis of Early Twenty-First Century Areal Changes in the Kabul River Basin Cryosphere, Earth Syst. Environ., 2(3), 563–571, 2018.

McCaffery, J.: Animals That Have Gone Extinct in the Last 100 Years, Read. Dig. [online] Available from: https://www.rd.com/culture/animals-extinct-last-100-years/ (Accessed 29 November 2018), 2015.

McCaffrey, S. C.: The harmon doctrine one hundred years later: buried, not praised, Nat. Resour. J., 965–1007, 1996.

McCaffrey, S. C.: Water, water everywhere, but too few drops to drink: the coming fresh water crisis and international environmental law, Denv J Intl Pol, 28, 325, 1999.

McCarthy, J. and Mustafa, D.: Despite the best intentions? Experiences with water resource management in northern Afghanistan, Water Post-Confl. Peacebuilding, 2014.

McGregor, J. A. and Pouw, N.: Towards an economics of well-being, Camb. J. Econ., 41(4), 1123–1142, 2016.

McIntosh, M. A.: Early Civilizations of the Indian Subcontinent, [online] Available from: https://brewminate.com/early-civilizations-of-the-indian-subcontinent/ (Accessed 1 May 2019), 2018.

McIntyre, O.: Environmental Protection of International Watercourses under International Law, 1st ed., Routledge. 2007.

McIntyre, O.: International Water R esources Law and the International Law Commission Draft Articles on Transboundary Aquifers: A Missed Opportunity for Cross-Fertilisation? Int. Community Law Rev., 13(3), 237–254, doi:10.1163/187197311X582386, 2011.

McKinney, D. C.: Transboundary water challenges: case studies, Cent. Res. Water Resour. Univ. Tex. Austin Google Sch., 2011.

McLellan, M. and Porter, T.: News, improved: How America's newsrooms are learning to change, CQ Press Washington, DC. 2007.

McMichael, A., Scholes, R., Hefny, M., Pereira, E., Palm, C. and Foale, S.: Linking ecosystem services and human well-being, Island Press., 2005.

McMichael, A. J., Woodruff, R. E. and Hales, S.: Climate change and human health: present and future risks, The Lancet, 367(9513), 859–869, 2006.

McMurray, J. C. and Tarlock, A. D.: The law of later-developing riparian states: the case of Afghanistan, NYU Envtl LJ, 12, 711, 2003.

MEA, M. E. A.: Ecosystems and human well-being, Island Press, Washington, DC. 2005.

Meagher, T. M.: Jefferson, Madison, Gallatin and the Resourcing of the War of 1812, in Financing Armed Conflict, Volume 1, pp. 105–184, Springer., 2017.

Mearsheimer, J.: Anarchy and the Struggle for Power, Realism Read. 179, 2014.

Mearsheimer, J. J.: The tragedy of great power politics, WW Norton & Company., 2001.

Mearsheimer, J. J.: Structural realism, Oxford: Oxford University Press., 2007.

Mearsheimer, J. J.: Hollow Victory, Foreign Policy, 2009.

Mechlem, K.: Water as a vehicle for inter-state cooperation: A Legal Perspective, Food and Agriculture Organization of the United Nations (FAO). Legal Office. 2003.

Mehari, A., Van Steenbergen, F. and Schultz, B.: Modernization of spate irrigated agriculture: A new approach, Irrig. Drain. 60(2), 163–173, 2011.

Mehsud, R.: Understanding _Pashtunwali,' Express Trib., 28th November [online] Available from: https://tribune.com.pk/story/999733/understanding-pashtunwali/ (Accessed 15 June 2018), 2015.

Mehta, P.: CROSSING THE LIMIT: E-WASTE DYNAMICS, IMPACTS, AND LEGISLATION IN INDIA, EVERYMAN'S Sci., 380, 2016.

Meir, P., Mencuccini, M., Binks, O., da Costa, A. L., Ferreira, L. and Rowland, L.: Short-term effects of drought on tropical forest do not fully predict impacts of repeated or long-term drought: gas exchange versus growth, Philos. Trans. R. Soc. B Biol. Sci., 373(1760), 20170311, 2018.

Mekonnen, M. M. and Hoekstra, A. Y.: The green, blue and grey water footprint of crops and derived crop products, Hydrol. Earth Syst. Sci., 15(5), 1577–1600, 2011.

Menga, F. and Mirumachi, N.: Fostering Tajik hydraulic development: Examining the role of soft power in the case of the Rogun Dam, Water Altern., 9(2), 373–388, 2016.

Menon, J.: Unit-9 Case Study of a Major Urban Centre: Mohenjodaro, IGNOU., 2018.

Merz, S. K., Evans, R. and Clifton, C. A.: Environmental water requirements to maintain groundwater dependent ecosystems, Environment Australia., 2001.

Mgquba, S. K. and Majozi, S.: Climate change and its impacts on hydro-politics in transboundary basins: a case study of the Orange-Senqu River basin, J. Water Clim. Change, 2018.

Milligan, S. R., Holt, W. V. and Lloyd, R.: Impacts of climate change and environmental factors on reproduction and development in wildlife, Philos. Trans. R. Soc. B Biol. Sci., 364(1534), 3313–3319, doi:10.1098/rstb.2009.0175, 2009.

Moench, M.: In Intensive Use of Groundwater: Challenges and Opportunities, Ab Ingdon UK Balkema, 2002.

Mollinga, P. P.: On the waterfront: Water distribution, technology and agrarian change in a South Indian canal irrigation system, Orient Blackswan., 2003.

Montaldo, C. R. B.: Sustainable development approaches for rural development and poverty alleviation & community capacity building for rural development and poverty alleviation., 2013.

Morris, K.-A.: Oil, power, and global hegemony, Stellenbosch: Stellenbosch University., 2015.

Morris, M. and de Loë, R. C.: Cooperative and adaptive transboundary water governance in Canada's Mackenzie River Basin: status and prospects, Ecol. Soc., 21(1), 2016.

Mosello, B.: How to Deal with Climate Change? Springer., 2015.

Mosse, D.: Rule and representation: Transformations in the governance of the water commons in British South India, J. Asian Stud., 65(1), 61–90, 2006.

Mosse, D.: A relational approach to durable poverty, inequality and power, J. Dev. Stud., 46(7), 1156–1178, 2010.

Mostert, E.: How can international donors promote transboundary water management?, German Development Institute, Bonn., 2005.

Moulherat, C., Tengberg, M., Haquet, J.-F. and Mille, B.: First evidence of cotton at Neolithic Mehrgarh, Pakistan: analysis of mineralized fibres from a copper bead, J. Archaeol. Sci., 29(12), 1393–1401, 2002.

Mukherjee, A.: Empire: How Colonial India Made Modern Britain, Econ. Polit. Wkly., 73–82, 2010.

Mukhtarov, F. and Gerlak, A. K.: Epistemic forms of integrated water resources management: towards knowledge versatility, Policy Sci., 47(2), 101–120, 2014.

Munia, H., Guillaume, J. H. A., Mirumachi, N., Porkka, M., Wada, Y. and Kummu, M.: Water stress in global transboundary river basins: significance of upstream water use on downstream stress, Environ. Res. Lett., 11(1), 014002, 2016.

Murgai, R., Ali, M. and Byerlee, D.: Productivity growth and sustainability in post–Green Revolution agriculture: the case of the Indian and Pakistan Punjabs, World Bank Res. Obs., 16(2), 199–218, 2001.

Murphy, W. F.: Constitutional democracy: creating and maintaining a just political order, JHU Press., 2007.

Murtaza, G. and Zia, M. H.: Wastewater production, treatment and use in Pakistan, in Second Regional Workshop of the Project _Safe Use of Wastewater in Agriculture, pp. 16–18., 2012.

Mustafa, D.: Colonial law, contemporary water issues in Pakistan, Polit. Geogr., 20(7), 817–837, 2001.

Nafees, M.: Environmental Study of Kabul River and its Tributaries in NWFP, Pakistan, M. Phil Thesis, Department of Environmental Sciences, University of Peshawar …., 2004.

Nafees, M., Ahmad, T. and Arshad, M.: A review of Kabul River uses and its impact on fish and fisherman, J. Humanit. Soc. Sci. Univ. Peshawar, 19(2), 73–84, 2011.

Nafees, M., Shabir, A. and Zahid, U.: Construction of dam on Kabul River and its socio-economic implication for Khyber Pukhtunkhwa, Pakistan, in seminar on ─Pak–Afghan Water Sharing Issue, Feburary, vol. 23, p. 2016., 2016.

Nafees, M., Ahmad, F., Butt, M. N. and Khurshed, M.: Effects of water shortage in Kabul River network on the plain areas of Khyber Pakhtunkhwa, Pakistan, Environ. Monit. Assess., 190(6), 359, 2018.

Naff, T.: Islamic law and the politics of water, in The evolution of the law and politics of water, pp. 37–52, Springer. 2009.

Naff, T. and Dellapenna, J.: Can there be confluence? A comparative consideration of Western and Islamic fresh water law, Water Policy, 4(6), 465–489, 2002.

Naidoo, S. and Olaniran, A.: Treated wastewater effluent as a source of microbial pollution of surface water resources, Int. J. Environ. Res. Public. Health, 11(1), 249–270, 2014.

Najmuddin, O., Deng, X. and Bhattacharya, R.: The Dynamics of Land Use/Cover and the Statistical Assessment of Cropland Change Drivers in the Kabul River Basin, Afghanistan, Sustainability, 10(2), 423, 2018.

National Water Policy of Pakistan: [online] Available from: http://waterbeyondborders.net/wp-content/uploads/2018/07/Pakistan-National-Water-Policy-2018.pdf (Accessed 15 July 2018), 2018.

NEA, U.: UK National Ecosystem Assessment, 2011, UK Natl. Ecosyst. Assess. Synth. Key Find. UNEP-WCMC Camb. 2011.

Neal, M. J., Greco, F., Connell, D. and Conrad, J.: The Social-Environmental Justice of Groundwater Governance, in Integrated Groundwater Management, pp. 253–272, Springer. 2016.

Nellemann, C.: The environmental food crisis: the environment's role in averting future food crises: a UNEP rapid response assessment, UNEP/Earthprint., 2009.

NEPA-Afghanistan: Afghanistan's environment, 2008, National Environmental Protection Agency of the Republic of Afghanistan : United Nations Environment Programme, Kabul. [online] Available from: http://mom.gov.af/Content/files/UNEP%20Environment%202008.pdf (Accessed 13 April 2018), 2008.

Ngan, T. T. T.: Neo-realism and the Balance of Power in Southeast Asia, Pap. Dipresentasikan Untuk Lokakarya Cent. East Eur. Int. Stud. Assoc. Stud. Assoc. CEEISA-ISA, 2016.

Nijssen, S.: From Dispute to Resolution: Managing Land in Afghanistan, Civ. Mil. Fusion Cent., 2011.

Nobles, R. and Schiff, D.: Using systems theory to study legal pluralism: what could be gained?, Law Soc. Rev., 46(2), 265–296, 2012.

Nolan, C. and Trew, A.: Transaction costs and institutions: investments in exchange, BE J. Theor. Econ., 15(2), 391–432, 2015.

Nordhaus, W. D.: Global Melting? The Economics of Disintegration of the Greenland Ice Sheet, National Bureau of Economic Research. 2018.

North, D. C.: A transaction cost theory of politics, J. Theor. Polit., 2(4), 355–367, 1990.

Notezai, M. A.: The lost civilisation of Mehrgarh: A treasure in ruins, DAWN.COM [online] Available from: https://www.dawn.com/news/1316715 (Accessed 28 July 2019), 2017.

Nye, J. S.: Soft power: The means to success in world politics, Public affairs., 2004.

Nye Jr, J. S.: The changing nature of world power, in Power in the Global Information Age, pp. 61–75, Routledge. 2004.

Omrani, B.: The Durand Line: History and Problems of the Afghan-Pakistan Border, Asian Aff., 40(2), 177–195, doi:10.1080/03068370902871508, 2009.

Omrani, B.: The Durand Line: Analysis of the Legal Status of the Disputed Afghanistan-Pakistan Frontier, , 26, 53, 2018.

Onifade, O. A., Adio-Moses, R., Adigun, J. O., Oguntunji, I. O. and Ogungboye, R. O.: Impacts of flood disaster on sustainable national development in Ibadan North local government, Oyo State, Arab. J. Bus. Manag. Rev. OMAN Chapter, 4(2), 139–147, 2014.

O'Riordan, T. and Jordan, A.: Institutions, climate change and cultural theory: towards a common analytical framework, Glob. Environ. Change, 9(2), 81–93, 1999.

Orme, M., Cuthbert, Z., Sindico, F., Gibson, J. and Bostic, R.: Good transboundary water governance in the 2015 Sustainable Development Goals: a legal perspective, Water Int., 40(7), 969–983, 2015.

Ostfeld, A., Barchiesi, S., Bonte, M., Collier, C. R., Cross, K., Darch, G., Farrell, T. A., Smith, M., Vicory, A. and Weyand, M.: Climate change impacts on river basin and freshwater ecosystems: some observations on challenges and emerging solutions, J. Water Clim. Change, 3(3), 171–184, 2012.

Ostrom, E.: The governance challenge: matching institutions to the structure of social-ecological systems, 2007.

Oteng-Peprah, M. and Acheampong, M. A.: Greywater Characteristics, Treatment Systems, Reuse Strategies and User Perception—a Review, Water. Air. Soil Pollut., 229(8), 255, 2018.

Owa, F. W.: Water pollution: sources, effects, control and management, Int. Lett. Nat. Sci., 3, 2014.

Ozkan, E. and Cetin, H. C.: The Realist and Liberal Positions on the Role of International Organizations in Maintaining World Order, Eur. Sci. J. ESJ, 12(17), 2016.

Pahl-Wostl, C., Craps, M., Dewulf, A., Mostert, E., Tabara, D. and Taillieu, T.: Social learning and water resources management, Ecol. Soc., 12(2), 2007.

Pai, S. and Sharma, P. K.: New institutionalism and legislative governance in the Indian states: A comparative study of West Bengal and Uttar Pradesh, Centre for the Study of Law and Governance, Jawaharlal Nehru University., 2005.

Palaniappan, M., Gleick, P. H., Allen, L., Cohen, M. J., Christian-Smith, J., Smith, C. and Ross, N.: Clearing the waters: a focus on water quality solutions, 2010.

Palka, E. J.: Afghanistan: A Regional Geography. 2001.

Palthe, J.: Regulative, normative, and cognitive elements of organizations: Implications for managing change, Manag. Organ. Stud., 1(2), 59, 2014a.

Palthe, J.: Regulative, normative, and cognitive elements of organizations: Implications for managing change, Manag. Organ. Stud., 1(2), 59, 2014b.

Panara, C.: The sub-national dimension of the EU: A legal study of multilevel governance, Springer., 2015.

Pascual, U., Balvanera, P., Díaz, S., Pataki, G., Roth, E., Stenseke, M., Watson, R. T., Dessane, E. B., Islar, M. and Kelemen, E.: Valuing nature's contributions to people: the IPBES approach, Curr. Opin. Environ. Sustain. 26, 7–16, 2017.

Pashakhanlou, A. H.: Comparing and contrasting classical realism and Neorealism, Saatavilla Osoitteessa Httpwww E-Ir Info20090723comparing--Contrasting-Class.--Neo-Realism Luettu, 10, 2014, 2009.

Pattberg, P. and Stripple, J.: Beyond the public and private divide: remapping transnational climate governance in the 21st century, Int. Environ. Agreem. Polit. Law Econ., 8(4), 367–388, 2008.

Pawari, M. J. and Gawande, S.: Ground water pollution & its consequence, Int. J. Eng. Res. Gen. Sci., 3(4), 773–76, 2015.

Pease, K.-K. S.: International organizations: [perspective on governance in the twenty-first century][eng], Boston, Mass.: Longman., 2012.

Pendleton, L., Donato, D. C., Murray, B. C., Crooks, S., Jenkins, W. A., Sifleet, S., Craft, C., Fourqurean, J. W., Kauffman, J. B. and Marbà, N.: Estimating global ‒blue carbon" emissions from conversion and degradation of vegetated coastal ecosystems, PloS One, 7(9), e43542, 2012.

Penuel, K. B., Statler, M. and Hagen, R.: Encyclopedia of crisis management, Sage Publications., 2013.

Perrings, C., Duraiappah, A., Larigauderie, A. and Mooney, H.: The biodiversity and ecosystem services science-policy interface, Science, 331(6021), 1139–1140, 2011.

Perry, M.: Afghanistan water crisis, prezi.com [online] Available from: https://prezi.com/ufz5odlced8o/afghanistan-water-crisis/ (Accessed 27 January 2019), 2015.

Person, R.: Balance of threat: The domestic insecurity of Vladimir Putin, J. Eurasian Stud., 8(1), 44–59, 2017.

Pervaz, I. and Khan, S.: Brewing Conflict over Kabul River; Policy Options for Legal Framework, [online] Available from: https://ndu.edu.pk/issra/issra_pub/articles/issra-paper/ISSRA_Papers_Vol6_IssueII_2014/03-Brewing-Conflict-over-Kabul-River.pdf (Accessed 26 January 2019), 2014.

Peters, B. G.: Globalization, institutions and governance, Gov. Twenty-First Century Revital. Public Serv., 29–57, 2000.

Peters, B. G. and Pierre, J.: Multi-level governance and democracy: a Faustian bargain?, Multi-Level Gov., 75–89, 2004.

Petersen-Perlman, J. D. and Fischhendler, I.: The weakness of the strong: re-examining power in transboundary water dynamics, Int. Environ. Agreem. Polit. Law Econ., 18(2), 275–294, 2018.

Petrie, C. A.: Diversity, variability, adaptation and _fagility'in the Indus Civilization, McDonald Institute for Archaeological Research., 2019.

Pfeiffer, J.: Traditional Dispute Resolution Mechanisms in Afghanistan and their Relationship to the National Justice Sector, Verfass. Recht Übersee Law Polit. Afr. Asia Lat. Am., 44(1), 81–98, 2011.

Piattoni, S.: Multi- level governance: a historical and conceptual analysis, Eur. Integr., 31(2), 163–180, 2009.

Piattoni, S.: The theory of multi-level governance: conceptual, empirical, and normative challenges, Oxford University Press., 2010.

Pidwirny, M.: The hydrologic cycle, Fundam. Phys. Geogr., 2006.

Pierre, J. and Peters, G. B.: Governance, politics and the state, 2000.

Piesse, M.: The Indus River and Agriculture in Pakistan, Future Dir. Int. [online] Available from: http://www.futuredirections.org.au/publication/the-indus-river-and-agriculture-in-pakistan/ (Accessed 27 January 2019), 2015.

Pingali, P. L.: Green revolution: impacts, limits, and the path ahead, Proc. Natl. Acad. Sci., 109(31), 12302–12308, 2012.

Pittock, J.: National climate change policies and sustainable water management: conflicts and synergies, Ecol. Soc., 16(2), 2011.

Possehl, G. L.: The Indus civilization: a contemporary perspective, Rowman Altamira., 2002.

Postel, S. and Richter, B.: Rivers for life: managing water for people and nature, Island Press., 2012.

Postel, S. L., Daily, G. C. and Ehrlich, P. R.: Human appropriation of renewable fresh water, Science, 271(5250), 785–788, 1996.

Powell, E. J.: Islamic law states and peaceful resolution of territorial disputes, Int. Organ., 69(4), 777–807, 2015.

Powell, E. J. and Staton, J. K.: Domestic judicial institutions and human rights treaty violation, Int. Stud. Q., 53(1), 149–174, 2009.

Powell, R.: Anarchy in international relations theory: the neorealist-neoliberal debate, Int. Organ., 48(2), 313–344, 1994.

Pressey, R. and Middleton, M.: Impacts of flood mitigation works on coastal wetlands in New South Wales., Wetl. Aust., 2(1), 2009.

Price, G.: Attitudes to water in South Asia, The Royal Institute of International Affairs, London. [online] Available from: https://www.chathamhouse.org/sites/default/files/field/field_document/20140627WaterSouthAsia.pdf (Accessed 15 December 2018), 2014.

Puffer, S. M., McCarthy, D. J. and Jaeger, A. M.: Institution building and institutional voids: can Poland's experience inform Russia and Brazil? Int. J. Emerg. Mark. 11(1), 18–41, 2016.

Qasem, J. R.: PROSPECTS OF WILD MEDICINAL AND INDUSTRIAL PLANTS OF SALINE HABITATS IN THE JORDAN VALLEY, , 20, 2015.

Qureshi, A. S.: Water resources management in Afghanistan: the issues and options, Iwmi., 2002.

Qureshi, A. S.: Water management in the Indus basin in Pakistan: challenges and opportunities, Mt. Res. Dev., 31(3), 252–260, 2011.

Qureshi, A. S.: Water crisis and food security, The Nation, 8th July [online] Available from: https://nation.com.pk/08-Jul-2018/water-crisis-and-food-security (Accessed 15 December 2018), 2018.

Raadgever, G. T. and Mostert, E.: Transboundary River Basin Management-State-of-the-art review on transboundary regimes and information management in the context of adaptive management, NeWater Rep. Ser., (10), 2005.

Radić, V. and Hock, R.: Glaciers in the Earth's hydrological cycle: assessments of glacier mass and runoff changes on global and regional scales, Surv. Geophys., 35(3), 813–837, 2014.

Rahaman, M. M.: Principles of Transboundary Water Resources Management and Ganges Treaties: An Analysis, Int. J. Water Resour. Dev., 25(1), 159–173, doi:10.1080/07900620802517574, 2009.

Rahman, A., Ali, M. and Kahn, S.: The British Art of Colonialism in India: Subjugation and Division, Peace Confl. Stud., 25(1), 5, 2018.

Rajput, M. I.: Inter-Provincial Water Issues in Pakistan, Pak. Inst. Legis. Dev. Transpar. PILDAT Retrieved 28th May, 2014.

Ramachandran, S.: India's Controversial Afghanistan Dams, The Diplomat [online] Available from: https://thediplomat.com/2018/08/indias-controversial-afghanistan-dams/ (Accessed 26 January 2019), 2018.

Ramay, S. A.: CPEC and the sustainable development it brings to Pakistan, 2018.

Ranjan, A.: Inter-Provincial water sharing conflicts in Pakistan, Pak. J. Pak. Stud., 4(2), 102–122, 2012a.

Ranjan, A.: Inter-Provincial water sharing conflicts in Pakistan, Pak. J. Pak. Stud., 4(2), 102–122, 2012b.

Rao, C. S., Gopinath, K. A., Rao, C. R., Raju, B. M. K., Rejani, R., Venkatesh, G. and Kumari, V. V.: Dryland agriculture in South Asia: Experiences, challenges and opportunities, in Innovations in Dryland Agriculture, pp. 345–392, Springer., 2016.

Rasouli, H., Kayastha, R. B., Bhattarai, B. C., Shrestha, A., Arian, H. and Armstrong, R.: Estimation of Discharge From Upper Kabul River Basin, Afghanistan Using the Snowmelt Runoff Model, J. Hydrol. Meteorol. 9(1), 85–94, 2015.

Rasul, G. and Sharma, B.: The nexus approach to water–energy–food security: an option for adaptation to climate change, Clim. Policy, 16(6), 682–702, 2016.

Ratnagar, S.: A critical view of Marshall's Mother Goddess at Mohenjo-Daro, Stud. People's Hist., 3(2), 113–127, 2016.

Rauniyar, G. P. and Kanbur, R.: Inclusive development: Two papers on conceptualization, application, and the ADB perspective, 2010.

Raza, S. A., Ali, Y. and Mehboob, F.: Role of agriculture in economic growth of Pakistan, 2012.

Razzaq, A.: POLICY: WATER SCARCITY MAY DISRUPT PAK-AFGHAN RELATIONS, DAWN.COM, 25th November [online] Available from: https://www.dawn.com/news/1447512 (Accessed 27 January 2019), 2018.

Rees, G.: The Role of Power and Institutions in Hydrodiplomacy, [online] Available from: https://transboundarywaters.science.oregonstate.edu/sites/transboundarywaters.science.oreg onstate.edu/files/Publications/Rees%2C%20G%202010%20- %20The%20Role%20of%20Power%20and%20Institutions%20in%20Hydrodiplomacy.pdf (Accessed 15 July 2016), 2010.

Rehman, A., Jingdong, L., Shahzad, B., Chandio, A. A., Hussain, I., Nabi, G. and Iqbal, M. S.: Economic perspectives of major field crops of Pakistan: An empirical study, Pac. Sci. Rev. B Humanit. Soc. Sci., 1(3), 145–158, 2015.

Rehman, F., Muhammad, S., Ashraf, I. and Ruby, T.: Effect of farmers' socioeconomic characteristics on access to agricultural information: Empirical evidence from Pakistan, Young 35, 52, 21–67, 2013.

Reis, J.: The state and the market: an institutionalist and relational take, RCCS Annu. Rev. Sel. Port. J. Rev. Crítica Ciênc. Sociais, (4), 2012.

Renner, M.: Water Challenges in Central-South Asia, South Asia, (4), 10, 2009.

Renner, M.: Troubled waters: Central and South Asia exemplify some of the planet's looming water shortages, World Watch, 23, 14–20, 2010.

Renner, M.: Water as a Trans-border Problem of Afghanistan, in Partners for Stability, pp. 151–166, Nomos Verlagsgesellschaft mbH & Co. KG., 2013.

Richter, W. L.: The political dynamics of Islamic resurgence in Pakistan, Asian Surv., 19(6), 547–557, 1979.

Rieu-Clarke, A.: International law and sustainable development, IWA Publishing., 2005.

Rignot, E., Mouginot, J. and Scheuchl, B.: Ice flow of the Antarctic ice sheet, Science, 333(6048), 1427–1430, 2011.

Rivera, J.: Institutional pressures and voluntary environmental behavior in developing countries: Evidence from the Costa Rican hotel industry, Soc. Nat. Resour., 17(9), 779–797, 2004.

Roberts, B. H.: Changes in urban density: its implications on the sustainable development of Australian cities, in Proceedings of the State of Australian Cities National Conference, pp. 720–739. 2007.

Robinson, M. and Ward, R.: Principles of hydrology, McGraw-Hill., 1990.

Rodríguez-Pose, A.: Do institutions matter for regional development?, Reg. Stud., 47(7), 1034–1047, 2013.

Rodwan, J. G.: Bottled water 2013: sustaining vitality, Bottled Water Report., 12–22, 2014.

Rogers, P. and Hall, A. W.: Effective water governance, Global water partnership Stockholm. 2003.

Roy, T.: Inequality in colonial India, 2018.

Ruddell, M. and Sanchez, R.: Comparison of the Development of Early Civilizations of Mesopotamia, Egypt, India, China, and Mesoamerica, [online] Available from: https://rudyruddell.wordpress.com/2012/12/28/comparison-of-the-first-civilizations/ (Accessed 15 September 2018), 2012.

Russell, R.: American diplomatic realism: A tradition practised and preached by George F. Kennan, Dipl. Statecraft, 11(3), 159–182, 2000.

Rutherford, M.: The old and the new institutionalism: can bridges be built?, J. Econ. Issues, 29(2), 443–451, 1995.

Rutherford, M.: Institutional economics: then and now, J. Econ. Perspect. 15(3), 173–194, 2001.

Rutten, M. and Mwangi, M.: How natural is natural? Seeking conceptual clarity over natural resources and conflicts, Confl. Nat. Resour. Glob. South–Conceptual Approaches, 51, 2014.

Saba, D. S.: Afghanistan: environmental degradation in a fragile ecological setting, Int. J. Sustain. Dev. World Ecol., 8(4), 279–289, 2001.

Sadeqinazhad, F., Atef, S. S. and Amatya, D.: Benefit-sharing framework in transboundary river basins: the case of the Eastern Kabul River Basin-Afghanistan, Cent. Asian J. Water Res., 4(1), 1–18, 2018.

Saeed, A.: Kabul River's famous Sher Mahi fish in peril, Third Pole [online] Available from: https://www.thethirdpole.net/en/2018/05/01/kabul-rivers-famous-sher-mahi-fish-in-peril/ (Accessed 25 January 2019), 2018.

Saeed, B. A., Hassani, S. and Malyar, I.: Transboundary Basin Management under conditions of Latent Conflict: A Multi-Sectoral and Multi-Disciplinary Approach towards the Kabul River Basin.   [online]   Available   from: https://pk.boell.org/sites/default/files/transboundary_basin_managment_kabul_river.pdf (Accessed 26 January 2019), 2016.

Salahuddin, S.: With war and neglect, Afghans face water shortage, Reuters, 24th March [online] Available   from: https://www.reuters.com/article/us-afghanistan-water-idUSTRE62N19Q20100324 (Accessed 27 January 2019), 2010.

Saleth, R. M. and Dinar, A.: Institutional changes in global water sector: trends, patterns, and implications, Water Policy, 2(3), 175–199, 2000.

Saleth, R. M. and Dinar, A.: Water institutions and sector performance: a quantitative analysis with cross-country data, World Bank Wash. DC, 2003.

Saleth, R. M. and Dinar, A.: The institutional economics of water: a cross-country analysis of institutions and performance, The World Bank., 2004a.

Saleth, R. M. and Dinar, A.: The institutional economics of water: a cross-country analysis of institutions and performance, The World Bank., 2004b.

Saleth, R. M. and Dinar, A.: Water institutional reforms in developing countries: Insights, evidences, and case studies, Econ. Dev. Environ. Sustain. New Policy Options, 2006.

Salman, M. A. S. and Uprety, K.: Conflict and Cooperation on South Asia's International Rivers: A Legal Perspective (Washington DC: The World Bank), 2002.

Salman, S. M.: The Helsinki Rules, the UN Watercourses Convention and the Berlin Rules: perspectives on international water law, Water Resour. Dev., 23(4), 625–640, 2007.

Salman, S. M.: The human right to water and sanitation: is the obligation deliverable?, Water Int., 39(7), 969–982, 2012.

Salman, S. M. and Bradlow, D. D.: Regulatory frameworks for water resources management: A comparative study, The World Bank., 2006.

Saltmarshe, D.: Local Governance in Afghanistan A View from the Ground. [online] Available from: https://reliefweb.int/sites/reliefweb.int/files/resources/Full_Report_1240.pdf, 2011.

Samim, M.: Afghanistan's Addiction to Foreign Aid, The Diplomat, 19, 2016.

Samim, M. S.: the impact of official development assistance on Agricultural Growth and agricultural Imports in developing countries, 2017.

Sanchez, J. C. and Roberts, J.: Transboundary Water Governance: Adaptation to Climate Change, IUCN. 2014.

Sanchez, N. and Gupta, J.: Recent changes in the Nile Region may create an opportunity for a more equitable sharing of the Nile River Waters, Neth. Int. Law Rev., 58(3), 363–385, 2011.

Sand, P.: The Effectiveness of Multilateral Environmental Agreements: Theory and Practice. 2016.

Santos, C., Matos, C. and Taveira-Pinto, F.: A comparative study of greywater from domestic and public buildings, Water Sci. Technol. Water Supply, 14(1), 135–141, 2014.

Sarfaraz, S.: The sub-regional classification of Pakistan's winter precipitation based on principal components analysis, Pak. J. Meteorol. Vol, 10(20), 2014.

Saruchera, D. and Lautze, J.: Measuring transboundary water cooperation: learning from the past to inform the sustainable development goals, International Water Management Institute (IWMI)., 2015.

Sarvilinna, A., Hjerppe, T., Arola, M., Hämäläinen, L. and Jormola, J.: Kaupunkipuron kunnostaminen (Restoring urban brooks). Environment Guide, Finn. Environ. Inst., 76, 2012.

Sattar, E., Robison, J. A. and McCool, D.: Evolution of Water Institutions in the Indus River basin: Reflections from the Law of the Colorado River, 2017.

Sattar, N.: Land rights in Islam, DAWN.COM, 2nd January [online] Available from: http://www.dawn.com/news/1154522 (Accessed 7 March 2018), 2015.

Savage, M., Dougherty, B., Hamza, M., Butterfield, R. and Bharwani, S.: Socio-economic impacts of climate change in Afghanistan, Stockh. Environ. Inst. Oxf. UK, 2009.

Schindler, S., O'Neill, F. H., Biró, M., Damm, C., Gasso, V., Kanka, R., van der Sluis, T., Krug, A., Lauwaars, S. G. and Sebesvari, Z.: Multifunctional floodplain management and biodiversity effects: a knowledge synthesis for six European countries, Biodivers. Conserv. 25(7), 1349–1382, 2016.

Schmeier, S.: The Institutional Design of River Basin Organizations – Introducing the RBO Institutional Design Database and its main Findings, , 34, 2012.

Schmeier, S.: The institutional design of river basin organizations–empirical findings from around the world, Int. J. River Basin Manag., 13(1), 51–72, 2015.

Schmeier, S., Gerlak, A. K. and Blumstein, S.: Clearing the muddy waters of shared watercourses governance: conceptualizing international River Basin Organizations, Int. Environ. Agreem. Polit. Law Econ., 16(4), 597–619, 2016.

Schmidt, S. K.: Mutual recognition as a new mode of governance, J. Eur. Public Policy, 14(5), 667–681, 2007.

Schmitter, P. C. and Kim, S.: The experience of European integration and the potential for Northeast Asian integration, 2005.

Schofer, E., Hironaka, A., Frank, D. J. and Longhofer, W.: Sociological institutionalism and world society, Wiley-Blackwell Companion Polit. Sociol., 33, 57, 2012.

Schroeder-Wildberg, E.: The 1997 International Watercourses Convention–Background and Negotiations, Work. Pap. Manag. Environ. Plan. Berl. Tech. Univ., 2002.

Schulze, S.: Public Participation in the Governance of Transboundary Water Resources–Mechanisms provided by River Basin Organizations, Eur. En Form. (3), 49–68, 2012.

Schwarzenbach, R. P., Egli, T., Hofstetter, T. B., Von Gunten, U. and Wehrli, B.: Global water pollution and human health, Annu. Rev. Environ. Resour. 35, 109–136, 2010.

Scott, W. R.: Institutional theory: Contributing to a theoretical research program, Gt. Minds Manag. Process Theory Dev., 37, 460–484, 2005.

Scott, W. R.: Approaching adulthood: the maturing of institutional theory, Theory Soc., 37(5), 427, 2008a.

Scott, W. R.: Institutions and organizations: Ideas and interests, Sage., 2008b.

Sedeqinazhad, F., Atef, S. S. and Amatya, D. M.: Benefit-sharing framework in transboundary river basins: the case of the Eastern Kabul River Basin-Afghanistan, , 18, 2018.

Seegert, J., Berendonk, T. U., Bernhofer, C., Blumensaat, F., Dombrowsky, I., Fuehner, C., Grundmann, J., Hagemann, N., Kalbacher, T. and Kopinke, F.-D.: Integrated water resources management under different hydrological, climatic and socio-economic conditions: results and lessons learned from a transdisciplinary IWRM project IWAS, Environ. Earth Sci., 72(12), 4677–4687, 2014.

Sehring, J.: Path Dependencies and Institutional Bricolage in Post-Soviet Water Governance. Water Altern., 2(1), 2009.

Selby, J.: Joint mismanagement: reappraising the Oslo water regime, in Water Resources in the Middle East, pp. 203–212, Springer. 2007.

Selby, J., Dahi, O. S., Fröhlich, C. and Hulme, M.: Climate change and the Syrian civil war revisited, Polit. Geogr., 60, 232–244, 2017.

Senier, A.: Rebuilding the judicial sector in Afghanistan: the role of customary law, Al Nakhalh Tufts Univ., 2006.

Shaarawy, S.: The role of customary international water law in settling water disputes by mediation: An examination of the Indus River and Renaissance dam disputes, 2016.

Shackleton, C., Shackleton, S., Gambiza, J., Nel, E., Rowntree, K. and Urquhart, P.: Links between Ecosystem Services and Poverty Alleviation, Situat. Anal. Arid Semi-Arid Lands South. Afr. Consort. Ecosyst. Poverty Sub-Sahar. Afr. 200p, 2008.

Shaffer, J. G. and Thapar, B. K.: Pre-Indus and early Indus cultures of Pakistan and India, Hist. Civiliz. Cent. Asia, 1, 247–281, 1992.

Shah, S. A.: Valuation of freshwater resources and sustainable management in poverty dominated areas, Colorado State University. Libraries. 2014.

Shah, S. A., Hoag, D. L. and Loomis, J.: Is willingness to pay for freshwater quality improvement in Pakistan affected by payment vehicle? Donations, mandatory government payments, or donations to NGO's, Environ. Econ. Policy Stud., 19(4), 807–818, 2017.

Shaheen, S., Bibi, Y., Hussain, M., Iqbal, M., Saira, H., Safdar, I., Mehboob, H., Ain, Q. T., Naseem, K. and Laraib, S.: A Review on Geranium wallichianum D-Don Ex-Sweet: An Endangered Medicinal Herb from Himalaya Region, Med. Aromat. Plants, 06(02), doi:10.4172/2167-0412.1000288, 2017.

Sharif, H.: Inter Province water distribution conflict in Pakistan, Rep. Intermedia Part Of, 2010.

Sharifi, A., Reshtin, A. and Sijapati, S.: ENSURING SUSTAINABLE FOOD PRODUCTION IN AFGHANISTAN IN SPITE OF CLIMATE CHANGE THROUGH SYSTEMATIC WATER MANAGEMENT, p. 5., 2016.

Sharma, A. B.: The Indian Forest Rights Act (2006): A Gender Perspective, ANTYAJAA Indian J. Women Soc. Change, 2(1), 2017.

Sheil, D.: Forests, atmospheric water and an uncertain future: the new biology of the global water cycle, For. Ecosyst. 5(1), 19, 2018.

Sheldon, G. W.: The history of political theory: ancient Greece to modern America, Peter Lang., 2003.

Shiklomanov, I. A.: Assessment of water resources and water availability in the world, Compr. Assess. Freshw. Re-Sources World, 1997.

Shrestha, R. P.: Land degradation in Afghanistan, in Unpublished. [online] Available from: http://rgdoi.net/10.13140/RG.2.2.19592.06400 (Accessed 28 January 2019), 2007.

Shrestha, S. and Kazama, F.: Assessment of surface water quality using multivariate statistical techniques: A case study of the Fuji river basin, Japan, Environ. Model. Softw. 22(4), 464–475, 2007.

Shroder, J. F.: Afghanistan, [online] Available from: http://www.iranica.com/newsite/ index.isc?Article=http://www.iranica.com/newsite/articles/v1f5/v1f5a040a.html (Accessed 17 January 2019), 2006.

Shroder, J. F. and Ahmadzai, S. J.: Transboundary water resources in Afghanistan: Climate change and land-use implications, Elsevier. 2016.

Siegert, M. J.: Role of Glaciers and Ice Sheets in Climate and the Global Water Cycle, Encycl. Hydrol. Sci., 2006.

Sigurdsson, H., Stix, J., Houghton, B., McNutt, S. R. and Rymer, H.: Encyclopedia of Volcanoes Academic Press, Lond. UK, 186–188, 2000.

Silsbe, G. and Hecky, R.: Are the Lake Victoria fisheries threatened by exploitation or eutrophication? Towards an ecosystem-based approach to management, Ecosyst. Approach Fish. 309, 2008.

Silvia, S. J. and Stanaitis, M.: Is Economic Hegemony Necessary for Maintaining Open Trade? An Empirical Challenge to the Hegemonic Stability Theory, 2013.

Singh, D.: Explaining varieties of corruption in the Afghan Justice Sector, J. Interv. Statebuilding, 9(2), 231–255, doi:10.1080/17502977.2015.1033093, 2015.

SIWI: MiCT_SIWI_Orphan-River_Final.pdf, [online] Available from: https://mict-international.org/wp-content/uploads/2016/01/MiCT_SIWI_Orphan-River_Final.pdf (Accessed 26 January 2019), 2015.

Smith, C.: A History of Water Rights at Common Law, JSTOR. 2005.

Snidal, D.: The limits of hegemonic stability theory, Int. Organ., 39(4), 579–614, 1985.

Snidal, D.: Relative gains and the pattern of international cooperation, Am. Polit. Sci. Rev., 85(3), 701–726, 1991.

Söderbaum, P. and Tortajada, C.: Perspectives for water management within the context of sustainable development, Water Int., 36(7), 812–827, 2011.

Sokile, C. S. and Van Koppen, B.: Local water rights and local water user entities: the unsung heroines of water resource management in Tanzania, Phys. Chem. Earth Parts ABC, 29(15–18), 1349–1356, 2004.

Song, J. and Whittington, D.: Why have some countries on international rivers been successful negotiating treaties? A global perspective, Water Resour. Res., 40(5), 2004.

Sophocleous, M.: Interactions between groundwater and surface water: the state of the science, Hydrogeol. J., 10(1), 52–67, 2002.

South China Morning Post: China and Pakistan pave way for ‚economic corridor' | South China Morning Post, [online] Available from: https://www.scmp.com/news/china/article/1431218/china-and-pakistan-pave-way-economic-corridor (Accessed 29 January 2019), 2014.

Steinmo, S.: Historical institutionalism, Approaches Methodol. Soc. Sci., 118, 2008.

Steinmo, S., Thelen, K. and Longstreth, F.: Structuring politics: historical institutionalism in comparative analysis, Cambridge University Press., 1992.

Stevens, C. J., Murphy, C., Roberts, R., Lucas, L., Silva, F. and Fuller, D. Q.: Between China and South Asia: A Middle Asian corridor of crop dispersal and agricultural innovation in the Bronze Age, The Holocene, 26(10), 1541–1555, 2016.

Stoa, R. B.: The United Nations Watercourses Convention on the Dawn of Entry Into Force, , 47, 50, 2014.

Stocker, T.: Climate change 2013: the physical science basis: Working Group I contribution to the Fifth assessment report of the Intergovernmental Panel on Climate Change, Cambridge University Press., 2014.

Stoecker, R.: Evaluating and rethinking the case study, Sociol. Rev., 39(1), 88–112, 1991.

Strand, A., Borchgrevink, K. and Harpviken, K. B.: Afghanistan: A Political Economy Analysis: [online] Available from: https://brage.bibsys.no/xmlui/bitstream/handle/11250/2470515/NUPI_rapport_Afghanistan_ Strand_Borchgrevink_BergHarpviken.pdf?sequence=2 (Accessed 29 January 2019), 2017.

Strange, S.: The persistent myth of lost hegemony, Int. Organ., 41(4), 551–574, 1987.

Strange, S.: Who governs? Networks of power in world society, Hitotsubashi J. Law Polit., 22, 5–17, 1994.

Stuchtey, B.: Colonialism and imperialism, 1450–1950, Notes, 2, 6, 2017.

Subrahmanyam, G.: Ruling continuities: colonial rule, social forces and path dependence in British India and Africa, Commonw. Comp. Polit., 44(1), 84–117, 2006.

Suddaby, R., Seidl, D. and Lê, J. K.: Strategy-as-practice meets neo-institutional theory, Sage Publications Sage UK: London, England. 2013.

Suich, H., Howe, C. and Mace, G.: Ecosystem services and poverty alleviation: a review of the empirical links, Ecosyst. Serv., 12, 137–147, 2015.

Swyngedouw, E.: Governance innovation and the citizen: the Janus face of governance-beyond-the-state, Urban Stud., 42(11), 1991–2006, 2005.

Tamanaha, B. Z.: The rule of law and legal pluralism in development, Hague J. Rule Law, 3(1), 1–17, 2011.

Tami, F.: Afghanistan and climate change in the Hindu Kush-Himalayan region - Norwegian Afghanistan Committee, [online] Available from: http://www.afghanistan.no/english/sectors/afghanistan_and_climate_change/index.html (Accessed 20 April 2018), 2013.

Tariq, S., Ahmad, M. and Mahmood, I.: Climate change impacts on Chitral-Kabul trans-boundary rivers, Northern Pakistan, , 2, 2014.

Tasan-Kok, T. and Vranken, J.: Handbook for Multilevel Urban Governance in Europe, Hague Eur. Urban Knowl. Netw. EUKN, 2011.

Team, C. W., Pachauri, R. K. and Meyer, L. A.: IPCC, 2014: climate change 2014: synthesis report. Contribution of Working Groups I, II III Fifth Assess. Rep. Intergov. Panel Clim. Change IPCC Geneva Switz., 151, 2014.

Teufel, N., Markemann, A., Kaufmann, B., Zárate, A. V. and Otte, J.: Livestock production systems in South Asia and the Greater Mekong sub-region, PPLPI Working Paper., 2010.

The Convention on Biological Diversity, I.: Convention on biological diversity, 1992.

The DAWN: How the British influenced Indian culture, DAWN.COM [online] Available from: http://www.dawn.com/news/881307 (Accessed 25 March 2018), 2010.

The DAWN: Soan River — witness to rise and fall of many civilisations, DAWN.COM, 31st March [online] Available from: http://www.dawn.com/news/707009 (Accessed 15 July 2018), 2012.

The DAWN: CPEC and water, DAWN.COM, 21st March [online] Available from: http://www.dawn.com/news/1246949 (Accessed 29 January 2019), 2016.

The DAWN: Lack of drinking water, DAWN.COM [online] Available from: https://www.dawn.com/news/1374724 (Accessed 8 September 2018), 2017.

The DAWN: Pakistan, Afghanistan urged to sign treaty on Kabul River water, DAWN.COM, 11th October [online] Available from: https://www.dawn.com/news/1438211 (Accessed 26 January 2019), 2018.

The Diplomat: The Diplomat, The Diplomat [online] Available from: https://thediplomat.com/2014/02/%20china-pakistan-flesh-out-new-economic-corridor (Accessed 29 January 2019), 2014.

The Express Tribune: Pakistan second-worst on tackling under-five's deaths, Express Trib., 1st February [online] Available from: https://tribune.com.pk/story/1623402/1-pakistan-second-worst-tackling-fives-deaths/ (Accessed 1 May 2019), 2018.

The Express Tribune: _9000-year-old Mehrgarh needs to be preserved' | The Express Tribune, , 5th June [online] Available from: https://tribune.com.pk/story/1966418/1-9000-year-old-mehrgarh-needs-preserved/ (Accessed 28 July 2019), 2019.

The Government of Afghanistan: Fifth National Report to the United Nation's Convention on Biological Diversity. [online] Available from: https://www.cbd.int/doc/world/af/af-nr-05-en.pdf (Accessed 27 January 2019), 2014.

The Rio Declaration: Rio declaration on environment and development. 1992.

The Seoul Rules: Seoul Rules on International Groundwaters, in Seoul, Korea: Report of the Sixty-Second Conference Held at Seoul. 1986.

The Stockholm Declaration: Stockholm declaration on the human environment, Int Leg. Mater, 11, 1416, 1972.

The World Bank: Scoping Strategic Options for Development of the Kabul River Basin A Multisectoral Decision Support System Approach. [online] Available from: http://documents.worldbank.org/curated/en/319391468185978566/pdf/522110ESW0Whit1anistan0Final0Report.pdf (Accessed 27 January 2019), 2010.

The World Bank: The World Bank 2014.pdf, [online] Available from: http://documents.worldbank.org/curated/en/245541467973233146/pdf/AUS9779-REVISED-WP-PUBLIC-Box391431B-Final-Afghanistan-ASR-web-October-31-2014.pdf (Accessed 26 January 2019), 2014.

The World Bank: Afghanistan - Irrigation Restoration and Development Project Proposal to Restructure. [online] Available from: http://documents.worldbank.org/curated/en/963441467992055040/pdf/PAD1670-REVISED-OUO-9-IDA-R2016-0100-1.pdf (Accessed 26 January 2019a), 2016.

The World Bank: AFGHANISTAN TO 2030 PRIORITIES FOR ECONOMIC DEVELOPMENT UNDER FRAGILITY, [online] Available from: http://documents.worldbank.org/curated/en/156881533220723730/pdf/129161-WP-P157288-Afghanistan-to-2030-PUBLIC.pdf (Accessed 27 January 2019b), 2016.

The World Bank: Disaster Risk Profile-Afghanistan. [online] Available from: https://www.gfdrr.org/sites/default/files/afghanistan_low_FINAL.pdf (Accessed 15 January 2018), 2017.

The World Bank: STRENGTHENING HYDROMET AND EARLY WARNING SERVICES IN AFGHANISTAN: A ROAD MAP. [online] Available from: http://documents.worldbank.org/curated/en/976021545165642530/pdf/133069-WP-P168141-PUBLIC-44684-Roadmap-Afghanistan-Report-Dec17-Digital.pdf (Accessed 28 February 2019), 2018.

Thomas, V.: Afghanistan and Pakistan: a decade of unproductive interactions over the Kabul-Indus basin, Third Pole [online] Available from: https://www.thethirdpole.net/en/2014/07/07/afghanistan-and-pakistan-a-decade-of-unproductive-interactions-over-the-kabul-indus-basin/ (Accessed 27 January 2019), 2014.

Thomas, V., Azizi, M. A. and Behzad, K.: Developing transboundary water resources: What perspectives for cooperation between Afghanistan, Iran and Pakistan?, , 111, 2016.

Thomson, E. F., Chabot, P. and Wright, I. A.: Production and marketing of red meat, wool, skins and hides in Afghanistan, Davis CA USA Macauley Res. Consult. Serv. Mercy Corps, 2005.

Thornton, J. A.: Global International Waters Assessment: Project Number GF/1100-99-01: Terminal Evaluation, UNEP. 2006.

Thu, H. N. and Wehn, U.: Data sharing in international transboundary contexts: the Vietnamese perspective on data sharing in the Lower Mekong Basin, J. Hydrol., 536, 351–364, 2016.

Tilly, C.: Big structures, large processes, huge comparisons, Russell Sage Foundation., 1984.

Tir, J. and M. Stinnett, D.: The Institutional Design of Riparian Treaties: The Role of River Issues. 2011.

Toft, P.: John J. Mearsheimer: an offensive realist between geopolitics and power, J. Int. Relat. Dev., 8(4), 381–408, 2005a.

Toft, P.: John J. Mearsheimer: an offensive realist between geopolitics and power, J. Int. Relat. Dev., 8(4), 381–408, 2005b.

Tokaranyaset, C.: Institution-based resource: Concept and cases, City University London., 2013.

Toynbee, A. J.: Study of history, Oxford University Press, London. 1946.

Trigueros, A.: The Human Right to Water: Will Its Fulfillment Contribute to Environmental Degradation? Indiana J. Glob. Leg. Stud., 19(2), 599–625, 2012.

Tsakatika, M.: The weakness of the neo-institutionalist approaches: how political institutions change, Sci. Soc. Rev. Polit. Moral Theory, 13, 135–166, 2004.

Tsuzuki, Y., Fujii, M., Mochihara, Y., Matsuda, K. and Yoneda, M.: Natural purification effects in the river in consideration with domestic wastewater pollutant discharge reduction effects, J. Environ. Sci., 22(6), 892–897, 2010.

Tunnermeier, T. and Himmelsbach, T.: Hydrogeology of the Kabul Basin. 52, 2005.

Tuomela, R.: The philosophy of sociality: The shared point of view, Oxford University Press., 2007.

Turton, A. and Funke, N.: Hydro-hegemony in the context of the Orange River Basin, Water Policy, 10(S2), 51–69, 2008.

Turton, A. R., Patrick, M. J. and Julien, F.: Transboundary water resources in Southern Africa: conflict or cooperation? Development, 49(3), 22–31, 2006.

Ukkola, A. M. and Prentice, I. C.: A worldwide analysis of trends in water-balance evapotranspiration, Hydrol. Earth Syst. Sci., 17(10), 4177–4187, 2013.

Ullah, S., Javed, M. W., Shafique, M. and Khan, S. F.: An integrated approach for quality assessment of drinking water using GIS: A case study of Lower Dir., J. Himal. Earth Sci., 47(2), 2014.

Umar, M., Hussain, M., Murtaza, G., Shaheen, F. A. and Zafar, F.: Ecological Concerns of Migratory Birds in Pakistan: A Review, Punjab Univ. J. Zool., 33(1), 69–76, 2018.

Umer, K.: Besides infrastructure, manufacturing industry needs big investment, Express Trib. [online] Available from: https://tribune.com.pk/story/1863482/2-besides-infrastructure-manufacturing-industry-needs-big-investment/ (Accessed 29 December 2018), 2018.

UN Millennium Declaration: United Nations millennium declaration, United Nations, Department of Public Information., 2000.

UN Water: Coping with water scarcity: challenge of the twenty-first century, Prep. World Water Day, 2007.

UNAMA: WATER RIGHTS - An Assessment of Afghanistan's Legal Framework Governing Water for Agriculture, [online] Available from: https://unama.unmissions.org/sites/default/files/2016_19_10_water_rights_final_v2.pdf (Accessed 26 January 2019), 2016.

UNCBD: UNITED NATIONS CONVENTION ON BIOLOGICAL DIVERSITY. [online] Available from: https://www.cbd.int/doc/legal/cbd-en.pdf (Accessed 26 January 2019), 1992.

UNCCD: UNITED NATIONS CONVENTION TO COMBAT DESERTIFICATION, [online] Available from: http://catalogue.unccd.int/936_UNCCD_Convention_ENG.pdf (Accessed 26 January 2019), 1994.

UNCCD: United Nations Convention to Combat Desertification in Those Countries Experiencing Serious Drought And/or Desertification, Particulary in Africa, Secretariat of the United Nations Convention to Combat Desertification. 1999.

UNDP, U. N. D.: Beyond scarcity: Power, poverty and the global water crisis, Palgrave Macmillan., 2006.

UNECE: Convention on the Protection and Use of Transboundary Watercourses and International Lakes, United Nations Economic Commission for Europe Geneva. 1992.

UNECE: CONVENTION ON ACCESS TO INFORMATION, PUBLIC PARTICIPATION IN DECISION-MAKING AND ACCESS TO JUSTICE IN ENVIRONMENTAL MATTERS, [online] Available from: https://www.unece.org/fileadmin/DAM/env/pp/documents/cep43e.pdf (Accessed 8 May 2018), 1998.

UNECE: ECONOMIC COMMISSION FOR EUROPE. 1999.

UNECE: AMENDMENT TO ARTICLES 25 AND 26 OF THE CONVENTION, [online] Available from: https://www.unece.org/fileadmin/DAM/env/documents/2004/wat/ece.mp.wat.14.e.pdf (Accessed 15 August 2018), 2003.

UNECE: THE GLOBAL OPENING of the 1992 Water Convention. [online] Available from: https://www.unece.org/fileadmin/DAM/env/water/publications/WAT_The_global_opening_of_the_1992_UNECE_Water_Convention/ECE_MP.WAT_43_Rev1_ENGLISH_WEB.pdf (Accessed 7 October 2018), 2016.

UNEP: Biodiversity Profile of Afghanistan. [online] Available from: https://postconflict.unep.ch/publications/afg_tech/theme_02/afg_biodiv.pdf (Accessed 19 January 2019), 2008.

UNEP, U. N. E. P. G. P.: Water quality for ecosystem and human health, UNEP GEMSWater Programme Ont., 2006.

UNESCO: Convention on Wetlands of International Importance especially as Waterfowl Habitat, UNESCO Ramsar, Iran. 1971.

UNESCO, W.: Water a shared responsibility. The United Nations, World Water Development Report 2, UN-WATER/WWAP/2006/3. Available at: http://unesdoc.unesco.org/images/0014. 2006.

UNFCCC: United Nations framework convention on climate change, United Nations New York., 1992.

UNGA: Resolution adopted by the General Assembly on 9 September 2014, 2010a.

UNGA: The human right to water and sanitation, [online] Available from: http://www.cawater-info.net/bk/water_law/pdf/n0947935.pdf (Accessed 13 September 2018b), 2010.

UNGA: Transforming our world: The 2030 agenda for sustainable development, Resolut. Adopt. Gen. Assem., 2015.

UNGA-HRC: Human rights and access to safe drinking water and sanitation, [online] Available from: http://www.cawater-info.net/bk/water_law/pdf/g1016309.pdf (Accessed 9 October 2018), 2010.

UN-Water: Managing water under uncertainty and risk, The United Nations world water development report 4, UN Water Reports, World Water Assessment Programme, UNESCO, Paris, France., 2012.

UN-WWAP, U. N. W. W. A. P.: The United Nations world water development report 2015: water for a sustainable world, U. N. World Water Assess. Programme, 2015.

USAID: NEEDS ASSESSMENT ON SOIL AND WATER IN AFGHANISTAN - Future Harvest Consortium to Rebuild Agriculture in Afghanistan. [online] Available from: https://afghanag.ucdavis.edu/irrigation-natural-resource/files/soil-access-water.pdf (Accessed 27 January 2019), 2002.

USAID: A River Runs Through It: Scientific Border Tales from Afghanistan and Pakistan | FrontLines September/October 2017 | U.S. Agency for International Development, [online] Available from: https://www.usaid.gov/news-information/frontlines/september-october-2017/river-runs-through-it-scientific-border (Accessed 10 May 2018a), 2017.

USAID: AGRICULTURE CONSOLIDATED PROJECT APPRAISAL DOCUMENT (PAD), [online] Available from: https://www.usaid.gov/sites/default/files/documents/1871/USAID_Afghanistan_Project_Appraisal_Document_-_Public_Version_2017-04-03.pdf (Accessed 26 January 2019b), 2017.

USGS, U. S. G. S.: Ice, Snow, and Glaciers: The Water Cycle, [online] Available from: https://water.usgs.gov/edu/watercycleice.html (Accessed 25 March 2017), 2018.

Valli, J.: CSR: Towards a Definition, XinXii. 2015.

Valters, C.: Theories of change in international development: communication, learning, or accountability, JSRP Pap., 17, 2014.

Van Noordwijk, M., Namirembe, S., Catacutan, D., Williamson, D. and Gebrekirstos, A.: Pricing rainbow, green, blue and grey water: tree cover and geopolitics of climatic teleconnections, Curr. Opin. Environ. Sustain. 6, 41–47, 2014.

Van Steenbergen, F.: Promoting local management in groundwater, Hydrogeol. J., 14(3), 380–391, 2006.

Van Weert, F. and van der Gun, J.: Saline and brackish groundwater at shallow and intermediate depths: genesis and world-wide occurrence, 39th Int. Assoc. Hydrol. Niagara F. Int. Assoc. Hydrol., 2012.

Vanham, D.: A holistic water balance of Austria–how does the quantitative proportion of urban water requirements relate to other users? Water Sci. Technol., 66(3), 549–555, 2012.

Vaux, H.: Water for agriculture and the environment: the ultimate trade-off, Water Policy, 14(S1), 136–146, 2012.

Vedrine, H. and Moisi, D.: France in an Age of Globalization. Washington, DC Brook, 2001.

Vedung, E. and Van der Doelen, F. C.: The sermon: Information programs in the public policy process: Choice, effects and evaluation, 1998.

Veeman, T. S. and Politylo, J.: The role of institutions and policy in enhancing sustainable development and conserving natural capital, Environ. Dev. Sustain., 5(3–4), 317–332, 2003.

Vella, P. S.: A theoretical model for inclusive economic growth in Indian Context, Int. J. Humanit. Soc. Sci., 4(13), 229, 2014.

Vick, M. J.: The Senegal River Basin: A retrospective and prospective look at the legal regime, Nat. Resour. J., 211–243, 2006.

Vick, M. J.: Steps towards an Afghanistan–Pakistan water-sharing agreement, Int. J. Water Resour. Dev., 30(2), 224–229, 2014a.

Vick, M. J.: Steps towards an Afghanistan–Pakistan water-sharing agreement, Int. J. Water Resour. Dev., 30(2), 224–229, 2014b.

Vihervaara, P., Kumpula, T., Tanskanen, A. and Burkhard, B.: Ecosystem services–A tool for sustainable management of human–environment systems. Case study Finnish Forest Lapland, Ecol. Complex., 7(3), 410–420, 2010.

Vijge, M. J.: The promise of new institutionalism: explaining the absence of a World or United Nations Environment Organisation, Int. Environ. Agreem. Polit. Law Econ., 13(2), 153–176, 2013.

Villholth, K. G., Tøttrup, C., Stendel, M. and Maherry, A.: Integrated mapping of groundwater drought risk in the Southern African Development Community (SADC) region, Hydrogeol. J., 21(4), 863–885, 2013.

Violatti, C.: Indus Valley Civilization, Anc. Hist. Encycl. [online] Available from: https://www.ancient.eu/Indus_Valley_Civilization/ (Accessed 13 May 2018), 2013.

Visalli, D.: Afghanistan: the legacy of the British Empire. A brief history, Glob. Res., 2013.

Vlek, C. and Steg, L.: ? Human Behavior and Environmental Sustainability: Problems, Driving Forces, and Research Topics, J. Soc. Issues, 63(1), 1–19, doi:10.1111/j.1540-4560.2007.00493.x, 2007.

Vogtmann, H. and Dobretsov, N.: Environmental Security and Sustainable Land Use - with Special Reference to Central Asia, Springer Science & Business Media., 2006.

Voinea, E.: Realism Today, E-Int. Relat. Stud., 2013.

Vörösmarty, C. J. and Sahagian, D.: Anthropogenic disturbance of the terrestrial water cycle, AIBS Bull., 50(9), 753–765, 2000.

Vörösmarty, C. J., McIntyre, P. B., Gessner, M. O., Dudgeon, D., Prusevich, A., Green, P., Glidden, S., Bunn, S. E., Sullivan, C. A. and Liermann, C. R.: Global threats to human water security and river biodiversity, Nature, 467(7315), 555, 2010.

Voudouris, K., Valipour, M., Kaiafa, A., Zheng, X. Y., Kumar, R., Zanier, K., Kolokytha, E. and Angelakis, A.: Evolution of water wells focusing on Balkan and Asian civilizations, Water Supply, 19(2), 347–364, 2019.

Walker, P.: The World's Most Dangerous Borders, Foreign Policy [online] Available from: https://foreignpolicy.com/2011/06/24/the-worlds-most-dangerous-borders/ (Accessed 11 December 2018), 2011.

Wall, D. H. and Nielsen, U. N.: Biodiversity and ecosystem services: is it the same below ground, Nat. Educ. Knowl., 3(12), 8, 2012.

Waltz, K. N.: Theory of international politics, Waveland Press., 2010.

Wang, P., Yu, J., Zhang, Y. and Liu, C.: Groundwater recharge and hydrogeochemical evolution in the Ejina Basin, northwest China, J. Hydrol., 476, 72–86, 2013.

Wang, W., Lu, H., Yang, D., Sothea, K., Jiao, Y., Gao, B., Peng, X. and Pang, Z.: Modelling hydrologic processes in the Mekong River Basin using a distributed model driven by satellite precipitation and rain gauge observations, PloS One, 11(3), e0152229, 2016.

Waqas, U., Malik, M. I. and Khokhar, L. A.: Conservation of Indus River Dolphin (Platanista gangetica minor) in the Indus River system, Pakistan: an overview, Rec Zool Surv Pak, 21, 82–85, 2012.

Wardak, A.: Building a post-war justice system in Afghanistan, Crime Law Soc. Change, 41(4), 319–341, 2004.

Warner, J.: More sustainable participation? Multi-stakeholder platforms for integrated catchment management, Water Resour. Dev., 22(1), 15–35, 2006.

Warner, J. and Zawahri, N.: Hegemony and asymmetry: Multiple-chessboard games on transboundary rivers, Int. Environ. Agreem. Polit. Law Econ., 12(3), 215–229, 2012.

Warner, J. and Zeitoun, M.: International relations theory and water do mix: A response to Furlong's troubled waters, hydro-hegemony and international water relations, Polit. Geogr., 27(7), 802–810, 2008.

Warner, J., Sebastian, A. and Empinotti, V.: Claiming (back) the land: the geopolitics of Egyptian and South African land and water grabs, Ambiente Soc., 16(2), 1–24, 2013.

Warner, J., Zeitoun, M. and Mirumachi, N.: 10. How _soft'power shapes transboundary water interaction, Glob. Water Issues Insights, 51, 2014.

Warner, J., Mirumachi, N., Farnum, R. L., Grandi, M., Menga, F. and Zeitoun, M.: Transboundary _hydro- hegemony': 10 years later, Wiley Interdiscip. Rev. Water, 4(6), e1242, 2017.

Warner, J. F.: Contested hydrohegemony: hydraulic control and security in Turkey, Water Altern., 1(2), 271–288, 2008.

Warraich, A. N.: Durand Line – A Binding International Border, , 19th February [online] Available from: http://courtingthelaw.com/2016/02/19/commentary/durand-line-a-binding-international-border/ (Accessed 26 January 2019), 2016.

Waterbury, J. and Whittington, D.: Playing chicken on the Nile? The implications of microdam development in the Ethiopian highlands and Egypt's New Valley Project, in Natural Resources Forum, vol. 22, pp. 155–163, Wiley Online Library., 1998.

Wegerich, K.: Hydro-hegemony in the Amu Darya basin, Water Policy, 10(S2), 71–88, 2008.

Weingast, B. R.: Constitutions as governance structures: The political foundations of secure markets, J. Institutional Theor. Econ. JITEZeitschrift Für Gesamte Staatswiss., 286–311, 1993.

Weinthal, E., Troell, J. J. and Nakayama, M.: Water and Post-Conflict Peacebuilding, Routledge., 2014.

Weiss, E. and Zohary, D.: The Neolithic Southwest Asian founder crops: their biology and archaeobotany, Curr. Anthropol. 52(S4), S237–S254, 2011.

Weiss, E. B.: International Environmental Law: Contemporary Issues and the Enmergence of a New World Order, Geo LJ, 81, 675, 1992.

WHO, W. H. O.: Water quality and health strategy 2013–2020, World Health Organ. Geneva Switz., 2013.

WHO/UNICEF, J. W.: Progress on sanitation and drinking water: 2015 update and MDG assessment, World Health Organization., 2015.

Wilhite, D. A., Svoboda, M. D. and Hayes, M. J.: Understanding the complex impacts of drought: A key to enhancing drought mitigation and preparedness, Water Resour. Manag. 21(5), 763–774, 2007.

Wilkinson, J. C.: Muslim land and water law, J. Islam. Stud., 1, 54–72, 1990.

Williamson, O. E.: The new institutional economics: taking stock, looking ahead, J. Econ. Lit., 38(3), 595–613, 2000.

Wilson III, E. J.: Hard power, soft power, smart power, Ann. Am. Acad. Pol. Soc. Sci., 616(1), 110–124, 2008.

Wingqvist, G. O. and Nilsson, A.: Effectiveness of River Basin Organisations–an institutional review of three African RBOs, Sida's Helpdesk for Environment and Climate Change., 2015.

Winter, T. C.: Relation of streams, lakes, and wetlands to groundwater flow systems, Hydrogeol. J., 7(1), 28–45, 1999.

Wittfogel, K. A.: The hydraulic civilizations, University of Chicago Press Chicago, IL., 1956.

Wolf, A. T.: Atlas of international freshwater agreements, UNEP/Earthprint., 2002.

Wolf, A. T.: Regional water cooperation as confidence building: water management as a strategy for peace, Citeseer., 2004.

Wolf, A. T., Yoffe, S. B. and Giordano, M.: International waters: identifying basins at risk, Water Policy, 5(1), 29–60, 2003.

Woodhouse, M. and Zeitoun, M.: Hydro-hegemony and international water law: grappling with the gaps of power and law, Water Policy, 10(S2), 103–119, 2008.

Worthington, E. B.: United Nations water conference, held at Mar Del Plata, Argentina, 14–25 March 1977, Environ. Conserv. 4(2), 153–154, 1977.

Wouters, P.: The relevance and role of water law in the sustainable development of freshwater: from —ḥdrosovereignty" to —ḥdrosolidarity," Water Int., 25(2), 202–207, 2000.

Wouters, P.: International law–facilitating transboundary water cooperation, 2013.

WRI, W. R. I.: World Resources: The Wealth Of The Poor: Managing Ecosystems To Fight Poverty, World Resources Institute (WRI)., 2005.

Yadav, N., Joglekar, H., Rao, R. P. N., Vahia, M. N., Adhikari, R. and Mahadevan, I.: Statistical Analysis of the Indus Script Using n-Grams, edited by F. Rapallo, PLoS ONE, 5(3), e9506, doi:10.1371/journal.pone.0009506, 2010.

Yang, W., Dietz, T., Liu, W., Luo, J. and Liu, J.: Going beyond the Millennium Ecosystem Assessment: an index system of human dependence on ecosystem services, PloS One, 8(5), e64581, 2013.

Yang, Y.-C. E., Brown, C., Yu, W., Wescoat Jr, J. and Ringler, C.: Water governance and adaptation to climate change in the Indus River Basin, J. Hydrol., 519, 2527–2537, 2014.

Yasuda, Y., Schillinger, J., Huntjens, P., Alofs, C. and de Man, R.: Transboundary Water Cooperation over the lower part of the Jordan River Basin: Legal Political Economy Analysis of Current and Future Potential Cooperation. The Hague Institute for Global Justice. [online] Available from: http://www.siwi.org/wp-content/uploads/2018/01/Jordan-Basin-Report_design.pdf. (Accessed 13 December 2018), 2017.

Yilmaz, S.: State, power, and hegemony, Int. J. Bus. Soc. Sci., 1(3), 2010.

Yin, R. K.: Validity and generalization in future case study evaluations, Evaluation, 19(3), 321–332, 2013.

Yıldız, D.: Afghanistan's Transboundary Rivers and Regional Security, World Sci. News, 16, 40–52, 2015.

Yoffe, S., Wolf, A. T. and Giordano, M.: CONFLICT AND COOPERATION OVER INTERNATIONAL FRESHWATER RESOURCES: INDICATORS OF BASINS AT RISR 1, JAWRA J. Am. Water Resour. Assoc., 39(5), 1109–1126, 2003.

Young et al., O.: INSTITUTIONAL DIMENSIONS OF GLOBAL ENVIRONMENTAL CHANGE. 2005.

Young, M.: Climate change implications on transboundary water management in the Jordan River Basin: A Case Study of the Jordan River Basin and the transboundary agreements between riparians Israel, Palestine and Jordan. 2015.

Young, O. R.: Institutions and the growth of knowledge: Evidence from international environmental regimes, Int. Environ. Agreem. 4(2), 215–228, 2004.

Young, O. R.: The effectiveness of international environmental regimes: Existing knowledge, cutting-edge themes, and research strategies, in Advances in International Environmental Politics, pp. 273–299, Springer. 2014.

Young, W. J., Anwar, A., Bhatti, T., Borgomeo, E., Davies, S., Garthwaite III, W. R., Gilmont, E. M., Leb, C., Lytton, L., Makin, I. and Saeed, B.: Pakistan: Getting More from Water, World Bank. 2019.

Yousaf, S.: KABUL RIVER AND PAK-AFGHAN RELATIONS, , 16, 2017.

Yousafzai, A. M., Khan, A. R. and Shakoori, A. R.: An assessment of chemical pollution in River Kabul and its possible impacts on fisheries, Pak. J. Zool., 40(3), 199, 2008a.

Yousafzai, A. M., Khan, A. R. and Shakoori, A. R.: Heavy Metal Pollution in River Kabul Affecting the Inhabitant Fish Population. 9, 2008b.

Yousafzai, A. M., Khan, A. R. and Shakoori, A. R.: Pollution of Large, Subtropical Rivers-River Kabul, Khyber-Pakhtun Khwa Province, Pakistan): Physico-Chemical Indicators, Pak. J. Zool., 42(6), 2010.

Yousafzai, I. A. and Yaqubi, H.: THE DURAND LINE: ITS HISTORICAL, LEGAL AND POLITICAL STATUS. 54(1), 20, 2017.

Yu, W., Yang, Y.-C., Savitsky, A., Alford, D., Brown, C., Wescoat, J., Debowicz, D. and Robinson, S.: The Indus basin of Pakistan: The impacts of climate risks on water and agriculture, The World Bank., 2013.

Zahraa, M. and Mahmor, S. M.: Definition and Scope of the Islamic Concept of Sale of Goods'(2001), Arab Law Q., 16, 215, 218, 2001.

Zang, C., Liu, J., Gerten, D. and Jiang, L.: Influence of human activities and climate variability on green and blue water provision in the Heihe River Basin, NW China, J. Water Clim. Change, jwc2015194, 2015.

Zaryab, A., Noori, A. R., Wegerich, K. and Kløve, B.: Assessment of Water Quality and Quantity trends in Kabul Aquifers with an outline for future water supplies, Cent. Asian J. Water Res. CAJWR Центральноазиатский Журнал Исследований Водных Ресурсов, 3(2), 1925, 2017.

Zawahri, N. A.: Designing river commissions to implement treaties and manage water disputes: the story of the Joint Water Committee and Permanent Indus Commission, Water Int., 33(4), 464–474, 2008.

Zeder, M. A.: The domestication of animals, J. Anthropol. Res., 68(2), 161–190, 2012.

Zeitoun, M.: Global environmental justice and international transboundary waters: an initial exploration, Geogr. J., 179(2), 141–149, 2013.

Zeitoun, M.: The relevance of international water law to later-developing upstream states, Water Int., 40(7), 949–968, 2015.

Zeitoun, M. and Allan, J. A.: Applying hegemony and power theory to transboundary water analysis, Water Policy, 10(S2), 3–12, 2008.

Zeitoun, M. and Mirumachi, N.: Transboundary water interaction I: Reconsidering conflict and cooperation, Int. Environ. Agreem. Polit. Law Econ., 8(4), 297, 2008.

Zeitoun, M. and Warner, J.: Hydro-hegemony–a framework for analysis of trans-boundary water conflicts, Water Policy, 8(5), 435–460, 2006.

Zeitoun, M., Eid-Sabbagh, K., Talhami, M. and Dajani, M.: Hydro-hegemony in the Upper Jordan waterscape: control and use of the flows, Water Altern., 6(1), 86, 2013.

Zhao, L., Li, Y., Jiang, F., Wang, H., Ren, S., Liu, Y. and Ouyang, Z.: Comparative advantage for the areas irrigated with underground blue water in North China Plain, Water Policy, wp2015114, 2015.

Zhu, Y. and Newell, R. E.: A proposed algorithm for moisture fluxes from atmospheric rivers, Mon. Weather Rev., 126(3), 725–735, 1998.

Zwarteveen, M. and Boelens, R.: Defining, researching and struggling for water justice: Some conceptual building blocks for research and action, Water Int., 39(2), 143–158, 2014.

263

ANNEX A: Thesis Log Frame

ANNEX A: Thesis Log Frame

Main Research Question

How can regional hydro-politics and institutions be transformed at multiple levels of governance through inclusive development objectives and incorporate the relationships with non-water sectors in addressing issues of water quality, quantity and climate change?

GAPS IN SCIENTIFIC KNOWLEDGE			SUB-RESEARCH QUESTIONS
Gap 1	Transboundary water governance (TWG) literature rarely combines the role of institutions in dealing with hydro hegemony at multiple geographic levels	Sub-Research Question 1	1) How do power politics and institutions influence water governance in transboundary river basins at multiple geographic levels?
Gap 2	TWG literature scarcely links international relations (IR) scholarship with multilevel governance scholarship promoted by European Union scholars	Sub-Research Question 2	2) How can the concept of biodiversity and ESS be incorporated in a framework to analyse the effectiveness of institutions, and the role of power, in governing transboundary water resources?
Gap 3	TWG literature insufficiently highlights the role of water outside the basin (e.g. rainwater, snow, greenwater), the ecosystem services of water, and non-water related issues and actors in transboundary water research and policies.	Sub-Research Question 3	3) Which principles and instruments address the causes/drivers of freshwater problems in transboundary river basins at multiple geographic levels?
Gap 4	TWG literature scarcely focuses on including the inclusive development approach of international development studies which prioritizes inequality and focuses on socio-relational and ecological aspects.	Sub-Research Question 4	4) How does legal pluralism affect transboundary water cooperation?

Chapters	Research Question	Research Sub-Question
1- Introduction	To highlight the real life and theoretical gaps, research questions, focus and limits as well as structure of the whole thesis.	---
2- Methodology and Analytical Framework	To situate the research in terms of current theoretical debates and elaborate the methodological approach.	---
3- Approaches to Transboundary Water Governance	How does power shape institutions and how do institutions limit the role of power in transboundary water governance at multiple geographic levels?	1) How do realist and institutionalist perspectives differ in international relations as well as in transboundary water governance? 2) How does power influence freshwater governance institutions at multiple geographic levels? 3) How does a combined approach of water governance and institutions help understand the influence of power at multiple geographic levels?
4- Ecosystem Services and Human	How can the various drivers of freshwater problems	1) What causes problems to ESS of freshwater?

Well-being	affect the ecosystem services (ESS) of different kinds of freshwater and how does this, in turn, affect human well-being?	2) How different types of freshwater have distinct ecosystem services? 3) How protection and enhancement of ESS can lead to improved human well-being and achieving inclusive and sustainable development?
5- Global Water Institutions & Its Relationship with Inclusive & Sustainable Development	How have global institutions for transboundary water management evolved and what are the implications of these institutions for governing transboundary river basins without a regulatory framework?	1) How have the key global institutions for governing transboundary water resources evolved? 2) Which governance instruments (principles and instruments) are included and which are excluded? 3) How can the establishment of RBO's promote cooperation and reduce conflict in a transboundary river basin? 4) How has power influenced the inclusion/exclusion of instruments (principles and instruments)?
6- Analysis of International Relations in the Kabul River Basin	--	1) How are various characteristics including ESS and drivers of freshwater problems taken into account at transboundary level in the KRB? 2) How have freshwater governance frameworks evolved at transboundary level in the KRB? 3) Which governance instruments address the drivers of freshwater problems at transboundary level in the KRB? 4) How does legal pluralism occur at transboundary level in the KRB? 5) How do power and institutions influence freshwater governance frameworks at transboundary level in the KRB?
7- Analysis of Water Governance in Afghanistan	--	1) How are various characteristics including ESS and drivers of freshwater problems taken into account at transboundary level in Afghanistan 2) How have freshwater governance frameworks evolved at transboundary level in Afghanistan? 3) Which governance instruments address the drivers of freshwater problems at transboundary level in Afghanistan? 4) How does legal pluralism occur at transboundary level in Afghanistan? 5) How do power and institutions influence freshwater governance frameworks at transboundary level in Afghanistan?
8- Analysis of Multilevel Freshwater Governance in Pakistan		1) How are various characteristics including ESS and drivers of freshwater problems taken into account at transboundary level in

ANNEX A: Thesis Log Frame

---	---	Pakistan? 2) How have freshwater governance frameworks evolved at transboundary level in Pakistan? 3) Which governance instruments address the drivers of freshwater problems at transboundary level in Pakistan? 4) How does legal pluralism occur at transboundary level in Pakistan? 5) How do power and institutions influence freshwater governance frameworks at transboundary level in Pakistan?
9- Multi-Level Integrated Analysis Focusing on Issues for Re-design	How do power and institutions influence multilevel freshwater governance in the KRB and the achievement of inclusive and sustainable development?	1) How are various characteristics including biodiversity, ESS and drivers of freshwater problems taken into account at multiple levels of governance in the KRB? 2) How have freshwater governance frameworks evolved at multiple levels of governance in the KRB? 3) Which governance instruments address the drivers of freshwater problems at multiple levels of governance in the KRB? 4) How does legal pluralism occur at multiple levels of governance in the KRB? 5) How do power and institutions influence water sharing at multiple level of governance in the KRB? 6) How can the current designs of the KRB multilevel institutional architecture become consistent with the key global institutions to achieve inclusive and sustainable development?
10-Conclusion of Thesis	How can regional hydro-politics and institutions be transformed at multiple levels of governance through inclusive development objectives and incorporate the relationships with non-water sectors in addressing issues of water quality, quantity and climate change?	1) How can the concept of biodiversity and ESS be incorporated in a framework to analyse the effectiveness of institutions, and the role of power, in governing transboundary water resources? 2) Which principles and instruments address the causes/drivers of freshwater problems in transboundary river basins at multiple geographic levels? 3) How does legal pluralism affect transboundary water cooperation? 4) How do power politics and institutions influence water governance in transboundary river basins at multiple geographic levels?

ANNEX B: Keywords searched in scientific databases

Freshwater Resources	Transboundary Water Resources
Freshwater Governance	Transboundary Water Governance
Freshwater Resources Depletion	Transboundary Water Depletion
Freshwater Resources Contamination	Transboundary Water Contamination
Freshwater Resources Pollution	Transboundary Water Pollution
Freshwater Resources Quality	Transboundary Water Quality
Freshwater Resources Quantity	Transboundary Water Quantity
Freshwater Ecosystem Services	Transboundary Ecosystem Services
Blue Surface Water Resources	Issues in Transboundary River Basins
Blue Groundwater Resources	Challenges in Transboundary River Basins
White Water Resources	Transboundary Rivers in South Asia
Frozen water Resources/Frozen-water Resources	Transboundary Water Governance in South Asia
Greywater Resources/Grey-Water Resources	Water Governance in Transboundary River Basins
Blackwater Resources/Black-Water Resources	Water Governance in Transboundary Rivers
Atmospheric Moisture	Governance in Transboundary River Basins
Climate Change and Freshwater Resources	Climate Change and Transboundary Water Resources
Ecosystems Services and Freshwater Resources	Ecosystems Services and Transboundary Water Resources
Energy Production and Freshwater Resources	Energy Production and Transboundary Water Resources
Agriculture and Freshwater Resources	Agriculture and Transboundary Water Resources
Food and Freshwater Resources	Food and Transboundary Water Resources
Governance of Freshwater Resources	Governance of Transboundary Water Resources
Conflict over Freshwater Resources	Conflict over Transboundary water resource
Freshwater Conflict	Transboundary Water Conflict
Freshwater Politics	Transboundary Water Politics

ANNEX C: List of people interviewed

PROFESSIONAL BACKGROUND	Organization	COUNTRY
Activist	Kissan Board	Pakistan
Activist	PNRDP	Pakistan
Agriculturalist	Duran	Afghanistan
Agriculture Officer	Agricultural Research Institute	Pakistan
Anthropologist	Jalalabad State University	Afghanistan
Bureaucrat- Water Wing	Ministry of Water & Power	Pakistan
Businessman (Fruits and Vegetables)	Self-Employed	Afghanistan
Businessman (Mineral Water and Recycling)	Self-Employed	Afghanistan
Climate change Expert	NEPA	Afghanistan
Deputy Commissioner	IWC	Pakistan
Development Practitioner / Activist	NRSP	Pakistan
DRR Expert	Meher Foundation	Pakistan
Environmental Lawyer	Self-Employed	Afghanistan
Environmental Management/ Geosciences	Kabul University	Afghanistan
Farmer	WUA	Afghanistan
Farmer / Activist	WUA	Pakistan
Farmer / Development Practitioner	Kissan Board	Pakistan
Farmer / Water Expert	WUA	Pakistan
Farmer/ Activist	FO	Pakistan
Fishermen/Farmers/Activists/Boat Drivers	Local CSO	Afghanistan/Pakistan
Fishermen/Farmers/Activists/Boat Drivers	Local CSO	Afghanistan/Pakistan
Former Chief Engineer	WAPDA	Pakistan
Freelance Environmentalist	Self-employed	Afghanistan
Freelance Journalist & Water Policy Expert	DAWN/NEWS/Reuters	Pakistan
Hydrogeologist	AREU	Afghanistan
Hydrogeologist	Kabul University	Afghanistan
Hydrogeologist	AREU	Afghanistan
Hydrogeologist	AREU	Afghanistan
Hydrogeologist	Duran	Afghanistan
Hydrologist	WAPDA	Pakistan
Hydrologist/Lecturer	Kabul University	Afghanistan
Irrigation and Hydrology	FATA-Secretariat	Pakistan
Journalist / Water Expert	The Kabul Times	Afghanistan
Journalist and Water Policy Expert	The DAWN News	Pakistan
Journalist and Water Policy Expert	Kawish Media Group	Pakistan
Journalist/ Practitioner	Kabul Weekly	Afghanistan
Journalist/ Practitioner	Kabul Weekly	Afghanistan
Lawyer (Basic Human Rights)	Self-Employed	Afghanistan
Lawyer (environment & other legal matters)	Self-employed	Afghanistan
Manager Special Programmes	RSPN	Pakistan
Manager Special Water Project for GB	RSPN	Pakistan
Monitoring Officer	OFWAM	Pakistan
Public Health Officer	Public Health Department	Pakistan
Researcher- Irrigation and Water Institutions	IWMI	Afghanistan
Researcher- Irrigation and Water Institutions	IWMI	Afghanistan
Researcher/Hydrologist	Freelancer	Afghanistan

Researcher-Hydrology/Water Management	IWMI	Afghanistan
Research-Governance & Water Institutions	IWMI	Pakistan
Research-Governance & Water Institutions	IWMI	Afghanistan
Senator / Water Expert / Activist	Senate of Pakistan	Pakistan
Senator / Water Expert / Activist	Senate of Pakistan	Pakistan
Spate Irrigation Expert	SPO/MetaMeta	Pakistan
Structure Engineer	OFWM	Pakistan
Urban Development Department	Kabul Municipality	Afghanistan
WASH Expert	AFD	Pakistan
WASH Expert	Plan International	Pakistan
WASH Expert	NCA	Pakistan
Water and Agriculture Sciences Expert	CABI	Pakistan
Water and DRR Expert	UNDP	Pakistan
Water Expert	The World Bank	Pakistan
Water Expert	SWP	Pakistan
Water Expert	USAID	Pakistan
Water Expert/Professor	Agriculture University Peshawar	Pakistan
Water Management	AREU	Afghanistan
Water Management and Irrigation Expert	Agriculture University Peshawar	Pakistan
Water Policy	AREU	Afghanistan
Water Policy and Renewable Energy Expert	PPAF	Pakistan
Water Policy Expert	Duran	Afghanistan
Water Policy Researcher	IUCN	Pakistan
Water Policy Researcher	Water Aid	Pakistan
Water Policy Researcher and Practitioner	SPO	Pakistan
Water Quality Expert/Lecturer	Kabul Polytechnic Institute	Afghanistan

ANNEX D: List of constitutions, water laws & policies in Afghanistan

- ❖ The Government of Islamic Repulic of Afghanistan 1939. Law for the sale of land under dams and river. Available from: http://extwprlegs1.fao.org/docs/pdf/afg50109.pdf.
- ❖ The Government of Islamic Repulic of Afghanistan 1955. Commercial Law (Commercial Code) of Afghanistan - Usulnameh on the Commercial Law of Afghanistan. Available from: http://www.asianlii.org/af/legis/laws/clcoa1955uotcloa713/.
- ❖ The Government of Islamic Repulic of Afghanistan 1955. Law concerning ownership and tax matter and the price of water for lands below Kajakai and Arghandab dams. Available from: https://www.ecolex.org/details/legislation/law-concerning-ownership-and-tax-matter-and-the-price-of-water-for-lands-below-kajakai-and-arghandab-dams-lex-faoc050112/.
- ❖ The Government of Islamic Repulic of Afghanistan 1962. Law for fixing the price and sale of water below Kajakai and Arghandab dams. Available from: https://www.ecolex.org/details/legislation/law-fixing-the-price-and-sale-of-water-below-kajakai-and-arghandab-dams-lex-faoc039930/
- ❖ The Government of Islamic Repulic of Afghanistan 1971: The Law on Pasture & Grazing Land. Available from: https://www.ecolex.org/details/legislation/law-on-pasture-and-grazing-land-lex-faoc078116/.
- ❖ Government of Afghanistan, 1981. Water Law.
- ❖ Government of Afghanistan, 1991. Water Law.
- ❖ The Constitution of Islamic Republic of Afghanistan 2004. Available from: http://www.afghanembassy.com.pl/afg/images/pliki/TheConstitution.pdf
- ❖ The Government of Islamic Repulic of Afghanistan. Afghanistan Environmental Law 2007: Available from: http://extwprlegs1.fao.org/docs/pdf/afg63169E.pdf.
- ❖ Government of Afghanistan, November 2007. Water Law
- ❖ Government of Afghanistan, June 2008. Water Law.
- ❖ The Government of Islamic Republic of Afghanistan. Water Law. Unofficial English translation of the Water Law as published in the Ministry Justice Official Gazette No. (980), 26 April 2009. Available from: http://extwprlegs1.fao.org/docs/pdf/afg172372.pdf
- ❖ The Government of Islamic Republic of Afghanistan. Water Sector Strategies (WSS). 2007a. Islamic Republic of Afghanistan. Afghanistan national development strategy, July 2007.
- ❖ The Government of Islamic Republic of Afghanistan. Water Sector Strategies (WSS). 2007b. Afghanistan national development strategy, October 2007.
- ❖ The Government of Islamic Republic of Afghanistan. Water Sector Strategies (WSS). 2008. Afghanistan national development strategy, February 2008.
- ❖ The Government of Islamic Republic of Afghanistan. Afghanistan National Development Strategy (ANDS). Draft water sector strategy 2008–2013, Kabul

ANNEX E: List of constitutions, water laws and policies in Pakistan

❖ The Government of Islamic Repulic of Pakistan. 1860. Pakistan Penal Code, XLV OF1860, 6th October, 1860. Available from: https://www.ma-law.org.pk/pdflaw/PAKISTAN%20PENAL%20CODE.pdf.

❖ The Government of Islamic Repulic of Pakistan. 1873. Canal and Drainage Act. 1873. Available from: http://extwprlegs1.fao.org/docs/pdf/pak64507.pdf

❖ The Government of Islamic Repulic of Pakistan. 1927. The Forest Act, 1927. Available from: http://punjablaws.gov.pk/laws/40.html.

❖ The Government of Islamic Republic of Pakistan. 1927. The Pakistan Water and Power Development Authority Act, 1958. W.P. Act XXXI of 1958. Available from: http://punjablaws.gov.pk/laws/86.html.

❖ The Government of Islamic Republic of Pakistan. 1973. Constitution of the Islamic Republic of Pakistan 1973. Available from: http://www.commonlii.org/pk/legis/const/1973/.

❖ The Government of Islamic Republic of Pakistan. 1992. National Conservation Strategy (NCS) 1992. Available from: http://www.commonlii.org/pk/legis/const/1973/.

❖ The Government of Islamic Republic of Pakistan. 1992. Indus River System Authority (IRSA) Ordinance 1992. The Gazette of Pakistan December 10, 1992. Available from: http://www.na.gov.pk/uploads/documents/1334288668_263.pdf.

❖ The Government of Islamic Republic of Pakistan. 1997. North-West Frontier Province Irrigation & Drainage ACT (PIDA) 1997. ACT V OF 1997. Available from: http://extwprlegs1.fao.org/docs/pdf/pak67388.pdf.

❖ The Government of Islamic Republic of Pakistan. 1997. Pakistan Environmental Protection Act (PEPA) 1997. https://www.elaw.org/system/files/Law-PEPA-1997.pdf.

❖ The Government of Islamic Republic of Pakistan. 2001. National Environmental Action Plan (NEAP) 2001. Available from: http://www.finance.gov.pk/survey/chapters/16-Environment.PDF.

❖ The Government of Islamic Republic of Pakistan. 2002. Cantonments Ordinance 2002. Available from: http://cbwah.gov.pk/assets/media/the-cantonment-ordinacne-2002.pdf.

❖ The Government of Islamic Republic of Pakistan. 2005. National Energy Conservation Policy (NEC) 2005. National Energy Conservation Centre (ENERCON) Ministry of Water & Power Governmnet of Pakistan. Available from: http://climateinfo.pk/frontend/web/attachments/data-type/National%20Energy%20Conservation%20Policy.pdf.

❖ The Government of Islamic Republic of Pakistan. 2005. National Environment Policy (NEP) 2005. Government of Pakistan, Ministry of Environment. Available from: https://mowr.gov.pk/wp-content/uploads/2018/05/National-Environmental-Policy-2005.pdf.

❖ The Government of Islamic Republic of Pakistan. 2006. National Sanitation Policy September 2006 Government of the Islamic Republic of Pakistan, Ministry of Environment, Islamabad, Pakistan. Available from: http://extwprlegs1.fao.org/docs/pdf/pak182158.pdf.

❖ The Government of Islamic Republic of Pakistan. 2007. National Policy and Strategy for Fisheries and Aquaculture Development in Pakistan 2007. Ministry of Food, Agriculture, and Livestock, Government of Pakistan, Islamabad. Available from: http://extwprlegs1.fao.org/docs/pdf/pak150786.pdf.

❖ The Government of Islamic Republic of Pakistan. 2009. National Drinking Water Policy, Government of Pakistan Ministry of Environment 2009. Available from: https://waterinfo.net.pk/sites/default/files/knowledge/Pakistan%20National%20Drinking%20Water%20Policy%20-%202009.pdf.

❖ The Government of Islamic Republic of Pakistan. 2009. National Wetlands Policy, Government of Pakistan, Ministry of Environment's Pakistan Wetlands Programme December 2009. Available from: https://waterinfo.net.pk/sites/default/files/knowledge/National%20Wetlands%20Policy%202009.pdf.

❖ The Government of Islamic Republic of Pakistan. 2012. National Climate Change Policy. Government of Pakistan, Ministry of Climate Change, Islamabad, Pakistan, September 2012. Available from: http://www.nrsp.org.pk/gcf/docs/National-Climate-Change-Policy-of-Pakistan.pdf.

❖ The Government of Islamic Republic of Pakistan. 2015. Pakistan National Biodiversity Strategy and Action

Plan, Government of Pakistan 2015. Available from:
http://www.mocc.gov.pk/moclc/userfiles1/file/Draft%20NBSAP%20(29nov16).pdf.

❖ The Government of Islamic Republic of Pakistan. 2015. National Forest Policy, Government of Pakistan, Ministry Of Climate Change, Islamabad, Pakistan 2015. Available from:
http://www.mocc.gov.pk/moclc/userfiles1/file/National%20Forest%20Policy%202015%20(9-1-17).pdf.

❖ The Government of Islamic Republic of Pakistan. 2017. National Food Security Policy (Draft), Government of Pakistan, Ministry of National Food Security and Research Islamabad, June 2017. Available from:
http://www.mnfsr.gov.pk/mnfsr/userfiles1/file/12%20Revised%20Food%20Security%20Policy%2002%20June%202017.pdf.

❖ The Government of Islamic Republic of Pakistan. 2018. National Water Policy, Government of Pakistan, Ministry of Water Resources, National Water Policy, April, 2018. https://ffc.gov.pk/wp-content/uploads/2018/12/National-Water-Policy-April-2018-FINAL_3.pdf.

ANNEX F: List of transboundary laws, proposal and discussions in the KRB

- ❖ The Anglo-Russian Agreement of 1873. Available from: http://www.iranicaonline.org/articles/anglo-russian-agreement-of-1873.
- ❖ The Anglo-Afghan Treaty of 1921. Available from: http://www.iranicaonline.org/articles/anglo-afghan-treaty-of-1921.
- ❖ The 1933-34 Agreement between Afghan government and state government in Chitral on Navigation in Kunar River. Cited in: http://www.lead.org.pk/attachments/presentations/PAK-AFGHAN-stakeholder-consultation-meeting/Day1/Session%201%20B/Water%20Conflict%20Management%20and%20Cooperation%20between%20Afghanistan%20and%20Pakistan%20-%20Fahima%20&%20Shakib.pdf.
- ❖ Pakistan Federal Flood Commission proposal for ‗Kabul River Treaty' with Afghanistan 2003. Aviable from: https://www.dawn.com/news/351801/committee-to-finalize-draft-treaty-on-28th-riparian-rights-over-kabul-kunnar-rivers.
- ❖ Discussions between Pakistan's Water & Power Development Authority and Afghanistan's Khost Province for a joint hydro-electric project. 2006. Available from: http://www.lead.org.pk/attachments/presentations/PAK-AFGHAN-stakeholder-consultation-meeting/Day1/Session%201%20B/Water%20Conflict%20Management%20and%20Cooperation%20between%20Afghanistan%20and%20Pakistan%20-%20Fahima%20&%20Shakib.pdf.
- ❖ The 2006 World Bank's proposal to secure a transboundary riparian agreement. Available from: http://www.lead.org.pk/attachments/presentations/PAK-AFGHAN-stakeholder-consultation-meeting/Day1/Session%201%20B/Water%20Conflict%20Management%20and%20Cooperation%20between%20Afghanistan%20and%20Pakistan%20-%20Fahima%20&%20Shakib.pdf.
- ❖ The 2008 Kuner cascades of Dam Project. Available from: http://www.lead.org.pk/attachments/presentations/PAK-AFGHAN-stakeholder-consultation-meeting/Day%202/Session%202%20A/Kabul%20River%20Basins%20Challenges%20&%20Opportunities%20-%20Idrees%20Malyar.pdf.
- ❖ Islamabad Declaration for regional collaboration in various sectors including water. 2009. http://www.ead.gov.pk/ead/userfiles1/file/EAD/Information%20Services/ADDREESby%20PM%20at%20RECCA.doc.
- ❖ Kabul River Basin Management Commission (KRBMC). 2011. Available from: http://documents.worldbank.org/curated/en/319391468185978566/pdf/522110ESW0Whit1anistan0Final0Report.pdf.
- ❖ Discussion between Finance ministers of Afghanistan and Pakistan for joint hydro-power projects on Kabul River. 2013. Available from: https://www.dawn.com/news/1038435.
- ❖ The Joint Chamber of Commerce (APJCC) initiative for joint hydropower projects on the Kabul River. 2014. Available from: https://mict-international.org/wp-content/uploads/2016/01/MiCT_SIWI_Orphan-River_Final.pdf.
- ❖ Meeting between Afghanistan's Ministry of Foreign Affairs and MEW with the Pakistani counterparts and the World Bank Officials in Dubai for transboundary water management. 2014. Available from: http://www.lead.org.pk/attachments/Second-stakeholder-consultation-Dubai.pdf.
- ❖ Meeting of Afghanistan, India and Pakistan's key stakeholders, experts and engineers at a regional climate change conference in Dubai. 2015. Available from: http://www.lead.org.pk/attachments/presentations/PAK-AFGHAN-stakeholder-consultation-meeting/Day1/Session%201%20B/Water%20Conflict%20Management%20and%20Cooperation%20between%20Afghanistan%20and%20Pakistan%20-%20Fahima%20&%20Shakib.pdf.
- ❖ Government of Afghanistan, China, and Pakistan Proposal for a joint 1500 megawatt hydropower project. 2015. Available from: https://www.sciencedirect.com/science/article/pii/S2352484717303219.
- ❖ Statement of officials from MEW-Afghanistan, during GLOF Conference in Islamabad. 2015. Available from: https://pamirtimes.net/2015/10/20/international-conference-on-glofs-held/.

ANNEX G: Evolution of transboundary-level formal/informal water governance frameworks in the KRB

Period	Law/Policy/MoU/Official Press Briefing	Year of Adoption	Political	Environmental	Social	Economic
				Main Principles		
The Era Before 1947	The Frontier Agreement between British-administered Afghanistan and Russia	1873	• Sovereignty			
	Agreement between British Empire & Afghan government	1921	• Sovereignty	• Protection and Preservation of Ecosystems	• Human Right to Water	
	Agreement between Afghan government & state government in Chitral on Navigation in Kunar River	1933-34		• Protection and Preservation of Ecosystems		
The Era After 1947	Pakistan Federal Flood Commission proposal for _Kabul River Treaty' with Afghanistan	2003	• Exchange of Information			
	Discussions between Pakistan's WAPDA and Afghanistan's Khost Province for a joint hydro-electric project	2006	• Obligation to Cooperate			
	World Bank's proposal to secure a transboundary riparian agreement	2006	• Exchange of Information • Obligation to Cooperate			
	Kunar cascades of Dam Project– (2 on Pakistan side and 3 in Afghanistan)	2008	• Exchange of Information			
	Islamabad Declaration for regional collaboration in various sectors including water	2009	• Exchange of Information • Obligation to Cooperate			
	Kabul River Basin Management Commission (KRBMC)	2011	• Exchange of Information • Obligation to Cooperate		• Capacity Building	• Water as an Economic Good
	Discussion between Finance ministers of Afghanistan and Pakistan for joint hydro-power projects on Kabul River	2013	• Obligation to Cooperate			

ANNEX G: Evolution of transboundary-level formal/informal water governance frameworks in the KRB

The Joint Chamber of Commerce (APJCC) initiative for joint hydropower projects on the Kabul River	2014	• Exchange of Information • Obligation to Cooperate	
Meeting between Afghanistan's Ministry of Foreign Affairs and MEW with the Pakistani counterparts and the World Bank Officials in Dubai for transboundary water management	2014	• Exchange of Information	
Meeting of Afghanistan, India and Pakistan's key stakeholders, experts and engineers at a regional climate change conference in Dubai	2015	• Exchange of Information • Obligation to Cooperate	
Government of Afghanistan, China, and Pakistan Proposal for a joint 1500 megawatt hydropower project	2015	• Obligation to Cooperate • Peaceful Resolution of Disputes	• Capacity Building • Poverty Eradication
Statement of officials from MEW-Afghanistan, during GLOF Conference in Islamabad	2015	• Exchange of Information • Obligation to Cooperate	

ANNEX H: Evolution of water governance frameworks in Afghanistan

ANNEX H: Evolution of water governance frameworks in Afghanistan

Period	Laws	Year of Adoption	Implementation Level	Main Principles Included in the Law			
				Political	Environmental	Social	Economic
The Pre-Colonial Era	Ancient Customs	9000-1900 BCE	Local	• Notification of emergency Situations	• Monitoring • Pollution prevention	• Public Participation	
The Pre-Colonial Era	Islamic Law	652 CE	Local	• Notification of planned measures • Obligation to cooperate • Peaceful resolution of disputes	• Pollution prevention • Water as a finite resource	• Human Right to Sanitation • Human Right to Water • Priority of Use • Public Awareness and Education • Rights of Women, Youth, and Indigenous Peoples	
The Colonial Era	Canal and Drainage Act	1873	Provincial	• Exchange of information • Notification of emergency Situations • Notification of planned measures • Peaceful resolution of disputes	• Basin as the unit of management • Pollution prevention • Precautionary principle • Protected areas for water	• Prior Informed Consent	• Polluter Pays
The Modern Era							
1939-1980	Law regulating the sale of land under dams and rivers	1939	National		• Water as a Finite Resource	• Capacity Building • Public Participation • Rights of Women, Youth, and Indigenous People	
1939-1980	Commercial Law (Commercial Code) of Afghanistan	1955	National		• Protection and Preservation of Ecosystems	• Priority of Use	• Water as an Economic Good

Period	Law	Year	Scope				
	Law concerning ownership, tax matter, and the price of water for lands below dams	1955	National	• Notification of Planned Measures • Peaceful Resolution of Disputes		• Priority of Use • Capacity Building • Rights of Women, Youth, and Indigenous People	• Water as an Economic Good
	Law fixing the price and sale of water	1962	National	• Notification of Planned Measures • Peaceful Resolution of Disputes	• Precautionary Principle	• Capacity Building • Priority of Use • Rights of Women, Youth, and Indigenous People	• Water as an Economic Good
	Law on Pasture and Grazing Land	1971	National	• Obligation to Cooperate • Peaceful Resolution of Disputes	• Protected Recharge and Discharge Zones • Protection and Preservation of Ecosystems	• Public Awareness and Education • Public Participation	
1980-1990	Afghan Water Law	1981	National	• Obligation to Cooperate • Peaceful Resolution of Disputes	• Monitoring • Water as a Finite Resource	• Capacity Building • Equitable and Reasonable Use • Priority of Use	• Water as an Economic Good
1990-2000 Post-2000	Afghan Water Law	1991	National	• Obligation to Cooperate • Peaceful Resolution of Disputes	• Monitoring • Water as a Finite Resource	• Equitable and Reasonable Use • Priority of Use • Public Participation	• Water as an Economic Good

ANNEX H: Evolution of water governance frameworks in Afghanistan

Constitution of Afghanistan	1931; 1964; 1977; 1980; 1987; 1992; 1994; 2004	National	• Obligation to cooperate	• Monitoring • Pollution Prevention • Protection and Preservation of Ecosystems	• Equitable and Reasonable Use • Human Right to Water • Priority of Use • Rights of Women, Youth, and Indigenous People	• Water as an Economic Good
Afghanistan Environmental Law	2007	National	• Exchange of Information • Obligate to Cooperate	• EIA • Monitoring • Pollution Prevention • Precautionary Principle • Protected Recharge and Discharge Zones • Protection and Preservation of Ecosystems	• Capacity Building • Equitable and Reasonable Use • Human Right to Sanitation • Human Right to Water • Intergenerational Equity • Priority of Use • Public Awareness and Education	• Water as an Economic Good
Afghanistan Water Sector Strategies	July 2007a Oct 2007b	National	• Exchange of Information • Notification of Planned Measures • Obligation to Cooperate • Peaceful Resolution of Disputes	• EIA • Monitoring • Pollution Prevention • Precautionary Principle • Protection and Preservation of Ecosystems	• Capacity Building • Equitable and Reasonable Use • Human Right to Sanitation • Human Right to Water • Intergenerational Equity • Public Awareness and Education • Public Participation • Rights of Women, Youth, and Indigenous People	• Water as an Economic Good

2000-Current

ANNEX H: Evolution of water governance frameworks in Afghanistan

Afghanistan Water Law	2009	National	• Exchange of Information • Obligation to Cooperate • Peaceful Resolution of Disputes	• EIA • Monitoring • Pollution Prevention • Protected Recharge and Discharge Zones • Protection and Preservation of Ecosystems	• Capacity Building • Equitable and Reasonable Use • Human Right to Sanitation • Human Right to Water • Priority of Use • Public Access to Information • Public Awareness and Education • Rights of Women, Youth, and Indigenous People	• Polluter Pays • Water as an Economic Good

ANNEX I: Evolution of freshwater governance frameworks from national to local level in Pakistan

ANNEX I: Evolution of freshwater governance frameworks from national to local level in Pakistan

Period	Laws/Policies	Year of Adoption	Implementation Level	Main Principles Included in the Law			
				Political	Environmental	Social	Economic
The Pre-Colonial Era	Ancient Customs	9000-1900 BCE	Local	• Notification of emergency Situations	• Monitoring • Pollution prevention	• Equitable and Reasonable Use • Human Right to Sanitation • Human Right to Water • Public Participation	
The Pre-Colonial Era	Islamic Law	652 CE	Local	• Notification of planned measures • Obligation to cooperate • Peaceful resolution of disputes	• Pollution prevention • Water as a finite resource	• Equitable and Reasonable Use • Human Right to Sanitation • Human Right to Water • Priority of Use • Public Awareness and Education • Rights of Women, Youth, and Indigenous Peoples	
The Colonial Era	Pakistan Penal Code	1860	Federal		• Pollution prevention • Protection and preservation of ecosystems		• Polluter Pays
The Colonial Era	Canal and Drainage Act	1873	Provincial	• Exchange of information • Notification of emergency Situations • Notification of planned measures • Peaceful resolution of disputes	• Basin as the unit of management • Pollution prevention • Precautionary principle • Protected areas for water	• Prior Informed Consent	• Polluter Pays

ANNEX I: Evolution of freshwater governance frameworks from national to local level in Pakistan

The Post-Colonial or Modern Era						
Forest Act	1927	Federal	• Notification of emergency Situations	• Monitoring • Pollution prevention • Precautionary principle		• Polluter Pays
First Period of Modern Era 1947 to 1977						
West Pakistan WAPDA Act	1958	Federal	• Peaceful resolution of disputes	• Pollution prevention • Protected areas for water	• Public Access to Information	• Polluter Pays
Constitution of the Islamic Republic of Pakistan	1973	Federal	• Peaceful resolution of disputes	• Water as a finite resource	• Equitable and Reasonable Use • Human Right to Sanitation • Human Right to Water • Public Access to Information • Public Awareness and Education • Public Participation • Rights of Women, Youth, and Indigenous Peoples	
Second Period of Modern Era 1977 to 1997						
National Conservation Strategy (NCS)	1992	Federal	• Exchange of information • Notification of emergency Situations • Obligation to cooperate • Peaceful resolution of disputes	• EIA • Monitoring • Pollution prevention • Protected areas for water • Protection and preservation of ecosystems	• Capacity Building • Human Right to Sanitation • Human Right to Water • Poverty Eradication • Public Access to Information • Public Awareness and Education • Public Participation	• Polluter Pays

ANNEX I: Evolution of freshwater governance frameworks from national to local level in Pakistan

	Year	Level				
Indus River System Authority (IRSA) Ordinance	1992	Provincial	• Peaceful resolution of disputes	• Monitoring	• Equitable and Reasonable Use	
North-West Frontier Province Irrigation & Drainage ACT (PIDA)	1997	Provincial	• Exchange of information	• Monitoring • Pollution prevention	• Equitable and Reasonable Use	• Water as an Economic Good
Pakistan Environmental Protection Act (PEPA)	1997	Federal	• Exchange of information	• EIA • Monitoring • Pollution prevention • Protected areas for water	• Human Right to Sanitation • Human Right to Water • Public Access to Information • Public Awareness and Education • Public Participation	• Polluter Pays
Third Period of Modern Era 1997 to 2007						
National Environmental Action Plan (NEAP)	2001	Federal		• EIA • Monitoring • Pollution prevention • Protected areas for water	• Human Right to Sanitation • Human Right to Water • Public Access to Information • Public Awareness and Education • Public Participation	• Polluter Pays
Cantonments Ordinance	2002	Federal			• Human Right to Sanitation • Human Right to Water	
National Energy Conservation Policy (NEC)	2005	Federal		• Pollution Prevention	• Human Right to Sanitation • Human Right to Water • Public Awareness and Education	

ANNEX I: Evolution of freshwater governance frameworks from national to local level in Pakistan

Policy	Year	Level				
National Environment Policy (NEP)	2005	Federal	• Exchange of Information • Notification of Emergency Situations • Obligation to Cooperate	• EIA • Invasive Species • Monitoring • Pollution Prevention • Protected Areas for water • Protected Recharge and Discharge Zones	• Capacity Building • Equitable and Reasonable Use • Human Right to Sanitation • Human Right to Water • Intergenerational Equity • Poverty Eradication • Public Access to Information • Public Awareness and Education • Public Participation	• Water as an Economic Good
National Fisheries Policy (NFP)	2006	Federal	• Exchange of Information • Obligation to Cooperate	• Monitoring	• Capacity Building • Poverty Eradication • Public Awareness and Education	
National Water Policy (approved by cabinet in 2009)	2006	Federal	• Exchange of Information • Obligation to Cooperate • Peaceful Resolution of Disputes	• Conjunctive Use • EIA • Monitoring • Pollution Prevention • Protected Areas for water • Protected Recharge and Discharge Zones • Protection and Preservation of Ecosystems • Subsidiarity • Water as a Finite Resource	• Capacity Building • Equitable and Reasonable Use • Human Right to Sanitation • Human Right to Water • Intergenerational Equity • Poverty Eradication • Priority of Use • Public Awareness and Education • Public Participation	• Polluter Pays • Water as an Economic Good

ANNEX I: Evolution of freshwater governance frameworks from national to local level in Pakistan

National Sanitation Policy (NSP)	2006	Federal	• Exchange of Information • Obligation to Cooperate	• Monitoring • Pollution Prevention	• Capacity Building • Equitable and Reasonable Use • Human Right to Sanitation • Human Right to Water • Poverty Eradication • Public Awareness and Education • Public Participation • Rights of Women, Youth, and Indigenous Peoples
Fourth Period of Modern Era 2007 to Present					
National Drinking Water Policy	2009	Federal	• Exchange of Information • Notification of Emergency Situations	• EIA • Monitoring • Pollution Prevention	• Capacity Building • Equitable and Reasonable Use • Human Right to Sanitation • Human Right to Water • Priority of Use • Public Awareness and Education • Public Participation • Rights of Women, Youth, and Indigenous Peoples

ANNEX I: Evolution of freshwater governance frameworks from national to local level in Pakistan

Framework	Year	Level				
Pakistan National Wetlands Policy	2009	Federal	• Exchange of Information • Peaceful Resolution of Disputes	• EIA • Invasive Species • Monitoring • Pollution Prevention • Precautionary Principle • Protected Areas for water • Protected Recharge and Discharge Zones • Water as a Finite Resource	• Capacity Building • Equitable and Reasonable Use • Human Right to Sanitation • Human Right to Water • Intergenerational Equity • Poverty Eradication • Public Access to Information • Public Awareness and Education • Public Participation	• Polluter Pays • Water as an Economic Good
National Climate Change Policy	2012	Federal	• Exchange of Information • Obligation to Cooperate	• EIA • Monitoring • Pollution Prevention • Precautionary Principle • Protected Areas for water • Protected Recharge and Discharge Zones • Protection and Preservation of Ecosystems • Water as a Finite Resource	• Capacity Building • Poverty Eradication • Public Awareness and Education • Public Participation • Rights of Women, Youth, and Indigenous Peoples	

ANNEX I: Evolution of freshwater governance frameworks from national to local level in Pakistan

Framework	Year	Level			
Biodiversity Action Plan	2015	Federal	• Exchange of Information • Obligation to Cooperate	• EIA • Invasive Species • Monitoring • Pollution Prevention • Precautionary Principle • Protection and Preservation of Ecosystems	• Capacity Building • Equitable and Reasonable Use • Poverty Eradication • Public Access to Information • Public Awareness and Education • Public Participation • Rights of Women, Youth, and Indigenous Peoples
National Forest Policy	2015	Federal	• Notification of Planned Measures	• Pollution Prevention • Protected Areas for water • Protection and Preservation of Ecosystems	
National Food Security Policy (Draft)	2017	Federal	• Peaceful Resolution of Disputes	• Monitoring • Pollution Prevention • Protected Areas for water • Water as a Finite Resource	• Human Right to Sanitation • Human Right to Water • Intergenerational Equity • Poverty Eradication • Public Awareness and Education

ANNEX J: Major principles & instruments in Pakistan's multilevel water governance fameworks

	Local Customs	Sharia Law	The 1860 Pakistan Panel Code	The 1873 Canal and drainage act	The 1927 forest act	The 1958 West Pakistan WAPDA act	The 1973 Constitution of the Islamic Republic of Pakistan	The 1992 National Conservation Strategy	The 1991 Interprovincial Water Apportionment Accord	The 1997 North-west frontier province irrigation and drainage	The 1997 Pakistan Environmental protection act (PEPA)	The 2001 National Environmental Plan (NEAP)	The 2002 Cantonments ordinance	The 2005 National Energy Conservation Policy (NEC)	The 2005 National Environment Policy	The 2006 National Fisheries Policy (NFP)	The 2018 National Water Policy (approved in 2009)	The 2006 National Sanitation Policy (NSP)	The 2009 National Drinking Water Policy	The 2009 Pakistan National Wetlands policy	The 2012 National Climate Change Policy	The 2015 Biodiversity Action Plan	The 2015 National Forest Policy	The 2017 National Food and Security Policy (draft)
Political Principles																								
Information exchange			■									■	■	■	■	■	■	■	■	■	■	■	■	■
Notification of Emergency Situations	■																							
Notification of Planned Measures			■				■						■											
Obligation to Cooperate			■						■	■	■	■	■	■	■	■	■	■	■	■	■	■	■	■
Peaceful settlement of Disputes			■						■	■	■	■	■	■	■	■	■	■	■	■	■	■	■	■
Limited Territorial Sovereignty/ Do Not Harm																								
Environmental Principles																								
Basin as the Unit of Management			■																					
BATT																								
Conjunctive Use												■												
EIA											■				■	■	■	■	■	■	■	■	■	■
Invasive Species																								
Monitoring								■	■	■	■	■	■	■	■	■	■	■	■	■	■	■	■	■
Prevention of Pollution	■									■	■	■	■	■	■	■	■	■	■	■	■	■	■	■

Precautionary Principle	
Protected Areas for Water	
Protected Recharge and Discharge Zones	
Protection and Preservation of Ecosystems	
Polluters Pay	
Water as a Finite Resource	
Social Principles	
Capacity Building	
Equitable & Reasonable Use	
Human Right to Water & Sanitation	
Intergenerational Equity	
Poverty Eradication	
Prior Informed Consent	
Priority of Use	
Public Access to Information	
Public Awareness & Education	
Public Participation	
Rights of Women, Youth, & Indigenous Peoples	
Food Security	
Human Well-being	
Quality Education	
Clean Energy	

Economic Growth																						
Infrastructure																						
Reduced Inequality																						
Sustainable Urbanisation																						
Responsible Consumption & Production																						

ANNEX K: Multilevel ecosystem services in the KRB

ESS	Transboundary	Afghanistan	Pakistan
Supporting	Soil formation (sediment retention and accumulation of organic matter) Habitat provision (provision of habitat for wildlife feeding, shelter and reproduction Nutrient cycling (storage, recycling, processing and acquisition of nutrients)	Habitat availability & genetic diversity ensure the functioning of all other services. Freshwater biodiversity of Koh-e Baba sustaining the provision of all ESS in the area Home to a large number of breeding and migratory birds Supporting diverse plant species & wildlife such as wolves, fox, wild cats, rabbit, deer, bats and numerous birds Genetic diversity of wild relatives of wheat and other flora provide genes of resilience and resistance to disease	Soil formation through sediment holding and accumulation of organic matter Nutrient storage Recycling, processing, and acquisition of nutrients Freshwater wetlands reduce the risk of pollution from agriculture runoff
Provisioning	Water collection from catchment areas through rainfall, snow melt & glacial melt; Water for irrigation, effluent & waste disposal, watering livestock, fishing, recreation, transportation, shelter, medicine, furniture, fuel washing and bathing; Water for livelihoods of many local people; Water for power supply system; Provisioning of groundwater through infiltration in muddy floodplains; Kabul River serves as an indicator of proper land use and land management	Natural storage facility Sustaining the vital flow in rivers through snow accumulation during winter, snow melt and rainfalls during spring, and release of frozen water from glaciers in summer Largest potential for irrigated agriculture and hydropower; Maintaining watersheds for the future prosperity; Freshwater in the KRB can enhance cross-border cooperation with neighbour countries	Production of food, fish, fruits, fibre, fuelwood and fodder; Freshwater storage and retention of water for domestic, industrial and agricultural use; Indus Flyway serve as migration route for ducks and wildfowl which provide local food sources for communities; Irrigates 17 million ha of farmland and provides employment to 40% population; The riverine forests and mangroves provide timber and fuelwood for local communities
Regulating	Hydrological regulation through flood water usage for irrigation in arid zones Groundwater recharge through infiltration of flood water into the ground Water quality improvement by removing nutrients and pollutants through soil layers Assimilation of organic waste by certain aquatic fauna such as fish and turtle Nutrients sequestration Forested floodplain prevent nonpoint source pollution from entering small streams Climate regulation through carbon sequestration in the floodplains and surrounding forests	Freshwater in Afghanistan naturally increases susceptibility to the processes of soil erosion Rainfed farming is especially detrimental to soil retention Preserve the natural vegetation and prevent soil erosion which enhance the productivity Pollination is indirect regulating service which provides an alternative source of income for local communities	Climate regulation by providing sinks for greenhouse gases; influencing local and regional temperature, precipitation and other climatic processes Water regulation (hydrological flows) and groundwater recharge/discharge Water purification by removing nutrients and other pollutants Natural hazard regulation, storm protection and pollination habitat for pollinators Kabul-Indus basin support three quarters of Pakistani population and irrigates 80% cropland Erosion regulation by slowing down the flow of water River beds and lakes provide support in flood control, provide an escape for floodwaters, holding them back and reducing the flood intensity

| Cultural | Recreational uses such as fishing, hunting waterfowl sports, canoeing, kayaking, white water rafting, sport fishing, swimming

Opportunity to worshipers to access water and clean themselves

Mass bathing and opportunity for social gathering mostly in summers

Riverside restaurants attract tourists and provide peace and quiet environment

Inspiring the imagination of the local people through songs | The snow line on the peak of the Koh-e Allah spell out the word 'Allah' is of cultural significance

Sacred natural hot springs attracted many tourists before the war

Provides opportunities for hiking and backcountry skiing as well as ecotourism to local communities and tourists

Many caves in the KRB on Afghanistan's side have ancient animal carvings that locals visit

Shrines in the KRB commemorate significant events

A mix of formal and traditional institutions in the KRB governs a patchy and uncertain land tenure system | Spiritual and religious freshwater attributes, with shrines of Pirs, Saints, Sufis and other religious leaders in some coastal and wetland areas

Shrines around Kalakahar Lake has archaeological and historical features that attract visitors

Mythological stories attract pilgrims and visitors during festivals

Lakes, beaches, rivers and streams carry recreational values which helped in establishing special tourism authorities, e.g., Tourism Corporation Development of Pakistan (TCDP), Tourism Corporation Khyber Pakhtunkhwa (TCKP), Tourism Development Corporation of Punjab (TDCP), Sindh Tourism Development Corporation (STDC) and The Department of Culture, Archives and Tourism in Balochistan

Rivers, lakes and streams are being used for educational purposes, creating awareness amongst local schools and colleges about the importance of water |
|---|---|---|---|

Source: Reproduced from Tables 6.1, 7.1 and 8.1

ABOUT THE AUTHOR

Author has diverse work experience in research, implementation, capacity building, monitoring, and overall projects management with a multitude of organizations including ADB, SANDEE, IHEID, and Higher Education Commission of Pakistan in sectors/themes such as, Transboundary Water Governance, Climate Change, Clean Energy, and DRR. He did his Graduate Studies in Climate Change and International Development from the School of International Development, University of East Anglia and wrote thesis on Power Politics and water governance in the Indus Basin. Since 2011 he has been associated with a Research Think-tank, Centre for Public Policy Research (CPPR) and also teach climate change and water governance in development studies faculty in a public sector university in Pakistan. He has contributed in four peer reviewed articles and various projects reports for different I/NGOs and Consultants. His PhD thesis topic at IHE Delft Institute for Water Education is Inclusive Development and Transboundary Water Governance in the Kabul River Basin. This PhD topic is a multi-disciplinary and revolving around institutional analysis of the Transboundary River Kabul by proposing a design/redesign of the governance principles and instruments to deal with status-quo and achieve inclusive and sustainable development.

LIST OF PUBLICATIONS

- Hayat, S., & Gupta, J. (2016). Kinds of freshwater and their relation to ecosystem services and human well-being. Water Policy, 18(5), 1229–1246.
- Jan, I., K. Humayun, and S. Hayat. 2012. Determinants of rural household energy choices: An example from Pakistan. Polish Journal of Environmental Studies. 21(3): 635-641.
- Jan, I., S. Hayat, and M.A. Khan. 2011. Analysis of rural household employment structure in northwest Pakistan. Spanish Journal of Rural Development. I I(4): 56-65.
- Jan, I., M.A. Khattak, S. Hayat and M.A. Khan. 2012. Factors affecting rural livelihood choices in northwest Pakistan. Sarhad Journal of Agriculture. 28(4): 617-626.

Netherlands Research School for the
Socio-Economic and Natural Sciences of the Environment

D I P L O M A

for specialised PhD training

The Netherlands research school for the
Socio-Economic and Natural Sciences of the Environment
(SENSE) declares that

Shakeel Hayat

born on 15 January 1980 in Swabi, Pakistan

has successfully fulfilled all requirements of the
educational PhD programme of SENSE.

Amsterdam, 5 March 2020

The Chairman of the SENSE board

Prof. dr. Martin Wassen

the SENSE Director of Education

Dr. Ad van Dommelen

K O N I N K L I J K E N E D E R L A N D S E

The SENSE Research School declares that Shakeel Hayat has successfully fulfilled all requirements of the educational PhD programme of SENSE with a work load of 30.8 EC, including the following activities:

SENSE PhD Courses

- ○ Environmental research in context (2014)
- ○ Research in context activity: 'Co-organizing UNESCO-IHE PhD Symposium on: From water scarcity to water security (Delft, 3-4 October 2016)'

External training at a foreign research institute

- ○ Research Leadership workshop for the Water Initiative South Asia (WISA), Imperial College London & Institute of Management Sciences, United Arab Emirates (2019)

Societal impact

- ○ Pakistan-Afghanistan stakeholders consultation: the way forward for benefit sharing in Kabul River Basin, LEAD-Pakistan and Institute of Management Sciences (2019)

Management and Didactic Skills Training

- ○ Member of PhD Association Board IHE (2014-2016)
- ○ Organising PhD week IHE (2016)
- ○ Teaching in a postgraduate course 'Lecture on international water law and transboundary water management', Institute of Management Sciences, Pakistan (2017)
- ○ Teaching in a postgraduate course 'Lecture on multilevel water governance' Institute of Management Sciences, Pakistan (2017)

Oral Presentation at international conference

- ○ *The use of socio-ecological information for enhancing transboundary water cooperation.* Water Initiative for the Future (WatIF), International Graduate Student Conference, 27-29 July 2016, Canada
- ○ *Enhancing transboundary water cooperation in the climate sensitive Kabul river by economic valuation of ecosystem services*, Pakistan-Afghanistan Stakeholders Consultation: The Way Forward for Benefit Sharing in Kabul River Basin, 16-17 October 2019, Dubai, United Arab Emirates

SENSE coordinator PhD education

Dr. ir. Peter Vermeulen

*For Product Safety Concerns and Information please contact
our EU representative GPSR@taylorandfrancis.com Taylor & Francis
Verlag GmbH, Kaufingerstraße 24, 80331 München, Germany*

T - #0136 - 160425 - C336 - 240/170/18 - PB - 9780367500740 - Gloss Lamination